THE CHEMISTRY OF
ORGANIC FILM FORMERS

THE CHEMISTRY OF ORGANIC FILM FORMERS

D. H. SOLOMON

CHIEF
DIVISION OF APPLIED ORGANIC CHEMISTRY
COMMONWEALTH SCIENTIFIC AND INDUSTRIAL
RESEARCH ORGANIZATION
AUSTRALIA

ROBERT E. KRIEGER PUBLISHING COMPANY
MALABAR, FLORIDA

Original Edition 1967
Second Edition 1977, 1982

Printed and Published by
ROBERT E. KRIEGER PUBLISHING CO., INC.
Krieger Drive
Malabar, Florida 32950

Library of Congress Cataloging in Publication Data

Solomon, David Henry.
   The chemistry of organic film formers.

   Reprint of the edition published by Wiley, New York.
   Includes bibliographies.
   Includes index.
   1. Polymers and polymerization.   I. Title.
[TP156.P6S58 1977]   547'.84   76-50117
ISBN 0-88275-165-4

PREFACE TO SECOND EDITION

During the last few years there have been rapid and significant changes in the methods relating to the application and curing of surface coatings. Both of these aspects of coating technology have been treated in this new edition, and the manner in which the chemistry of the polymer systems has been adjusted to cope with new requirements discussed. In particular, developments in powder coatings, radiation curing of coatings and in electrodeposition are discussed, with the broad principles of the chemistry of these systems being defined.

Melbourne, Australia
September 1976

D. H. Solomon

PREFACE

Until recently, the formulation of surface coatings was basically an art, but it is now rapidly changing to a science. This book is designed to provide the basic chemistry necessary to approach coating technology from a scientific point of view.

During my 20 years association with the surface-coating industry, I have noticed numerous instances where a single textbook covering the chemistry of surface coatings would have been invaluable for a number of reasons. First, the research findings in this field appear in a wide range of scientific publications, and the collation of this information in a single reference book would save much time and effort. Second, those involved in the more technological aspects of coating technology often require, and would benefit considerably from, an understanding of the chemistry of the systems involved. Third, while lecturing on this subject at the Royal Melbourne Institute of Technology, I became aware of the need for a single textbook concerned predominantly with the chemistry (as opposed to the technology) of coating compositions.

In the first chapter, I have discussed sufficient basic polymer science for the reader to be able to understand and control both stepwise and chain polymerizations. I have also brought together information that has not, to my knowledge, been presented in a single text, although this information is scattered throughout other books. For example, I have included a discussion of the methods of controlling the molecular weight of a polymer that is produced by a chain reaction. The remaining chapters deal specifically with the chemistry of the various types of vehicles used in coating composition.

Throughout the book, the more technological aspects of coating compositions have been deliberately kept to a minimum, because these are well covered in other textbooks and in the literature available from the suppliers of chemicals and polymers.

I have attempted to cover the literature up to the end of 1965 and to include existing and projected coating compositions. I hope this book will be of value to the technologists and scientists working in the industry, and to students concerned with organic polymers suitable for use as coating compositions, plastics, or rubbers.

The Appendix contains the formulae and manufacturing instructions for typical polymer compositions used in surface coatings. These have been included to relate the chemistry discussed in the text with the polymer systems used in practice.

Melbourne, Australia
January 1966

D. H. Solomon

ACKNOWLEDGMENTS

In writing this book I have been assisted greatly by many of my colleagues. In particular I am deeply indebted to Dr. George F. Walker for encouragement when it was most needed and for the great personal satisfaction I have gained from working with him in our present field of interest.

My thanks are also due to Mr. A. J. Gaskin, Chief of the Division of Applied Mineralogy in the C.S.I.R.O., for providing the environment which has made it possible for me to write this book and to work in the new and exciting field of mineral-organic systems.

I gratefully acknowledge the assistance given by Mrs. J. D. Swift and Mr. B. C. Loft in reading, checking, and commenting upon the manuscript, and in checking the galley and page proofs. I also thank Dr. D. Willis, Mr. R. Harris, Mr. P. J. Wigney, and Mr. M. Cochrane for their comments on selected chapters of the book.

Checking original references is not a very interesting pastime; the burden was lightened considerably by the assistance given by the staff of the C.S.I.R.O. Chemical Research Laboratories Library.

The manuscript was typed by Miss Freda Cheal who has earned my sincere thanks and admiration for the competent manner in which this task was performed.

Permissions granted to reproduce copyright material by the following companies, journals, and their publishers are gratefully acknowledged.

A.D.M. Chemicals (Archer Daniels Midland Company), Minneapolis, Minn.; E. I. du Pont de Nemours and Company (Elastomer Chemical Department), Wilmington, Del.; Novadel Ltd. (London); Shell Chemical (Australia), Pty. Ltd.; American Chemical Society, publishers of *Journal of the American Chemical Society, Industrial and Engineering Chemistry,* and *Chemical Reviews; Canadian Journal of Chemistry; Die Makromolekulare Chemie* (Hüthig und Wepf Verlag, Basel); *Fette, Seifen, Anstrichmittel* (Germany); *Industrial Chemist* (Tothill Press Ltd.); Interscience Publishers, New York, N.Y.; John Wiley and Sons, New York, N.Y.; *Journal American Oil Chemists' Society; Journal of Applied Chemistry* (Society of Chemical Industry); *Journal of Applied Polymer Science; Journal of the Oil and Colour Chemists' Association;*

Journal of Paint Technology; *Journal of Polymer Science*; *Proceedings of the Royal Australian Chemical Institute*; Reinhold Publishing Corporation, New York, N.Y.; *Reviews of Pure and Applied Chemistry*.

D.H.S.

ACKNOWLEDGMENTS TO SECOND EDITION

In preparing the Second Edition of this book I received valuable advice and assistance from Mr. K. F. Alcock, Research Manager of Dulux Paints and the members of his staff. In particular I wish to record my appreciation for the help given by Mr. D.W. Berryman, the late Dr. R. W. Kershaw, Mr. F. J. Lubbock and Mr. David Owen. Permission to use the Dulux Paints Library is gratefully acknowledged.

I thank Mr. B. C. Loft for his comments on the revised manuscript and Miss C. Polinelli, Mr. I. Lomas and Mr. R. Allen for their assistance in the preparation of the manuscript.

Permission to reproduce copyright material from the following publications is gratefully acknowledged: J. Paint Technology, Proceedings of the Royal Australian Chemical Institute. Proceedings of the International Symposium on Powder Coatings.

CONTENTS

CHAPTER 1

THE FORMATION, CHARACTERIZATION, AND PROPERTIES OF POLYMERS

Surface coatings generally consist of a polymeric resinous material dissolved or dispersed in suitable solvents or diluents; usually additives are also present. If required, color and/or opacity is provided by the incorporation of selected pigments, which are virtually insoluble in the other coating components. Inerts or fillers, materials with a refractive index approximately that of the resinous phase, are sometimes used to give filling and other properties. Formulations designed for surface preparation or undercoating generally contain fillers. The basic chemical reactions involved in the formation of the film from the solution or dispersion are related to the polymer system present, although the pigments and/or fillers will influence these film-forming reactions by inhibiting, retarding, or catalyzing them to various degrees. Therefore, the chemistry of organic film formers is basically concerned with the preparation of the polymers that are used to prepare the liquid coatings, and with the mechanisms by which these liquid coatings form dried films.

As a general rule, it is found that the film properties of a surface coating improve as the molecular weight or molecular complexity of the final polymer increase; from a chemical point of view, two distinct methods are used to arrive at films conforming to one or other of these requirements. In the first approach, the polymer used is of a sufficiently high molecular weight to give the desired film properties without further chemical or molecular weight changes. Such polymers are applied as solutions or dispersions, and film formation results from evaporation of the solvent or diluent. These systems are referred to as thermoplastic or lacquer-type formulations. The second method of obtaining a desirable film is to use a polymer system with a complex molecular structure. Since such a polymer would be insoluble in common solvents (if an attempt were made to achieve full molecular complexity initially), this complex structure must be formed from predominantly linear, but reactive, polymers during or after the time the solvent evaporates from the applied film. These polymer systems are of the thermosetting or enamel type.

Since the enamel-type polymer undergoes an increase in molecular

1

weight and complexity during film formation, the initial molecular weight of the polymer is lower than when a thermoplastic system is used. Consequently, higher solids contents are obtained in a coating at a given viscosity, since solution viscosity is a function of molecular weight; conversely at a given viscosity (for example, spraying), less high molecular weight polymer can be tolerated. Table 1.1 compares the properties of solutions of thermoplastic and thermosetting polymers of the type used in automobile topcoat formulations.

In many coating requirements the properties desired in the final film will dictate the general class of polymers to be considered. For example, if a solvent resistant finish is specified, then a thermoplastic system, which is soluble in the solvents used to deposit the film (and probably many other

TABLE I.I

COMPARISON OF THERMOPLASTIC AND THERMOSETTING POLYMER
SYSTEMS AS AUTOMOTIVE TOPCOAT FORMULATIONS

Property	Thermoplastic (Lacquer)	Thermosetting (Enamel)
Solids[a] at application viscosity (this includes polymer, pigment, and plasticizer)	20–30% (approx.)	50% (approx.)
Solvent composition	Esters, ketones (expensive)	Hydrocarbons (relatively cheap)
Drying conditions required	Very flexible, will air dry or can be force dried at elevated temperatures (i.e. $\frac{1}{2}$ hr. at 85°C)	Conditions more specific and critical than thermoplastic systems
Film properties	Film is sensitive to solvents and will re-dissolve. Damage is easy to repair. Requires polishing to develop full gloss.	Film becomes insoluble and is difficult to repair. It may require a specially catalyzed system. Full gloss is developed without polishing.
Cost per sq. ft. per mil. dry (ratio)	3–4	1

[a] The solids figures shown refer to systems applied by spray.
Reproduced by courtesy of the editor, *Rev. Pure Appl. Chem.* **13**, 171, (1963).

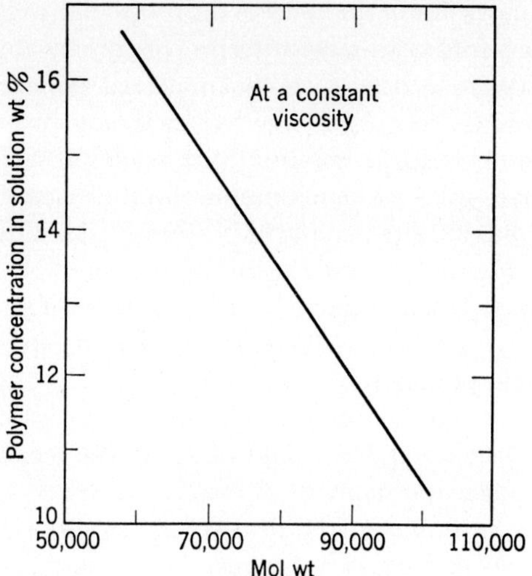

Figure 1.1. Relationship between solids content and the molecular weight of a polymer.

solvents also), would not generally be used. Once the general class of polymer required is known, the choice of the specific polymer type and composition is a more involved—and less definite—process. It is usually necessary to balance out a number of opposing factors (price, durability, application solids, etc.), and to arrive at a compromise formulation, which satisfies the specified properties to various degrees. In reaching this

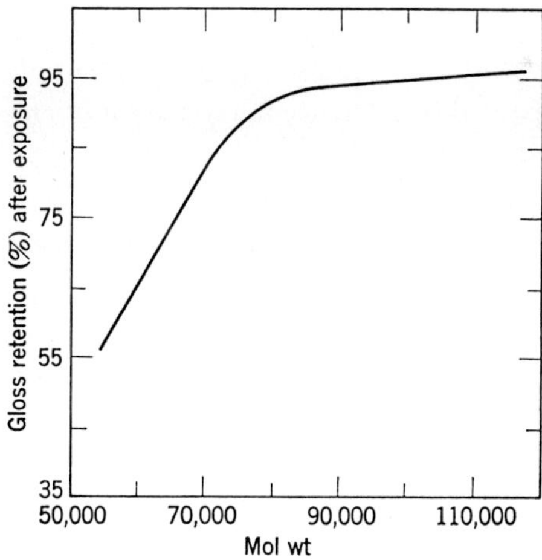

Figure 1.2. Relationship between durability and molecular weight of poly(methyl methacrylate).

compromise, a knowledge of the effect of the molecular weight and composition of the polymer on the various coating properties is invaluable, and enables the coatings scientist to decide on the preferred polymer. For example, in acrylic lacquers based on poly(methyl methacrylate) it is desirable to have the maximum solids in solution at application viscosity, and yet to use a polymer that will give the maximum durability on exterior exposure. Generally, the solids requirement would be best satisfied by a low molecular weight polymer (Figure 1.1), and the durability would be satisfied by a high molecular weight polymer (Figure 1.2). The compromise chosen takes into account the comparatively small increase in durability for molecular weights above 90,000 (approximately).

Satisfactory control of the molecular weight, composition, and reactivity of a polymer molecule requires a knowledge of the mechanisms by which polymers are formed, the methods of characterization of macromolecules, and the physical measurements indicative of film properties; these aspects are discussed below and in later chapters. For a more detailed coverage of these subjects, General References 1 to 8 should be consulted.

POLYMER PREPARATION

Polymers are derived from simple organic molecules, called monomers, by two distinct mechanisms; these involve stepwise (condensation) or chain (addition) reactions. Typical surface-coating polymers prepared by stepwise polymerization are the alkyd resins, phenol- and amine-formaldehyde condensates, epoxy-type polymers, and polyurethanes. Chain reactions are used to prepare polymers from acrylic or vinyl monomers.

The reactions occurring in a stepwise polymerization are analogous to those that take place in the formation of much simpler organic molecules. This may be illustrated by reference to the esterification of an alcohol with an acid to yield the ester and water.

$$ROH + R'COOH \rightleftharpoons R'COOR + H_2O$$

The reaction equilibrium can be shifted to the right by removal of the water and this is the usual approach adopted in polymer synthesis.

If a glycol and dibasic acid are reacted, esterification takes place and a linear polymer results.

$$n\ HO\!-\!R\!-\!OH + n\ HOOC\!-\!R'\!-\!COOH$$
$$\downarrow -H_2O$$
$$HO\!-\!R\!-\!O\!\!-\!\!\left[CO\!-\!R'\!-\!CO\!-\!O\!-\!R\!-\!O\right]_{n-1}\!\!CO\!-\!R'\!-\!COOH$$

Since the glycol and dibasic acid have two reactive groups, they are termed bifunctional. It is important to recognize that the buildup of the above polyester involves a number of individual esterification reactions, each similar to the simple esterification of an alcohol and monobasic acid. Consequently, the molecular weight of the polymer increases in a stepwise manner, and the number of ester links increases with a corresponding loss of hydroxyl and carboxyl groups.

In polyesterifications the rate of reaction decreases markedly with time and this observation was once taken as indicating a decrease in the reactivity of the functional group with increasing size of the molecule.

From a study of the kinetics of simple and polymeric esterification processes, Flory (Gen. Ref. 1) has concluded that the hydroxyl and carboxyl groups in a polymer molecule have the same chemical activity as those in a simple monomeric compound. Third-order kinetics apply over the latter part of esterification, and polyesterification; consequently,

$$-\frac{d[\text{COOH}]}{dt} = k[\text{COOH}]^2[\text{OH}]$$

Besides taking part in ester formation, the acid acts as a catalyst. Therefore, the slow rate of reaction toward the end of polyester formation is not a function of reduced activity because of molecular size, but is characteristic of a third-order reaction.

Most polyesters used in surface coatings require a low free acid-group content or acid value to minimize reaction with other polymers or pigments. In order to reach a low acid value in a reasonable processing time, more than the stoichiometric amount of glycol is used. In addition to increasing the rate of esterification, the use of excess glycol serves as a means of controlling the maximum average molecular weight by limiting the actual functionality of the glycol. For example, in the extreme case where two moles of glycol to one mole of dibasic acid are used, the average composition will be as shown:

$$2\ \text{HO—R—OH} + \text{HOOC—R'—COOH}$$
$$\downarrow -2\,\text{H}_2\text{O}$$
$$\text{HO—R—O—CO—R'—COO—R—OH}$$

Here the glycol is only monofunctional (that is, forms a mono ester with one free hydroxyl group per glycol residue). In practical linear polyester formulations, the glycol functionality will be slightly less than two. The theory of condensation polymerization is further developed in Chapter 3 on alkyd resins.

The molecular weight of a polyester, or other polymers formed by step-growth polymerization, can be measured and controlled by a number of methods. The course of a polysterification is most commonly followed by end-group analysis (usually carboxyl) and by viscosity. An increase in the temperature of the polymerization does not alter the possible molecular weight, but serves to increase the rate at which the polymer is formed.

The use of alcohols and acids with functionalities of more than two gives rise to polymers that can have branched and eventually three-dimensional structures. The polymerization of these more highly functional systems follows the same general principles as the simpler bifunctional compositions; the molecular weight of the polymer increases in a stepwise manner with increasing time of reaction. However significant deviations from the expected functional group reactivity have been noted between the linear system studied by Flory and the more highly functional systems, which are of technical importance in the design and use of polymers for surface coatings. It has been shown,[1] that in some polymers the functional groups, particularly the functional groups in alkyd resins, do not have the expected reactivity and they are certainly not comparable with those present in simple organic molecules or in linear polyesters. As a general observation the deviation between the calculated and measured hydroxyl values in alkyd resins is more pronounced in the highly branched, or crosslinked structures. Formulations which are essentially linear often behave as predicted from model compounds but the processing conditions used to prepare the polymer can alter the polymer structure and the availability of the functional groups. In some complex alkyd formulations it is doubtful if some groups are accessible to other reactive entities at all. These findings offer a possible explanation for the failure of, for example, the reaction between glycerol and phthalic anhydride[2] to follow third-order kinetics, since the number of hydroxyl groups present and the number available for esterification are not the same. This arises from the presence of primary and secondary hydroxyl groups, and from the formation of microgel particles in nonlinear systems (see page 113). This lack of availability or reactivity of some of the groups must also be considered where subsequent chemical cross-linking is essential for film formation. These aspects are considered further in Chapter 3.

The general features of stepwise, polymer-forming reactions are compared and contrasted with chain reactions in Table 1.2.

Chain or addition polymerization operates when monomers of the general type

$$
CH_2 = \underset{\underset{Y}{|}}{\overset{\overset{X}{|}}{C}}
$$

$$
(X = H, CH_3, \text{halogen})
$$
$$
(Y = H, -COOR, -OCOR, \text{halogen},
$$
$$
C_6H_5-, -C\equiv N, -COCH_3, -OR)
$$

are converted to polymers. These monomers are susceptible to chain polymerization by a number of mechanisms, but the one with by far the greatest significance to the coatings industry involves free radical intermediates and proceeds as shown in Scheme 1.

Scheme 1

Mechanisms for Free Radical Polymerization of Vinyl and/or Acrylic Monomers

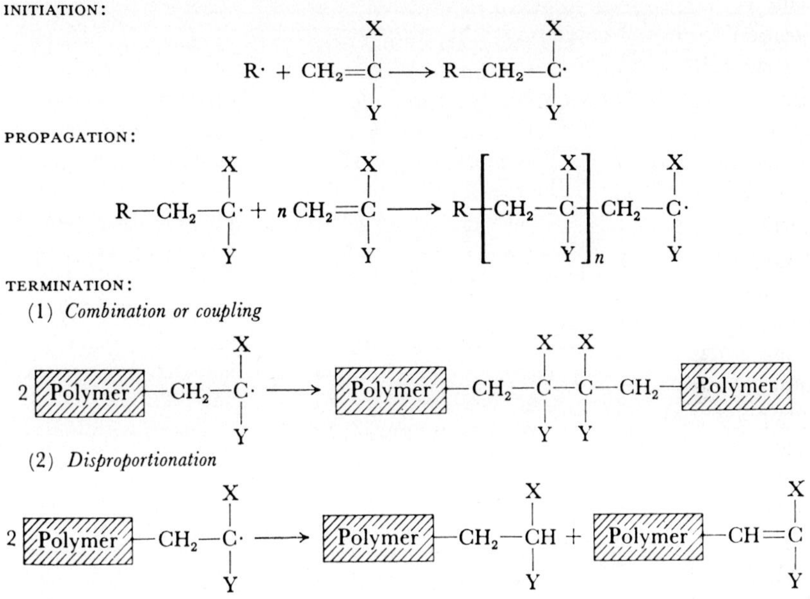

INITIATION:

$$R\cdot + CH_2{=}\underset{\underset{Y}{|}}{\overset{\overset{X}{|}}{C}} \longrightarrow R{-}CH_2{-}\underset{\underset{Y}{|}}{\overset{\overset{X}{|}}{C}}\cdot$$

PROPAGATION:

$$R{-}CH_2{-}\underset{\underset{Y}{|}}{\overset{\overset{X}{|}}{C}}\cdot + n\,CH_2{=}\underset{\underset{Y}{|}}{\overset{\overset{X}{|}}{C}} \longrightarrow R{\left[CH_2{-}\underset{\underset{Y}{|}}{\overset{\overset{X}{|}}{C}}\right]_n}CH_2{-}\underset{\underset{Y}{|}}{\overset{\overset{X}{|}}{C}}\cdot$$

TERMINATION:

(1) *Combination or coupling*

$$2\,\boxed{Polymer}{-}CH_2{-}\underset{\underset{Y}{|}}{\overset{\overset{X}{|}}{C}}\cdot \longrightarrow \boxed{Polymer}{-}CH_2{-}\underset{\underset{Y}{|}}{\overset{\overset{X}{|}}{C}}{-}\underset{\underset{Y}{|}}{\overset{\overset{X}{|}}{C}}{-}CH_2{-}\boxed{Polymer}$$

(2) *Disproportionation*

$$2\,\boxed{Polymer}{-}CH_2{-}\underset{\underset{Y}{|}}{\overset{\overset{X}{|}}{C}}\cdot \longrightarrow \boxed{Polymer}{-}CH_2{-}\underset{\underset{Y}{|}}{\overset{X}{C}}H + \boxed{Polymer}{-}CH{=}\underset{\underset{Y}{|}}{\overset{\overset{X}{|}}{C}}$$

Once a monomer unit is attacked by the initial free radical (initiation), the total time taken to complete the formation of the polymer molecule is of the order of a fraction of one second. Consequently, these polymerizations are characterized by the formation of polymer very early in the reaction and, except in special circumstances, this polymer is not further altered during the period of subsequent reaction. Monomer is present in these systems right up until the completion of the polymerization, and the main effect of time is to increase the yield of polymer. Also, the only entities capable of undergoing chain growth are the small number of radicals present, and not all molecular species as with a stepwise reaction. Consequently, the methods used to control the molecular weight of the polymers obtained by chain processes are quite distinct. The general characteristics of a chain polymer are summarized in Table 1.2.

The molecular weight of the polymer formed in a free radical chain polymerization is determined by the formulation and processing conditions

TABLE I.2

EFFECT OF PROCESSING VARIABLES ON A STEPWISE AND A CHAIN POLYMERIZATION

Processing Variables	Stepwise Polymerization	Chain Polymerization
Time	The monomeric units are nearly all incorporated into larger molecules early in the reaction. The molecular weight increases with time. The yield of polymeric species is not a function of time in the latter stages of reaction.	High molecular weight polymer is formed immediately the reaction starts. The monomer concentration decreases steadily throughout the polymerization. The yield of polymer will increase with time.
Temperature	The rate of polymerization increases with temperature. However, the molecular weight of the final polymer is not appreciably affected.	The rate of reaction increases, but the molecular weight decreases, with increasing temperature.
Concentration of polymer-forming entities	Diluents usually slow down the rate of polymerization; the final molecular weight will not be appreciably affected.	The molecular weight decreases with increasing concentration of diluents.

chosen, and it is not usually necessary to follow the course of the reaction by analytical procedures. The major factors that have a bearing on the molecular weight of a polymer formed in solution are the following:

1. Initiator concentration.
2. Temperature.
3. Monomer concentration.
4. Presence of chain-transfer agents.

The higher the initiator concentration, the lower the molecular weight of the polymer formed, since more chains will be growing at a given time and these will compete for the available monomer units. This is illustrated below:

$$R\cdot...X...X...X...X \qquad\qquad R\cdot...X...X...X\ ..X$$
$$X...X...X...X \qquad\qquad R\cdot...X...X...X...X$$

(X = monomer unit)

In practice, the relationship between initiator concentration and the molecular weight is slightly more involved than shown above, and it is necessary to allow for other factors which include the efficiency of the radical in initiating a polymer chain and, particularly at higher concentrations, the likelihood of two initiator radicals interacting without polymer initiation (cage effect). Consequently, a relationship of the type shown in Figure 1.3 is often observed and the molecular weight is inversely proportional to the square root of the initiator concentration. Over the range of A to B (Figure

POLYMER PREPARATION

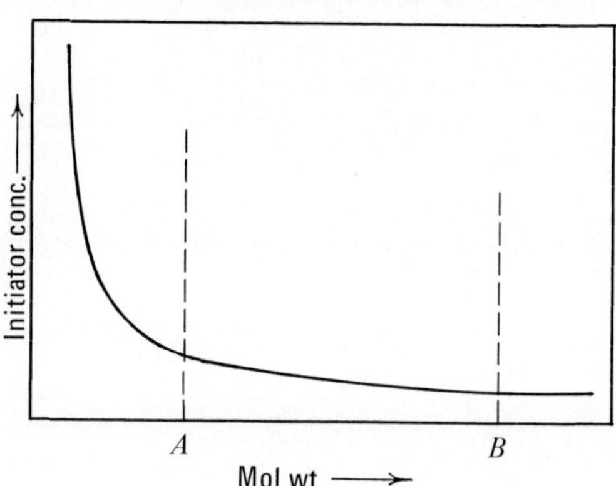

Figure 1.3. Relationship between initiator concentration and molecular weight.

1.3), the initiator concentration is a satisfactory method of controlling molecular weight, but for lower molecular weights the amount of initiator is excessive so that alternative, more economical, and controllable methods are used. As a general guide, initiator concentrations up to 2% (on the weight of monomer) are used in solution polymerizations to give polymers of molecular weight down to approximately 30,000.

Both the propagation and termination steps in the free-radical polymerization proceed at faster rates at higher temperatures, but, as pointed out by Flory (Gen. Ref. 1), the termination reaction increases more markedly than propagation. Consequently, termination is favored at higher temperatures. In other words, within certain limits, higher temperatures result in lower molecular weight polymers.

The monomer concentration also has a bearing on the molecular weight of the polymer formed since, in the presence of a diluent or solvent, the chances of a growing polymer radical finding a monomer unit will decrease with a decrease in monomer concentration. Therefore, termination is more likely to occur at lower monomer concentrations, and these give lower molecular weight polymers. The surface-coating polymers prepared by solution polymerization are normally used as such. Consequently, the lower limit of polymer solids is 30 to 40% (approximately), whereas the upper limit will be in the vicinity of 60% because of the need to maintain a reasonably low viscosity during manufacture.

Chain-transfer agents are used to lower further the molecular weight of the polymer being formed. It is usual to arrive at a balance between the concentrations of the chain-transfer agent, monomer, and initiator, and the temperature of the polymerization. This balance is chosen to give maximum control over the reaction, and to keep the formulation cost to a minimum. A chain-transfer agent functions by limiting the growth of one chain and starting another as shown below. Therefore, the rate of polymerization is maintained since, for every polymer radical lost, a new chain is started (Scheme 2).

The common transfer agents are compounds that readily lose a hydrogen atom, for example, the mercaptans. Some solvents also have a limited effect

Scheme 2

Mechanism of Chain Transfer

$$\text{Polymer}-CH_2-\underset{\underset{Y}{|}}{\overset{\overset{X}{|}}{C}}\cdot + RSH \longrightarrow \text{Polymer}-CH_2-\underset{\underset{Y}{|}}{\overset{\overset{X}{|}}{C}}H + RS\cdot$$

$$RS\cdot + CH_2{=}\underset{\underset{Y}{|}}{\overset{\overset{X}{|}}{C}} \longrightarrow RS-CH_2-\underset{\underset{Y}{|}}{\overset{\overset{X}{|}}{C}}\cdot$$

as transfer agents, and typical chain-transfer constants are given in Table 1.3.[3]

TABLE 1.3

CHAIN-TRANSFER CONSTANTS OF MERCAPTAINS AND SOLVENTS IN THE POLYMERIZATION OF METHYL METHACRYLATE[3]

Transfer Agent	Transfer Constant at 60°
Thio-β-naphthol	2.5
Thiophenol	2.2
n-Butyl mercaptan	0.52
Ethyl thioglycolate	0.50
t-Butyl mercaptan	0.14
Toluene	2.6×10^{-5}
	$\times 10^{-5}$ (at 80° C)
Carbon tetrachloride	24
Isopropyl benzene	19
Methyl ethyl ketone	7
Methyl isobutyl ketone	7
Toluene	5.3
n-Butyl alcohol	2.5
Acetone	2.25
Benzene	0.75

The chain transfer constant (C) is defined in terms of the concentration of monomer (M), the concentration of transfer agent (S) and the molecular weight of the polymer formed in the presence (P) and absence (P_o) of the transfer agent, i.e.

$$\frac{1}{P} = \frac{1}{P_o} + C\frac{S}{M}$$

The generation of the free radicals required to initiate polymerization may be accomplished in a number of ways, but the most important to the coatings industry include the following:

1. Action of heat.
2. Irradiation.
3. Thermal decomposition of thermolabile organic molecules.
4. Redox systems.

When vinyl and/or acrylic monomers are heated, free radicals are generated, although in many cases the actual structure of the initial product is not known with certainty. In the particular case of styrene, it has recently been proposed that the following takes place.[4]

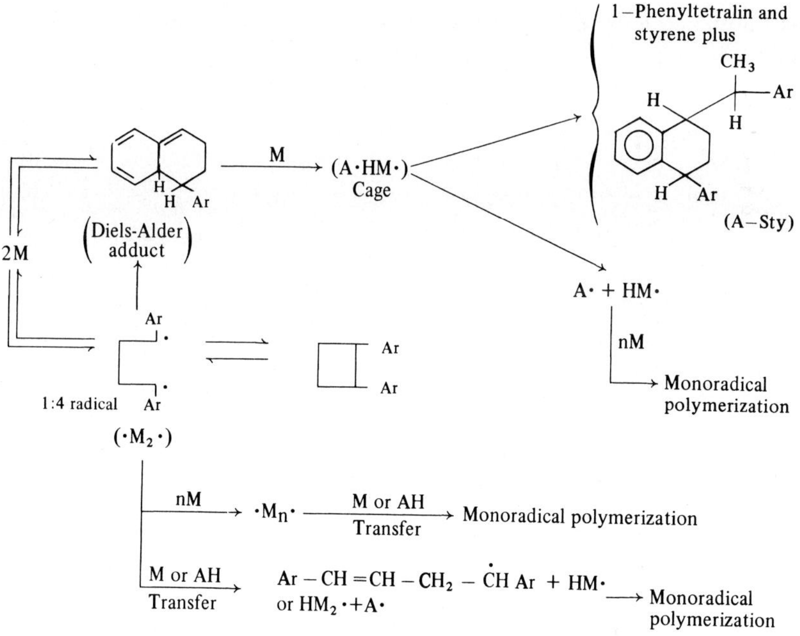

1—Phenyltetralin and styrene plus

(A—Sty)

$M \longrightarrow (A \cdot HM \cdot)$ Cage

$A \cdot + HM \cdot$

nM

Monoradical polymerization

(Diels-Alder adduct)

2M

1:4 radical

$(\cdot M_2 \cdot)$

$$\xrightarrow[\text{Transfer}]{\text{nM}} \cdot M_n \cdot \xrightarrow[\text{Transfer}]{\text{M or AH}} \text{Monoradical polymerization}$$

$$\xrightarrow[\text{Transfer}]{\text{M or AH}} Ar-CH=CH-CH_2-\overset{\cdot}{C}H\,Ar + HM\cdot$$
or $HM_2\cdot + A\cdot \longrightarrow$ Monoradical polymerization

The term molecule induced homolysis has been used by Pryor to describe reactions such as the thermal polymerization of styrene. The interaction of two styrene molecules results in the formation of radicals at a temperature far below that expected for a unimolecular homolytic process.

The existence and involvement of the Diels-Alder adduct in initiation has been demonstrated for styrene[4] by the isolation of 1-phenyltetralin. The relative importance of the various reaction paths shown for the thermal polymerization of styrene are discussed by Pryor[4].
The concentration of radicals present during thermal polymerization is very small. Therefore, high molecular weight polymers are formed and the concentration of monomer decreases slowly.

Irradiation of either the monomer or the monomer containing a photosensitizer is a method of generating free radicals[5] which could substantially increase in importance in the next few years (see page 145 on polyesters). For example, a mixture of methyl methacrylate, azodiisobutyronitrile, and benzophenone yields radicals when irradiated with light of 3000 Å, and polymerization results.

The most widely used method of forming free radicals is by the thermal decomposition of suitable organic compounds. For example, the breakdown of benzoyl peroxide, and of azodiisobutyronitrile (ADIB or AIBN), can be represented by the following equations.

$$\underset{\underset{CN}{|}}{\overset{\overset{CH_3}{|}}{CH_3-C-}}N=N\underset{\underset{CN}{|}}{\overset{\overset{CH_3}{|}}{-C-CH_3}}\ \xrightarrow{\Delta}\ 2\ \underset{\underset{CN}{|}}{\overset{\overset{CH_3}{|}}{CH_3-C\cdot}}\ +\ N_2\uparrow$$

$$C_6H_5-\overset{\overset{O}{\|}}{C}-O-O-\overset{\overset{O}{\|}}{C}-C_6H_5\ \xrightarrow{\Delta}\ C_6H_5\cdot\ +\ C_6H_5-COO\cdot\ +\ CO_2\uparrow$$

Each of these compounds is characterized by a critical temperature, which is defined as the temperature at which free radicals are produced by thermal dissociation,[6] and by a half-life period under specified conditions, as shown in Table 1.4.[6]

Solvents have an important bearing on the rate of homolytic fission of free radical initiators as shown by the figures in Table 1.5.[23] Similar effects would be expected in polymers or polymer solutions.

It is necessary to select the thermolabile organic compound that decomposes at an appropriate rate under the conditions of the polymerization. Also, the free radicals that are generated vary in their activity and the

TABLE I.4

**HALF-LIFE PERIODS AND CRITICAL TEMPERATURES OF THERMALLY
UNSTABLE ORGANIC COMPOUNDS[6]**

Compound	Critical Temperature, °C	Half-Life Time	
		Temperature, °C	Time, ⟨Hr.⟩
Tert-butyl hydroperoxide	110	100	165
		115	21.5
		130	3.2
Cumene hydroperoxide	100	115	472
		130	113
		145	29
Paramenthane hydroperoxide	100	130	12.5
		145	3.2
		160	0.9
Diisopropyl benzene hydro-peroxide	100		
Dicumyl peroxide	100	115	12.4
		130	1.84
		145	0.28
Ditertiary butyl peroxide	100	100	218
		115	34
		130	6.4
Lauroyl peroxide	65	50	54.2
		70	3.38
		85	0.5
Benzoyl peroxide	70	70	13.0
		85	2.15
		100	0.40
Parachloro benzoyl peroxide	75	50	310
		85	2.9
		100	0.5
Acetyl peroxide	75		
Methyl ethyl ketone peroxide	80	85	81.2
		100	16.2
		115	3.6
Cyclohexanone peroxide	90	85	20
		110	3.85
		115	1.01
		130	0.37

TABLE 1.5

THE EFFECT OF SOLVENT ON THE DECOMPOSITION OF
BENZOYL PEROXIDE AT 79.8°

| | Per cent. decomposition | |
Solvent	1 Hr.	4 Hr.
Tetrachloroethylene	13.0	35.0
Carbon tetrachloride	13.5	40.0
Cyclohexene	14.0	39.5
Methyl benzoate	14.5	41.4
Benzene	15.5	50.4
Toluene	17.4	49.5
Styrene	19.0	
Cumene	20.0	53.3
Methylene chloride	24.5	62.2
Ethyl chloride	26.0	64.7
Bromobenzene	26.3	69.0
t-Butylbenzene	28.5	69.5
Acetone	28.5	
Ethyl bromide	33.6	71.8
Allyl bromide	37.2	
Acetic anhydride	48.5	
Cyclohexane	51.0	84.3
Ethyl acetate	53.5	85.2
Acetic acid	59.3	87.4
Pyridine	77.3	
Dioxane	82.4	

Values selected from reference 23.

efficiency with which they will attack monomers to the exclusion of other
reactions. For example, it has been claimed that the radicals derived from
benzoyl peroxide are more likely to attack polymer molecules, and abstract
a hydrogen atom, than are those derived from azodiisobutyronitrile.[7]
Consequently, in some polymerizations, benzoyl peroxide may give rise to
chain branching at the newly created polymeric radical (see page 24 on
graft copolymers).

Redox-type initiating systems are of particular interest where polymerizations must be carried out below the critical temperature of the common
initiators; for example, when an unsaturated polyester-styrene composition
(see page 141) is applied to a heat-sensitive substrate it may be necessary to
initiate the polymerization at or near room temperature. This can be done by
the use of a compound which forms a Redox system with the peroxide and
enables free-radicals to form at a much lower temperature than by thermal
dissociation. N,N-dimethyl aniline is often used in conjunction with benzoyl
peroxide for this purpose. Also the polymer structure can be influenced by
the temperature, particularly where conjugated dienes, capable of either 1:2

or 1:4 addition, are involved. Under these circumstances, cobalt salts or aromatic tertiary amines are used to induce initiator decomposition. The detailed mechanisms are discussed further on page 142.

The arrangement of the individual monomer units in the polymer chain will be governed by the propagation steps in the polymerization. A growing free radical chain, or an initiating radical, can attack a monomer unit so as to give a radical of either structure **1** or **2**.

Addition to give structures of type **2** is preferred; this requires less energy and gives a more stable radical, since the odd electron can be distributed over the substituent group (Gen. Ref. 1). However the difference in energy and steric requirements between the head-tail and head-head addition depends on the monomer. With poly(styrene), poly(vinyl chloride) and poly(methyl vinyl ketone) the polymers are predominantly head-tail whereas substantial amounts of head-head arrangement are found in poly(vinylidene chloride) (10%), poly(vinylidene fluoride) (10%) and poly(vinyl fluoride) (up to 30%). (Gen. Ref. 7).

$$R\left[CH_2-\underset{Y}{\overset{X}{C}}\right]_n CH_2-\underset{Y}{\overset{X}{C}}\cdot \xrightarrow{CH_2=\underset{Y}{\overset{X}{C}}} R\left[CH_2-\underset{Y}{\overset{X}{C}}\right]_n CH_2-\underset{Y}{\overset{X}{C}}-\underset{Y}{\overset{X}{C}}-CH_2\cdot \qquad (1)$$

$$R\left[CH_2-\underset{Y}{\overset{X}{C}}\right]_n CH_2-\underset{Y}{\overset{X}{C}}-CH_2-\underset{Y}{\overset{X}{C}}\cdot \qquad (2)$$

In the particular case of styrene ($X = H$, $Y = C_6H_5$), the contributing structures of the radical are

Therefore, each monomer unit will tend to add to the chain in a regular and defined manner, and will give rise to the predominantly head-tail structure of most addition polymers.

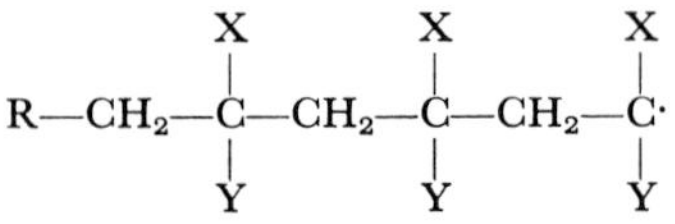

$$R-CH_2-\overset{\overset{\displaystyle X}{|}}{\underset{\underset{\displaystyle Y}{|}}{C}}-CH_2-\overset{\overset{\displaystyle X}{|}}{\underset{\underset{\displaystyle Y}{|}}{C}}-CH_2-\overset{\overset{\displaystyle X}{|}}{\underset{\underset{\displaystyle Y}{|}}{C}}\cdot$$

(head–tail) (head–tail) (head–tail)

A more detailed examination of this polymer structure shows that every second carbon atom is asymmetric (except those in the center of the molecule), and a predominantly D or L arrangement of the substituent groups is possible.

To distinguish the various possible polymer structures, Natta has introduced the term tacticity. Isotactic refers to polymers with the same configuration around the asymmetric carbon atom, syndiotactic to an alternating polymer configuration, and atactic denotes a random arrangement of D and L structures.

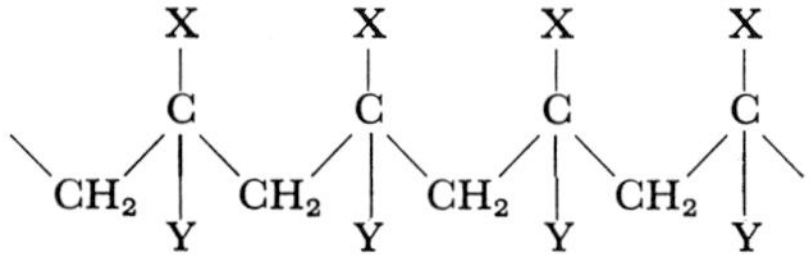

(Isotactic)

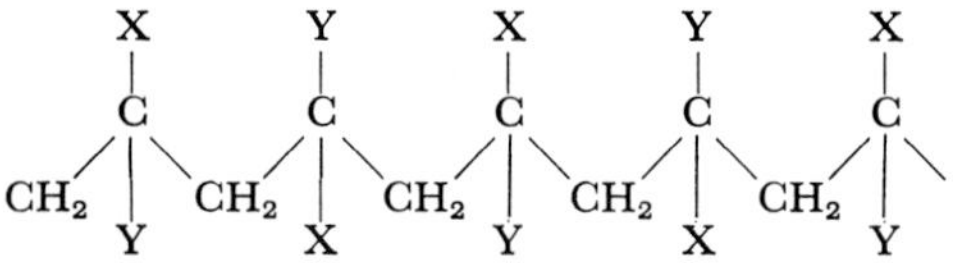

(Syndiotactic)

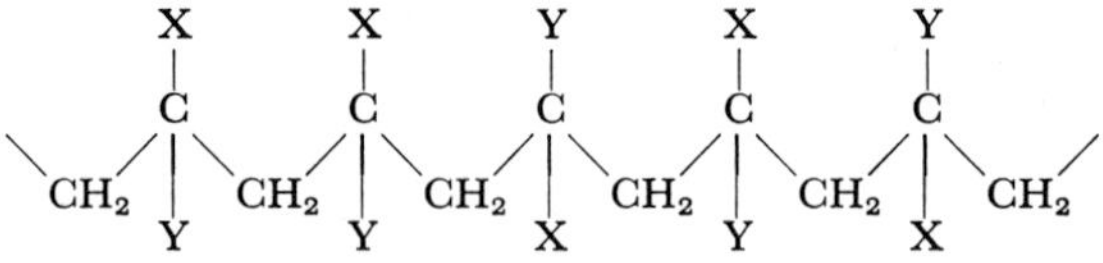

(Atactic)

The formation of stereospecific polymers (iso- and syndiotactic) is usually carried out in the presence of a heterogeneous catalyst system but, recently, examples of control of the propagation step have been noted in free radical polymerizations. These are often carried out at low temperatures and in the presence of solvents that can "complex" with the monomer; such conditions influence the addition of the monomer unit to the growing chain. The properties of stereoregular polymers differ markedly from those of an

TABLE 1.6

COMPARISON OF PROPERTIES OF POLYMERS OF DIFFERENT TACTICITIES[8]

(a) Poly(methyl methacrylate)

Tacticity	Glass Temperature, °C	Melting Temperature, °C
Isotactic	115	200
Syndiotactic	45	160
Atactic	104	—

(b) Poly(vinyl isobutyl ether)

Tacticity	Properties
Stereospecific (iso- or syndiotactic)	Crystalline, sharp X-ray diffraction Insoluble in methyl ethyl ketone and butanol
Atactic	Rubberlike, diffuse X-ray diffraction Soluble in methyl ethyl ketone and butanol

atactic polymer, and this is of significance to the design and preparation of polymers for use in surface coating. The differences are not such as would be expected by analogy with simple organic molecules. Whereas the absence of certain elements of symmetry in a small molecule produces optical activity, it has little effect on the capacity of a molecule to crystallize. In macromolecules, however, lack of symmetry has a profound effect on the ability to crystallize, on mechanical properties, softening point, and solubility, but it does not generally show up as optical rotatory power. The properties of various poly(methyl methacrylates) and poly(vinyl isobutyl ethers) are shown in Table 1.6.[8]

The reduced solubility of the more tactic polymers limits their use in solution-coating compositions where high solids at a given viscosity are normally required. Therefore it is usual to carry out the polymerization under conditions which minimize stereoregular growth.[9]

The termination of polymer growth can be either by combination (coupling, dimerization) or by disproportionation, and the relative importance of these possibilities is a function of the polymerization conditions, and the substituents X and Y (page 7) on the monomer. For example, at 60°C poly

TABLE 1.7

COMPARISON OF POLYMERIZATION METHODS

Polymerization Method	Characteristics of Technique
Bulk polymerization	Heat dissipation is difficult. The polymer is isolated as a solid that must be dissolved or dispersed before use. Not in general use for preparing surface-coating type polymers.
Solution polymerization	The polymerization temperature can be controlled by the refluxing solvent. The polymer is obtained as a solution that may be used as such, and the method is widely used to prepare solution vinyl and acrylic polymers.
Suspension polymerization	Heat of polymerization is easily controlled. The polymer is usually isolated by coagulation and is contaminated with stabilizers, etc. Sometimes used for production of vinyl polymers subsequently used as organosols.
Emulsion polymerization (a) Aqueous continuous phase	Rapid polymerization to polymers with a higher molecular weight than is possible in solution polymerization at a comparable rate. The molecular weight distribution is usually narrower than that obtained in solution polymerization. Copolymers are possible from monomers that do not copolymerize easily in solution. The emulsion is usable directly in some paint systems.
(b) Nonaqueous continuous phase	This method has the advantages listed for aqueous emulsion polymerization. In addition, the stabilizers and colloids are not water sensitive and the polymer is, therefore, obtained in a more hydrophobic form. The technique can be easily applied to the preparation of solutions of polymers by the addition of solvents for the polymer after the polymerization is complete. This method has advantages over solution polymerization—better viscosity control and heat dissipation, narrower molecular weight distribution, and control of the copolymerization tendency of monomers by the use of suitable monomeric or polymeric surfactants.

(styrene) terminates by dimerization whereas methyl methacrylate chains tend to disproportionate (Gen. Ref. 2). The presence of the double bond at the chain end of disproportionated molecules is a potential breakdown center which can have a disproportionately large effect on polymer stability. This arises because some polymers break down by chain reactions which are the reverse of the propagation step in polymer synthesis. Consequently a weak point at a chain end can lead to the formation of a reactive polymer chain which "unzips" back to monomer. Similar effects can arise from initiator residues. It has been shown by Grassie[10] that poly(methyl methacrylate) chains with unsaturated end groups break down more readily than polymer with saturated end groups. Furthermore, the double bond will induce activity into the adjacent methyl group and enhance reactions at this point. Such activity in the polymer molecule is generally undesirable; it can lead to premature weathering of a paint film, or breakdown under other conditions of use. By careful choice of polymerization conditions, termination by disproportionation can be minimized. One of the simplest means is to use a chain-transfer agent or solvent of activity such that each polymer molecule is terminated by chain transfer.[10]

$$\boxed{Polymer}-CH_2-\underset{Y}{\overset{X}{C}}\cdot \;+\; \underset{}{\overset{CH_3}{C_6H_5}} \longrightarrow \boxed{Polymer}-CH_2-\underset{Y}{\overset{X}{C}}H \;+\; \underset{}{\overset{CH_2\cdot}{C_6H_5}}$$

There are a number of methods or conditions by which monomeric compounds may be converted into polymers; these are compared in Table 1.7. All of these processes occur by free radical mechanisms but they differ in significant respects (Gen. Refs. 1 to 6).

The polymerization of vinyl or acrylic monomers can proceed by mechanisms other than those involving free radicals. Cationic polymerization is occasionally used to prepare hydrocarbon-type polymers and copolymers for use in varnishes, and the general mechanism, which is formally similar to the free radical process already discussed, is shown in Scheme 3.

Scheme 3

Mechanism for the Cationic Polymerization of Vinyl and/or Acrylic Monomers

---

INITIATION:

$$R^+ \;+\; CH_2{=}\underset{Y}{\overset{X}{C}} \longrightarrow R-CH_2-\underset{Y}{\overset{X}{C}}{}^+$$

PROPAGATION:

$$R-CH_2-\underset{Y}{\overset{X}{C}}{}^+ \;+\; n\,CH_2{=}\underset{Y}{\overset{X}{C}} \longrightarrow R{-}\!\left[CH_2-\underset{Y}{\overset{X}{C}}\right]_n\!\!CH_2-\underset{Y}{\overset{X}{C}}{}^+$$

TERMINATION:

$$R{-}\!\left[CH_2-\underset{Y}{\overset{X}{C}}\right]_n\!\!CH_2-\underset{Y}{\overset{X}{C}}{}^+ \longrightarrow H^+ \;+\; R{-}\!\left[CH_2-\underset{Y}{\overset{X}{C}}\right]_n\!\!CH{=}\underset{Y}{\overset{X}{C}}$$

Other mechanisms are discussed in detail in the General References 1 to 6.

TABLE 1.8

INCOMPATIBLE POLYMER MIXTURES[11]

PMA/PEA	PBA/PiBMA
PMA/PMMA	PsBA/PtBA
PMA/PBA	PiBMA/PMMA
PEA/PMMA	PMMA/PBzMA
PEA/PEMA	PMMA/PiBoMA
PEA/PBMA	PBzMA/PiBoMA
PEA/PiPA	PMA/PtBVE
PEA/PBA	PtBVE/PiBVE
PEA/PiBoMA	PMA/PS
PEMA/PMMA	PEA/PS
PiPA/PBA	PBA/PS
PiPA/PtBA	PBMA/PS
PBA/PMMA	PMMA/PS
PBA/PsBA	PiBoMA/PS
PBA/PtBA	PBzMA/PS

Where:

PMA	=	poly(methyl acrylate)
PMMA	=	poly(methyl methacrylate)
PBA	=	poly(butyl acrylate)
PEA	=	poly(ethyl acrylate)
PBMA	=	poly(butyl methacrylate)
PiPA	=	poly(isopropyl acrylate)
PiBoMA	=	poly(isobornyl methacrylate)
PEMA	=	poly(ethyl methacrylate)
PtBA	=	poly(*tert*-butyl acrylate)
PsBA	=	poly(*sec*-butyl acrylate)
PiBMA	=	poly(isobutyl methacrylate)
PBzMA	=	poly(benzy-methacrylate)
PtBVE	=	Poly(*tert*-butyl vinyl ether)
PiBVE	=	poly(isobutyl vinyl ether)
PS	=	polystyrene

The compatibility of the polymer mixtures was measured in the presence of chloroform.

COPOLYMERS

A given homopolymer often requires modification in order to give the
necessary balance of properties for the particular coating requirement. The
addition of simple low molecular weight compounds for this purpose will
be discussed later but, in a number of circumstances, it is necessary and
desirable to change chemically the original homopolymer. This alters the
properties and/or controls subsequent chemical cross-linking reactions.

The simplest approach to this problem of achieving the required balance
of properties would, at first sight, appear to be the blending of suitable
homopolymers to give a mixture of intermediate hardness and flexibility.
Unfortunately, this approach is limited since many homopolymers are incompatible with each other, as shown by the results in Table 1.8.[11]
Significantly, even polymers as similar as poly(ethyl methacrylate) and
poly(methyl methacrylate) are not compatible.

The theoretical requirements for polymer compatibility are that sufficient
polymer-polymer interaction occurs, so that ΔH, in Eq. 1.1, is negative

$$\Delta G = \Delta H - T \Delta S \qquad (1.1)$$

where ΔH = change in internal heat of the system that results from inter-
molecular interactions during mixing
T = absolute temperature
ΔS = entropy change during mixing
ΔG = free energy change during mixing

Polymer-polymer interaction is favored by the presence of polar groups in
the polymer molecules. For example, amide groups could be introduced into
one polymer and acid into the other to give, in some cases at least, compatibility. This approach is used in a number of coating applications, but it
suffers from some drawbacks, notably that the polar groups are derived from
monomers which are comparatively expensive, and it is necessary to carry
stocks of two polymer solutions.

An alternative and widely used method of modifying polymer properties
is to prepare copolymers that contain monomer units chosen to give the
desired physical properties. There are three main types of copolymer
possible from monomers A and B, and these are:

1. Random copolymer

A—A—B—A—B—A—A—A—A—B—A—A—B—B—A

2. Block copolymer

3. Graft copolymer

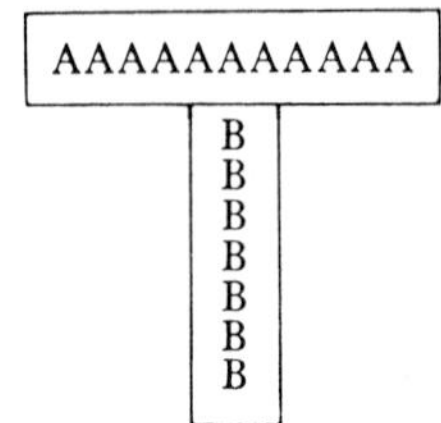

Random Copolymers.—When a mixture of two monomers is polymerized quantitatively to polymer the overall composition of the copolymer is necessarily that of the monomer mixture used initially. However the individual copolymer molecules are rarely of the same composition as that of the overall average mixture. Staudinger and Schneiders[12] have shown that a vinyl chloride-vinyl acetate copolymer made from equimolar quantities of the two monomers gave, on fractionation, a range of compositions varying from 75:25 down to 42:58 (vinyl chloride:vinyl acetate). Similarly, it has been shown that acrylic esters enter copolymer chains much faster than does vinyl chloride. Some unsaturated compounds that homopolymerize only with difficulty have been found to enter copolymer reactions readily; the classical example is maleic anhydride which slowly forms the homopolymer,[13] and yet readily gives a 1:1 copolymer with styrene. A large amount of experimental work has been carried out to develop a satisfactory theory to explain some of the phenomena of the type referred to above. This work has resulted in the development of the copolymer equation.[14] If we consider two monomer species M_1 and M_2 which give the radicals m_1 and m_2, then, under steady state conditions, there are four possible ways in which a monomer can add to the chain:

$$
\begin{array}{ll}
& \textit{Rate} \\
\sim\sim\sim m_1 + M_1 \longrightarrow \sim\sim\sim m_1 m_1 & k_{11} \\
\sim\sim\sim m_1 + M_2 \longrightarrow \sim\sim\sim m_1 m_2 & k_{12} \\
\sim\sim\sim m_2 + M_2 \longrightarrow \sim\sim\sim m_2 m_2 & k_{22} \\
\sim\sim\sim m_2 + M_1 \longrightarrow \sim\sim\sim m_2 m_1 & k_{21}
\end{array}
$$

The relative rate of reaction of radical m_1 with monomers M_1 and M_2 is called the reactivity ratio r_1 for monomer M_1 in this particular system. A similar ratio gives r_2 for the comonomer and its radical. Therefore

$$
r_1 = \frac{k_{11}}{k_{12}} \qquad r_2 = \frac{k_{22}}{k_{21}}
$$

The equation relating copolymer composition to monomer composition has been developed in detail elsewhere,[14] and is

$$
\frac{m_1}{m_2} = \frac{M_1}{M_2} \left[\frac{r_1 M_1 + M_2}{r_2 M_2 + M_1} \right]
$$

where m_1 and m_2 are the proportions of each unit in the copolymer and M_1 and M_2 are the monomer concentrations.

As an example of the use of this equation, let us consider the case mentioned above for an equimolecular mixture of vinyl chloride and vinyl acetate. The reactivity ratios determined for this system are $r_1 = 2.1$ and $r_2 = 0.3$ where vinyl chloride is M_1 and vinyl acetate M_2. Then, using the copolymer equation, we can calculate the initial copolymer composition as follows:

$$\frac{m_1}{m_2} = \frac{50}{50} \left[\frac{2.1 \times 50 + 50}{0.3 \times 50 + 50} \right]$$

$$= \frac{155}{65}$$

$$m_1 + m_2 = 100$$

whence $m_1 = 70\%$ and $m_2 = 30\%$. Therefore, the initial copolymer will be 70% vinyl chloride. It follows that vinyl chloride is being used up at a faster rate than the vinyl acetate, and so the composition of the remaining monomer mixture will change, and with it the composition of the next copolymer formed. Toward the end of the copolymerization (high conversion), the copolymer will be rich in vinyl acetate.

On the other hand, we can use the copolymer equation to calculate a monomer feed to give a desired copolymer composition. If we require a 50/50 copolymer of vinyl acetate/vinyl chloride, then, substituting in the equation we get

$$\frac{50}{50} = \frac{M_1}{M_2} \left[\frac{2.1M_1 + M_2}{0.3M_2 + M_1} \right]$$

Solving this equation, we find that a monomer composition of 27.4% vinyl chloride (M_1) and 72.6% vinyl acetate (M_2) will give an initial copolymer containing equal parts of the two components. Copolymers that are prepared by reacting the monomers without regard for the reactivity ratios are termed nondistributed, whereas control of the monomer composition throughout the reaction gives a distributed copolymer. While both copolymers will have the same average overall composition, the individual molecules in a distributed copolymer will vary less than those in the nondistributed one.

In some coating-type copolymers, the property differences between a distributed and a nondistributed copolymer are not sufficiently great to warrant the more elaborate manufacturing conditions necessary for the distributed polymer. However, the control of chemical distribution in copolymers is likely to become of increasing importance, particularly in thermosetting systems where one of the comonomers is essential for subsequent chemical reaction.

Polymer molecules which, because of the poor distribution of functional groups throughout the system, do not contain sufficient functional groups for chemical cross-linking will not enter the final polymer net-work and could be subsequently exuded from the composition.

Block and Graft Copolymers. Block and graft copolymers usually exhibit quite different properties from a random copolymer of the same composition. However, they are not widely used in surface coatings, although the general principles of arranging monomer units as a block or graft sequence are applicable; this applies particularly in systems where two different polymer species are chemically united either before or during film formation. For example, in the formation of a polyurethane from a polymeric glycol and a polymeric diisocyanate, a block-type copolymer structure is built up where the size of the units will determine, to some extent, the properties of the film.

$$\text{HO}-[\text{Polymer}^1]-\text{OH} + \text{OCN}-[\text{Polymer}^2]-\text{NCO}$$

$$\downarrow$$

$$-[\text{Polymer}^1]-\text{OCONH}-[\text{Polymer}^2]-\text{NHCOO}-[\text{Polymer}^1]-\text{OCONH}-[\text{Polymer}^2]-$$

Graft copolymers can also be formed by the generation of active free radical sites along a polymer molecule, either by hydrogen abstraction or by addition across any residual double bonds. Polymer chains subsequently grow from these active sites.

$$[\text{Polymer A}] + \text{R}\cdot \longrightarrow [\text{Polymer A}] + \text{RH} \longrightarrow [\text{Polymer A}]$$

This type of graft reaction is favored by using initiators that give active radicals, which are likely to attack polymer backbone chains. A possible use for this type of copolymer arises from indications that it gives more effective use of a plasticizing monomer than is obtained in a simple random copolymer. For example, a graft of styrene on to poly(ethyl acrylate) is more flexible, and softer, than a random copolymer of the same composition. Alternatively, at the same flexibility, the graft composition requires less ethyl acrylate and, hence, is cheaper. Other examples of block and graft copolymers will be dealt with under the individual chapters.

POLYMER CHARACTERIZATION

A polymer usually is a mixture of molecules of different sizes, and the molecular weight as determined is an average figure. The molecular weight as measured varies with the method used; some determinations rely on the number of molecules, whereas others are more dependent on the weight of the individual molecules. Consequently, the molecular weights that are obtained are referred to as number- and weight-average molecular weights. Typical methods that yield a number-average molecular weight are osmotic pressure, end-group analysis, ebulliometry, and cryoscopy; weight averages are given by light scattering and ultracentrifugation. The viscosity average usually lies between the other two averages, but approaches more closely the weight average. This difference in molecular weight averages can be illustrated by considering a polymer mixture containing an equal number of molecules of molecular weights 30,000 and 120,000.

Then
$$\overline{M}_n = 30,000 \times \tfrac{1}{2} + 120,000 \times \tfrac{1}{2}$$
$$= 75,000$$

$$\overline{M}_w = 30,000 \times \frac{30,000}{150,000} + 120,000 \times \frac{120,000}{150,000}$$

$$= 6000 + 96,000$$
$$= 102,000$$

The molecular weight of a polymer or copolymer may be determined by a number of methods (Gen. Refs. 1 to 6). The particular technique chosen will be related to the type of polymer and the use for which the polymer is designed. The methods most widely used to characterize polymers that are designed for use in surface coatings are end-group analysis and viscosity measurement in suitable solvents.

The determination of molecular weight by end-group analysis is particularly suitable for polymers prepared by stepwise or condensation mechanisms. In the formation of a polyester, for example, the number of hydroxyl and carboxyl groups decreases as the molecular weight increases. Therefore, a direct determination of the concentration of either of these functional groups is a measure of the number average molecular weight. It should be noted that these methods were originally developed and applied to linear polymers; caution is necessary in applying results to the more complex branched structures often used for surface coatings. The estimation of carboxyl end groups by titration with alcoholic sodium or potassium hydroxide solution appears to give an accurate measure of this end group, but it has been shown recently that the determination of hydroxyl groups by the standard chemical methods is open to doubt.[1] The use of physical methods, such as infrared spectroscopy, is another means of determining end groups. Often in describing a polymer, a molecular weight figure is not quoted as such, but the end-group concentration is given. In polyesters or alkyd resins it is conventional to quote an acid value or acid number (milligrams of KOH required to neutralize 1 g of resin) rather than a molecular weight figure. End-group analyses are usually limited to the lower molecular weight polymers, particularly those designed for use in thermosetting coatings. The accuracy of the method decreases with increasing molecular weight because of the lower concentration of end groups in the high molecular weight polymers.

The measurement of the molecular weight of a polymer by the viscosity of a dilute solution is a simple, rapid, and accurate method of characterizing a polymer. The viscosity of a dilute solution of the polymer is determined in a suitable viscometer (Figure 1.4) at a constant temperature. The intrinsic viscosity $[\eta]$ is determined by carrying out measurements at a number of

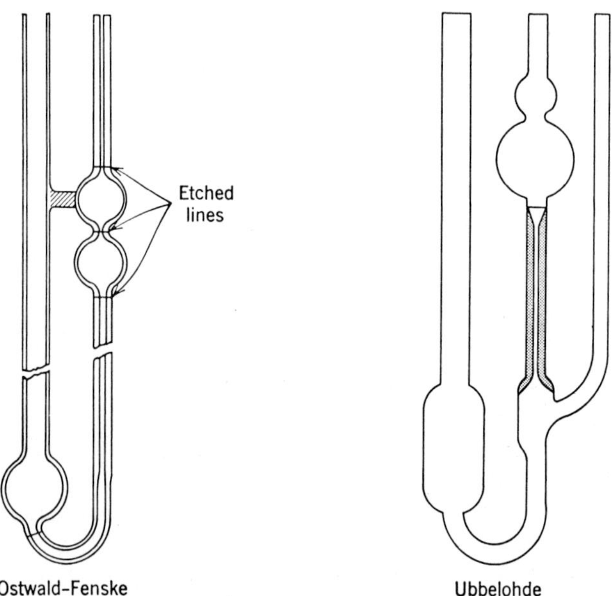

Figure 1.4. Capillary viscometers used to determine dilute polymer solution viscosity (Gen. ref. 2).

concentrations, and extrapolating to zero concentration by plotting concentration against $1/c \ln (t/t_0)$:

$$[\eta] = \lim_{c \to 0} \frac{1}{c} \ln \left(\frac{t}{t_0}\right)$$

$(t = \text{efflux time of polymer solution})$
$(t_0 = \text{efflux time of solvent})$

In practice, the determination of viscosity is often carried out at a fixed concentration (0.25 or 0.5 g per 100 ml) and the inherent viscosity used as an approximation for $[\eta]$.

Viscosity measurement does not yield a direct value for the molecular weight of a polymer, but the intrinsic viscosity has been related to molecular weight by the equation:

$$[\eta] = KM^a$$

The constants K and a are fortunately known for many polymer systems, and so the viscosity reading can be converted to a molecular weight value. The characterization and control of polymers by viscosity measurement at much higher concentrations (40 to 100%) is widely practiced in the coatings industry. Although these systems are difficult to analyze theoretically, the method is ideal for control purposes. Generally, the viscosity of the polymer solution is compared with a series of standard viscosity tubes (Gardner-Holdt) at a standard temperature, 25°C. These tubes cover a range of viscosities from 0.5 to 148 stokes, and this range is designated alphabetically from A through to Z_6. The polymer viscosity is then quoted as being equivalent to a particular Gardner-Holdt body (viscosity) tube. Another advantage of this method is that these tubes are available in most countries and thus, they enable ready comparisons to be made.

With polymers the weight-average molecular weight is always higher than the number-average, and the two values become identical only when all molecules are the same size. Therefore, the ratio $\overline{M}_w/\overline{M}_n$ is a measure of the distribution of the molecular weights of the individual molecules. This parameter is related to a number of polymer properties, including solution viscosity, sprayability, film strength, and drying time.

The measurement of molecular weight distributions can be a time consuming and tedious undertaking particularly where fractionation procedures are used. However, the invention of the technique of gel permeation chromatography (GPC) has overcome most of the problems of the older methods and usually this technique enables rapid measurement of the entire molecular weight distribution of a polymer mixture.

GPC makes use of a column packed with rigid porous beads, the pore size of which is of the same order as the polymer molecules to be separated. The polymer solution flows down the column of beads where the smaller molecules enter the internal structure of the beads. The larger polymer molecules spend less time in the pore structure and consequently they are the first to come through the column. Thus the molecules exit from the column in order of their size, the largest ones first (see Figure 1.5).

It should be noted that while GPC can give rapid information of the size distribution of the molecular species it does not give an absolute measure of the molecular weight of each fraction and it does not give information about changes in chemical constitution which would be expected with typical condensation polymers.

The effects of changes in the molecular weight distribution and the properties of polymers are further considered when discussing the individual polymer systems.

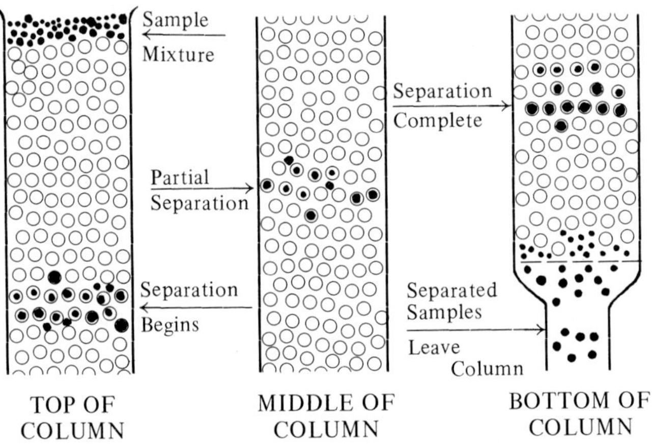

Figure 1.5.   The principle of separation of polymer molecules by Gel Permeation Chromatography.[24]

THE GLASS-TRANSITION TEMPERATURE RANGE

Some polymers are brittle or glassy, while others are tacky or rubbery at room temperature; for a given polymer, the change from a glass to a rubber-like state takes place at a temperature (or temperature range) which is called the glass temperature, T_g. Glass temperature corresponds to the abrupt change that shows in a plot of a physical property against temperature, and is a reflection of the temperature at, or above which, segmental chain motion in the polymer can occur.

The brittle point (B.P.) approximates to the glass temperature, but is more readily measured; one method is to flex strips of polymer and determine the temperature at which the polymer breaks.[15] Burrell[16] has shown that both the solution and film properties of a polymer are related to T_g. In particular, the flexibility, solution viscosity, solvent release, drying speed, hardness, adhesion, impact resistance, tensile strength, abrasion resistance, rate of cure, color, hiding, permeability, and electrical properties will depend on the relation between T_g and the service temperature of the coating composition.

The extent to which the T_g will affect each of these properties has been admirably elucidated by Burrell,[16] and the preferred T_g for a polymer system will depend on the end use of the coating. More recently, Tordella[17] has pointed out that it is more appropriate to consider a temperature range over which the change from a rubber to glasslike state occurs. The properties exhibited by the polymer depend on time factors as well as on temperature. For example, under different rates of loading, polyvinyl acetate films display hardness figures that may differ by several factors of ten.

The T_g of a polymer is related to its chemical composition, molecular weight, and the degree of cross-linking. With increasing molecular weight the T_g increases, but eventually becomes approximately constant over the range used for most practical formulations. The increase of T_g with cross-linking has been claimed in the acrylic series to be several degrees per mole percent of cross-links.[15] Large changes in T_g occur with changes in chemical composition as shown in Table 1.9.[15] The glass temperature of a copolymer is intermediate between those of the two homopolymers, and varies according to the equation

$$\frac{1}{T_g} = \frac{W_1}{T_{g_1}} + \frac{W_2}{T_{g_2}}$$

where W_1 and W_2 are the weight fractions of monomers 1 and 2 and T_{g_1} and T_{g_2} are the glass temperatures in °K for the homopolymers of monomers 1 and 2, respectively. Consequently, the composition of a copolymer necessary to satisfy a given T_g can be predicted.

TABLE 1.9

VARIATION OF T_g WITH POLYMER
COMPOSITION[15]

Polymer	T_g °C
Poly(methyl acrylate)	8
Poly(ethyl acrylate)	-22
Poly(n-propyl acrylate)	-55.5
Poly(n-butyl acrylate)	-54
Poly(methyl methacrylate)	105
Poly(ethyl methacrylate)	65
Poly(isobutyl methacrylate)	48
Poly(n-butyl methacrylate)	20
Poly(styrene)	100
Poly(vinyl chloride)	75
Poly(vinyl acetate)	30

The tendency of a polymer to crystallize is related to the regularity and symmetry of the polymer backbone and, by careful selection of the monomer units, polymers ranging from crystalline to amorphous are obtainable. For example, the polyesters derived from ethylene glycol and the phthalic acids differ in properties; the terephthalate has a crystalline melting point of 265°C, the isophthalate is amorphous and softens at 105°C, and the ortho-phthalate polymer softens well below room temperature.[18]

An alternative method of modifying the T_g is to add a low molecular weight compound as a plasticizer for the polymer. This method of plasticization is usually referred to as *external plasticization*, since the plasticizer is not chemically combined with the polymer, whereas a copolymer is stated to be *internally plasticized*. The terms are used in this sense throughout this book, although in other fields they sometimes have a different significance. The use of a low molecular weight external plasticizer is usually limited to thermoplastic polymers. The relative advantages—and disadvantages—of this method of lowering the T_g over chemical modification by copolymerization, are given in Table 1.10. In thermosetting polymer systems, the plasticizer

TABLE 1.10

COMPARISON OF INTERNAL AND EXTERNAL PLASTICIZATION OF A
POLYMER

Internal Plasticization	External Plasticization
The plasticizing component is chemically united into the film and is consequently not lost on exposure, etc.	The plasticizer may migrate from the film causing embrittlement on aging and/or poor adhesion by attacking the substrate. On the other hand, a temporary or fugitive plasticizer chosen so that it will volatilize after film formation can be an advantage—for example, in latex systems.
The plasticizing comonomer is sometimes more expensive than a simple external plasticizer.	Modification to a formulation is easily achieved by use of different amounts of the plasticizer, or by a combination of plasticizers.
Mechanical properties often suffer at high comonomer contents.	Solvent release is often favored, since the plasticizer keeps the film surface mobile and relieves stresses set up during drying.
	External plasticizer is often effective over a wider temperature range.
	Less external plasticizer (on a weight basis) is usually required, and the desirable properties of the original polymer are not diluted as much as with internal plasticization.

may be another polymer molecule that chemically combines with the other film formers during the curing reactions.

External plasticizers function by reducing polymer-polymer chain interactions, thereby increasing the segmental motion within the chains. The action of a plasticizer is determined by the presence (in the polymer and in the plasticizer) of groups capable of interaction, by the shape of the molecule, and by geometrical factors.[19] Consequently, a plasticizer is not uniformly distributed since a polar plasticizer will interact at polar areas of the polymer molecule. This effectively reduces the overall polarity of the polymer chain and results in a weakening of interchain forces. For example, it has been shown[20] that an alcohol plasticizer is held by hydrogen bonding to poly(methyl methacrylate).

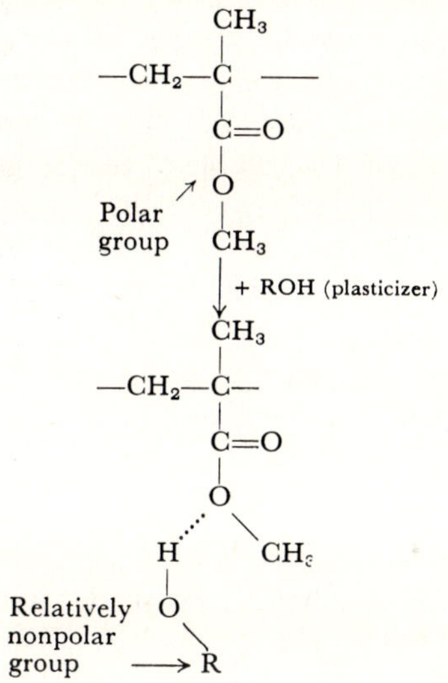

$$
\begin{array}{c}
CH_3 \\
| \\
-CH_2-C\ -\!\!-\!\!-\!\!- \\
| \\
C=O \\
| \\
O \\
| \\
CH_3
\end{array}
$$

Polar group

$+$ ROH (plasticizer)

$$
\begin{array}{c}
CH_3 \\
| \\
-CH_2-C- \\
| \\
C=O \\
| \\
O \\
\diagdown \\
H\cdots\quad CH_3 \\
| \\
O \\
\diagdown \\
R
\end{array}
$$

Relatively nonpolar group

The effectiveness of plasticizers of the same molecular size will be similar, provided the bonding tendency to the polymer molecules is also similar.

Solvents function in a similar manner to a plasticizer, but the molecules—being smaller—are more volatile, and are lost during film formation. A rough general guide to the choice of a solvent or solvent mixture for a particular polymer system can be gained from the old rule of "like dissolves like," especially if the need to have one solvent component capable of polymer bonding is taken into account.

There are two steps necessary to dissolve a solid polymer. First, the solvent molecule must diffuse between the polymer chains to give a swollen polymer gel; second, this gel must be dispersed to give the solution. Only the second of these steps is speeded up by agitation. It should also be noted that a kinetically good solvent, that is, one which diffuses rapidly to give a gel, may not be a thermodynamically good choice. This situation can partially account for the wide use of solvent mixtures, rather than individual solvents, for polymer solutions; (evaporation rates are also of considerable importance). The properties of a polymer film are also related to the solvent composition from which the film was deposited. In thermodynamically good solvents the polymer molecules are extended; polymer-polymer interaction between long segments of different molecules makes the film strong. On the other hand, in a poor solvent, the polymer molecules tend to exist as a coiled-up ball-like structure, and the films formed are weak.

The Concept of Solubility Parameters. A more detailed and scientific approach to solvency is, however, necessary to explain some of the solution properties of polymers. It is known that polymer solutions frequently tolerate large amounts of nonsolvents and, in some cases, these improve the solution properties. Also, there are examples where blends of two nonsolvents become a good solvent for a polymer. The classical case quoted is nitrocellulose (cellulose nitrate), which is soluble in diethylether-ethanol mixtures, but not in either solvent alone. Conversely, cellulose acetate is soluble in either aniline or acetic acid, but not in a mixture of the two. Burrell[21] has developed the concept of the solubility parameter to explain the above facts, and to provide a basis for predicting suitable solvents or solvent mixtures. The process of mixing or dissolving a solvent and polymer is governed by the free energy equation (page 20).

$$\Delta G = \Delta H - T\Delta S$$

If the free energy has a negative value, the process of solubilization will proceed. The more a molecule or segments of a molecule can move, the easier it will be for a solvent molecule to occupy sites associated with the polymer chain; crystalline polymers, where movement is restricted, are generally less soluble than amorphous polymers. The separation of the polymer chains by the solvent results in an increase in the total entropy. The entropy change is invariably large, but the heat term ΔH determines the sign of the free energy change.[22] Now

$$\Delta H_m = V_m \left[\left(\frac{\Delta E_1}{V_1} \right)^{\frac{1}{2}} - \left(\frac{\Delta E_2}{V_2} \right)^{\frac{1}{2}} \right]^2 \phi_1 \phi_2$$

where ΔH_m = overall heat of mixing (cal)
V_m = total volume of the mixture (cc)
ΔE = energy of vaporization of component 1 or 2 (cal)
V = molar volume of component 1 or 2 (cc)
ϕ = volume fraction of component 1 or 2 in the mixture

The term $\Delta E/V$ is the cohesive energy density, and represents the concentration of forces which cause the molecules to cohere. Then, rewriting the above equation, we have

$$\frac{\Delta H_m}{V_m \phi_1 \phi_2} = \left[\left(\frac{\Delta E_1}{V_1}\right)^{\frac{1}{2}} - \left(\frac{\Delta E_2}{V_2}\right)^{\frac{1}{2}}\right]^2$$

It can be seen that the heat of mixing is proportional to the square of the differences between the square roots of the cohesive energy density terms. A term δ, the solubility parameter, is used to denote $\delta = (\Delta E/V)^{\frac{1}{2}}$. Solution will occur when the heat of mixing is small, or when the value of δ for the polymer and solvent are similar.

Since Burrell's original development of the concept of solubility parameter it has become apparent that other factors, particularly the tendency to form hydrogen bonds and the dipole moment, also have an effect on the solubility of a polymer in a solvent or solvent mixture. These factors have been discussed at length in the recent literature and the reader is recommended to reference 25 to 30 for further reading.

In practice, it is not only necessary to adjust the solubility parameter of the solvent to the polymer, but to select a solvent with suitable hydrogen bonding characteristics. It is equally necessary to use a solvent mixture that gives acceptable viscosity, evaporation rate, and film properties. Detailed calculations of solvent mixtures and the further application of the concept of solubility parameter are given in Reference 21.

GENERAL REFERENCES

1. Flory, P. J., *Principles of Polymer Chemistry*, Cornell University Press, New York, 1953.
2. Billmeyer, F. W., *Textbook of Polymer Chemistry*, Interscience Publishers, Inc., New York, 1957; *Textbook Polymer Science*, Interscience Publishers, Inc., New York, 1962.
3. Stille, J. K., *Introduction to Polymer Chemistry*, John Wiley and Sons, New York, 1962.
4. Plesch, P. H., *The Chemistry of Cationic Polymerization*, Pergamon Press, Oxford, 1963.
5. Burnett, G. M., *Mechanism of Polymer Reactions*, Interscience Publishers, Inc., New York, 1954.
6. Solomon, D. H., "Polymerization Mechanisms," *Proc. R. Australian Chem. Inst.*, **32**, 117 (1965).
7. Lenz, R. W., *Organic Chemistry of Synthetic High Polymers*, Interscience Publishers, New York (1967).
8. *The Kinetics and Mechanisms of Step-Growth Polymerizations*, Ed. D. H. Solomon, Marcel Dekker, Inc., New York, 1967.

REFERENCES

1. Solomon, D. H. and J. J. Hopwood, *J. Appl. Polymer Sci.*, **10**, 981 (1966).
2. Kienle, R. H., P. A. Van der Meulen, and F. E. Petke, *J. Am. Chem. Soc.*, **61**, 2258 (1939).

3. Riddle, E. H., *Monomeric Acrylic Esters*, Reinhold Publishing Corp., New York, 1954, p. 57.
4. Pryor, W. A., *Polymer Preprints*, **12**, No. 1, 49, (March 1971).
5. Hrubesh, A., H. Craubner, and W. Luck, Badische Anilin-and Soda-Fabrik Akt.-Ges, Ger. Pat., 1099,166 (1961). Cooper, W. and M. Fielden, Dunlop Rubber Co. Ltd., Ger. Pat., 1055,814 (1959).
6. *The Role of Peroxides in Curing Polyester Resins and their Influence on the Physical Properties of Reinforced Plastics*, Novadel, Ltd., London, 1960, p. 96
7. Merrett, F. M., *Trans. Faraday Soc.*, **50**, 759 (1954).
8. Fox, T. G., B. S. Garrett, W. E. Goode, S. Gratch, J. F. Kincaid, A. Spell, and J. D. Stroupe, *J. Am. Chem. Soc.*, **80**, 1768 (1958); Schildknecht, C. E., S. T. Gross, H. R. Davidson, J. M. Lambert, and A. O. Zoss, *Ind. Eng. Chem.*, **40**, 2104 (1948).
9. Gaylord, N. G., and H. F. Mark, *Linear and Stereoregular Addition Polymers*, Interscience Publishers, Inc., New York, 1959, Ch. IV.
10. Grassie, N., *The Chemistry of High Polymer Degradation Processes*, Butterworth & Co., London, 1956, p. 31.
11. Hughes, L. J., and G. E. Britt, *J. Appl. Polymer Sci.*, **5**, 337 (1961).
12. Staudinger, H., and J. Schneiders, *Ann.*, **541**, 151 (1939); via *Chem. Abstr.*, **34**, 2322 (1940).
13. Lang, J. L., W. A. Pavelich, and H. D. Clarey, Am. Chem. Soc. Div. Polymer Chem., Abstr. 140th (Chicago) Meeting, 1961, 3U.
14. Ham, G. E. in G. E. Ham, Ed., *Copolymerization*, Interscience Publishers, Inc., New York, 1964, Ch. 1.
15. Brown, W. H., and T. J. Miranda, *Offic. Dig. Federation Soc. Paint Technol.*, **36**, 92 (1964).
16. Burrell, H., *Offic. Dig. Federation Soc. Paint Technol.*, **34**, 131 (1962).
17. Tordella, J. P., *Offic. Dig. Federation Soc. Paint Technol.*, **37**, 349 (1965).
18. Wilfong, R. E., *J. Polymer Sci.*, **54**, 385 (1961).
19. Voskresenskii, V. A., and E. M. Orlova, *Russ. Chem. Rev.*, **33**, 151 (1964).
20. Zhurkov, S. N., and B. Ya. Levin, Symposium, "Stroenie Veshchestva i Spectroskopiya" (Structure of Matter and Spectroscopy), *Izd. Akad. Nauk SSSR*, Moscow, 1960, via Ref. 19.
21. Burrell, H., *Offic. Dig. Federation Soc. Paint Technol.*, **27**, 726 (1955).
22. Hildebrand, J., and R. Scott, *The Solubility of Non-electrolytes*, 3rd ed., Reinhold Publishing Corp., New York, 1950.
23. Nozaki, K., and P. D. Bartlett, *J. Am. Chem. Soc.*, **68**, 1686 (1946).
24. Billmeyer, F. W., *J. Paint Technol.*, **41**, 2 (1969).
25. Brown, I., *Nature*, **188**, 1021 (1960).
26. Hansen, C. M., *J. Paint Technol.*, **39**, 104 (1967).
27. Crowley, J. D., G. S. Teague, and J. W. Lowe, *J. Paint Technol.*, **38**, 296 (1966).
28. Teas, J. P., *J. Paint Technol.*, **40**, 19 (1968).
29. Lieberman, E. P., *Offic. Dig. Federation Soc. Paint Technol.*, **34**, 30 (1962).
30. Lieng-Huang Lee, *J. Paint Technol.*, **42**, 365 (1970).

CHAPTER 2

GLYCERIDE OILS

Glyceride oils of either vegetable or marine origin are the glyceryl esters of fatty acids (**1**).

$$CH_2\text{—}OCOR^1$$
$$CH\text{—}OCOR^2$$
$$CH_2\text{—}OCOR^3$$

(**1**)

$(R^1, R^2, \text{ and } R^3 = \text{fatty acid residue generally a } C_{17}H_{35-x})$

Oils, in the raw and variously modified form, are the vehicles for a large volume of surface coatings[1] although the advent of the synthetic polymer systems based on vinyl and acrylic monomers could reduce the use of oil-based coatings in the future.

Raw oil is the basis of some roof and exterior household paints, but its poor drying characteristics and the comparatively ready attack of moisture on the dried film limit further application. Heat-treated, isomerized, and modified oils—including varnishes and alkyds—are widely used as marine coatings, as air-drying enamels for household or industrial use, and as baking or stoving automotive enamels (often in conjunction with other vehicles). The chemistry of these various systems is, at least in part, that of the fatty acids present, so that while the oils as such are not of great importance as vehicles for coatings, the chemistry of a large number of paint systems is, to a greater or lesser degree, that of glyceride oils. For example, similar chemical reactions are involved in the air drying of both an alkyd resin and a glyceride oil.

OCCURRENCE, ISOLATION, AND STRUCTURE OF GLYCERIDE OILS

The oils present in various plants and marine animals are usually isolated by either a pressure or solvent extraction process. This yields what is termed raw oil, which contains small quantities of impurities, particularly carbohydrates and phosphatides (**2**); these "break" and precipitate from the oil solution when it is heated.

35

$$CH_2—O—CO—R$$
$$CH—O—CO—R'$$
$$CH_2—O—\overset{O^-}{\underset{O^-}{P^+}}—O—CH_2—CH_2—\overset{+}{N}(CH_3)_2$$

$$(2)$$

Raw oil must be refined prior to its use in varnishes, alkyds, modified oils and alkyds, etc., where the oil is processed at elevated temperatures. If this is not done, the precipitated sludge can adversely affect the clarity, gloss, and drying time of the resin. The impurities can also interfere with certain chemical reactions, particularly the formation of a monoglyceride in alkyd preparation (page 86). Mechanical refining of raw oils involves heating in the presence of a filter aid, such as fullers earth, and then removing the decomposed products by filtration. Widely practiced alternatives are the use of acid or alkaline treatments; oils from which the "break" material has been removed by these methods are termed acid-refined or alkali-refined oils.

Oils are usually described as drying, semidrying, or nondrying, and these divisions reflect the influence of oxidation on the film-forming properties of the oil. Table 2.1 gives the classification of some of the more common oils,

TABLE 2.1

CLASSIFICATION OF VEGETABLE OILS

Drying Oils	Semidrying Oils	Nondrying Oils
Linseed oil	Dehydrated castor oil	Coconut oil
Tung or China-wood oil	Tall oil	Olive oil
Oiticica oil	Soya bean oil	Castor oil
Perilla oil	Safflower oil	Hydrogenated castor oil

although the division is arbitrary and some overlap exists. It should be noted that it is possible for a processed vehicle prepared from a semidrying oil to give good air-drying properties. Examples of these vehicles are to be found in alkyd resins (Chapter 3), where a 50% soya bean oil resin will air dry because its initial molecular weight is higher than the straight oil, and it requires fewer cross-links to reach a dry state.

The tendency for an oil to air dry is directly related to its fatty acid components; typical compositions for some oils are listed in Table 2.2.

The composition of a given oil may vary from year to year with changes in climatic conditions or variations in soil structure and, hence, it is usual to give a range of fatty acid contents.

The iodine value gives an indication of the degree of unsaturation and this is useful in predicting the properties of an oil. The drying index as defined by Greaves[2] (drying index = % linoleic + 2 × % linolenic acid) represents an attempt to put on a quantitative basis the classification of oils into drying, semidrying and nondrying. It has been suggested that an index of over 70 is indicative of a drying oil.

Oils vary in both the type and amount of carbon-carbon unsaturation present (Table 2.2) and these structural features influence subsequent chemical reactions. The nonconjugated or methylene-interrupted system of double bonds, which is present in linseed, soya bean, and safflower oils, is characterized by a *cis* configuration of each double bond; the conjugated unsaturation of wood and dehydrated castor oils has some unsaturation with a *trans* arrangement of the double bond substituents.

Oils are not pure compounds but are mixtures of fatty acids present as glycerides, and the actual structure of the individual glyceride molecules or, in other words, the distribution of the fatty acids is of considerable importance—and interest—in subsequent reactions and uses of the oil. The simplest distribution would be one in which the oil was a mixture of simple triglycerides of the individual fatty acid—that is, in a linoleic/oleic acid mixture, all the linoleic acid would be present as glyceryl trilinoleate, and the oleic acid would be present as glyceryl trioleate. Theories of the above type have now been generally discarded and replaced by a number of different hypotheses based on random distribution of the fatty acids.[3] The evidence for these various proposals has been reviewed by Gunstone[3] who has proposed an elegant theory satisfying the major experimental evidence on glyceride structure. Gunstone suggests that the fatty acids are distributed so that the secondary or β-hydroxyl group of glycerol is esterified by the unsaturated acids (oleic, linoleic, linolenic), and that the two remaining α- or primary hydroxyls are then acylated in a statistical manner by the remaining acids. Figure 2.1 shows the theoretically predicted structures of a mono-unsaturated glyceride and of a di-unsaturated glyceride.

In discussing the chemistry of glyceride oils, it is convenient to consider the reactions of the individual fatty acids or their esters with monohydric alcohols and, in fact, most of the experimental evidence has been gained on such systems. Certain advantages and disadvantages result from the use of the simple methyl or ethyl esters. First, it is possible to work with relatively pure materials, since the fatty acid or its ester can be distilled or separated by relatively simple techniques. Second, the reduced functionality, compared with that of a triglyceride, reduces problems associated with viscosity changes, and the reaction products are often more easily separated and identified. As against these advantages, it is necessary to extrapolate results from simple esters to triglycerides, and the different viscosity characteristics

$$CH_2—O—CO—(CH_2)_{16}—CH_3$$
$$CH—O—CO—(CH_2)_7—CH=CH—CH=CH—(CH_2)_5—CH_3$$
$$CH_2—O—CO—(CH_2)_{16}—CH_3$$

(mono-unsaturated glyceride)

$$CH_2—O—CO—(CH_2)_7—CH=CH—CH=CH—(CH_2)_5—CH_3$$
$$CH—O—CO—(CH_2)_7—CH=CH—CH=CH—(CH_2)_5—CH_3$$
$$CH_2—O—CO—(CH_2)_{16}—CH_3$$

(di-unsaturated glyceride)

Figure 2.1.   Predicted structures of unsaturated glycerides.

often modify the chemistry considerably. For example, oxygen and volatile decomposition products may diffuse through a liquid simple ester much more readily than through a semisolid, partially cross-linked, triglyceride. Another significant point to bear in mind when applying results to triglycerides is the presence of mixed fatty acids and the distribution of these among the glyceride molecules.

REFERENCES

1. *Chem. Eng. News,* **42**, **(41)**, 100 (1964).
2. Greaves, J. H., *Oil Colour Trades J.,* **113**, 949 (April 23, 1948).
3. Gunstone, F. D., *Chem. Ind. (London),* **1962**, 1214.

AUTOXIDATION

The behavior of fatty acids or their esters toward oxidation is very dependent on a number of structural features, which are related to the unsaturated portion of the molecule. The most important of these features are the following:

1. The number of carbon-carbon double bonds.
2. The presence of a conjugated or nonconjugated (methylene interrupted) system of double bonds, that is, 9:11 or 9:12 dienes.
3. The geometrical arrangement of the substituents about the double bonds, that is, *cis* or *trans*.

Even the fully saturated fatty acid esters are susceptible to oxidation, and it has been suggested by Skellon and Wharry[1] that the methylene group α-, or even β-, to the ester link can form a hydroperoxide.

$$R^1\text{—}CH_2\text{—}COOR \xrightarrow{O_2} R^1\text{—}CH\text{—}COOR$$
$$\qquad\qquad\qquad\qquad\qquad\quad \overset{|}{O}OH$$

This reaction is comparatively slow (approximately 1/100th rate of mono-olefin[2]), and is of no great significance for film formation; however, it is a potential point of attack for film breakdown or degradation.

Studies on the autoxidation of unsaturated fatty acids have shown marked differences in the behavior of the conjugated and nonconjugated isomers, and distinct theories have been developed to explain the reaction of the two types of unsaturation. These theories are presented here in a summarized form; for further details the reader should consult the General References.

Nonconjugated Fatty Acids and Esters. The drying of oil films based on non-conjugated fatty acids is characterized by a number of distinct stages. First, there is an induction period during which the antioxidants that are present in the oil are destroyed. Then follows the uptake of oxygen by the film and, finally, the polymerization stage involving cross-linking, the liberation of volatile oxidation products, and the formation of a dried film.

In developing a theory to explain the autoxidation of nonconjugated oils, considerable effort has been directed to simple oleate esters, principally the methyl and ethyl esters. The corresponding triglyceride ester, olive oil, is nonfilm-forming because it is a nondrying oil but, nevertheless, the oleates represent the simplest olefins directly related to the semidrying and drying oils. Consequently, oleates have been used to develop theories that have then been applied to the more unsaturated oils.

The observation by Criegee and co-workers[3] that the autoxidation of cyclohexene gave the hydroperoxide (**3**), and not the cyclic peroxide (**4**) expected from attack at the double bond, was used by Farmer and his colleagues[4] as a basis for the development of a general theory to explain the autoxidation of olefins.

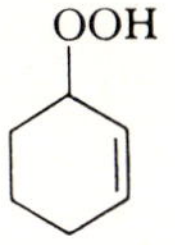

(3) 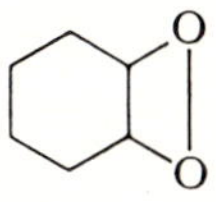   (4)

Farmer proposed that the methylene groups α- to the double bonds of oleate esters were "activated," and that removal of a hydrogen atom from

C_8 or C_{11} would give the resonance-stabilized allylic structures **5–8**. These radicals can then form C_8, C_9, C_{10}, and C_{11} hydroperoxides as shown.

$$CH_3-(CH_2)_7-\overset{cis}{CH}=CH-(CH_2)_7-COOR$$

The double bond shifts in structures **6** and **8**, and it changes from the original *cis* to the more stable *trans* configuration.[5] Recent evidence suggests that the hydroperoxides are largely derived from structures **6** and **8** with small amounts of other isomers.[6] Controversy still exists as to the initiating step in the reaction. Farmer and other investigators now seem to favor initial attack at the double bond to form sufficient peroxides to trigger subsequent attack at the α-methylene group. The presence of α-glycol structures strongly suggests at least some direct double bond attack.[7] The amount of double bond attack, compared to α-methylene attack, increases at higher temperatures.[1]

A more recent explanation for the formation of the initial hydroperoxides is that they are formed by the attack of singlet oxygen on the active methylene groups.[21] Non-conjugated hydroperoxides are formed. The singlet oxygen is said to form by a photosensitized reaction.

While it is clearly established that the major products in the early stages of reaction are hydroperoxides, other oxygenated compounds, including some dihydroperoxides, are also formed. The latter compounds are possible precursors of volatile lower aldehydes; the following mechanism has been suggested to account for the formation of dihydroperoxides.[8]

$$
\begin{array}{c}
\quad\quad\quad\quad CH_2\!-\!CH_2 \\
-CH\!=\!CH\!-\!CH \diagup \quad\quad \diagdown CH\!-\!CH_2- \\
\quad\quad\quad O \quad\quad\quad H \\
\quad\quad\quad\quad \diagdown O\!\cdot \\
\quad\quad\quad\quad \downarrow \\
\quad\quad\quad\quad CH_2\!-\!CH_2 \\
-CH\!=\!CH\!-\!CH \diagup \quad\quad \diagdown CH\!-\!CH_2- \\
\quad\quad\quad OOH \quad\quad\quad \cdot \\
\quad\quad\quad\quad \downarrow O_2 \\
-CH\!=\!CH\!-\!CH\!-\!CH_2\!-\!CH_2\!-\!CH\!-\!CH_2- \\
\quad\quad\quad\quad |\quad\quad\quad\quad\quad\quad | \\
\quad\quad\quad\quad OOH \quad\quad\quad OO\cdot \\
\quad\quad\quad\quad \downarrow RH \\
-CH\!=\!CH\!-\!CH\!-\!CH_2\!-\!CH_2\!-\!CH\!-\!CH_2- \\
\quad\quad\quad\quad |\quad\quad\quad\quad\quad\quad | \\
\quad\quad\quad\quad OOH \quad\quad\quad OOH
\end{array}
$$

The autoxidation of elaidic acid and its esters, which are the *trans* analogues of the oleates, is of considerable importance in relation to the theories on the mechanism of oxidation. The most significant point to note is that the *trans* elaidate oxidizes more slowly than the *cis* oleate (Gen. Ref. 2). Other work, however, suggests that this difference is only apparent when traces of oxidation catalysts are present.[9]

When the general theory of autoxidation developed for mono-olefins is applied to the linoleates and linolenates, a number of additional factors become apparent. Of prime importance to the surface-coating industry is the very much faster rate of oxidation and polymerization; it has been claimed that the ratios of the oxidation rates are triolein:trilinolein:trilinolenin = 1:120:330.[10] With the methyl esters of the acids, ratios of 1:12:25 and 1:2:4 have been quoted.[11] This greatly increased activity of the linoleates and linolenates over the oleates suggests that the methylene groups at C_{11} and C_{14} (for linolenates) are much more susceptible to oxidation; this can be related to the additional unsaturation located α- to these methylene groups. Arguments similar to those advanced for oleates have been used to explain the hydroperoxides formed by the autoxidation of linoleates and linolenates. The theory of Farmer et al.[4] and others,[12] suggests that with linoleates, hydrogen abstraction from C_{11} yields the allylic radical (**9**) that would be stabilized by conjugation in the form of structures **10** and **11**.

$$CH_3-(CH_2)_4-\overset{cis}{CH=CH}-CH_2-\overset{cis}{CH=CH}-(CH_2)_7-COOR$$

$$\downarrow O_2$$

$$CH_3-(CH_2)_4-\overset{cis}{CH=CH}-\overset{\bullet}{CH}-\overset{cis}{CH=CH}-(CH_2)_7-COOR$$

(9)

<table>
<tr><th>(10)</th><th>(9)</th><th>(11)</th></tr>
<tr><td>

CH$_3$
|
(CH$_2$)$_4$
|
·CH
|
CH
‖ *trans*
CH
|
CH
‖ *cis*
CH
|
(CH$_2$)$_7$
|
COOR
(10)

</td><td>

</td><td>

CH$_3$
|
(CH$_2$)$_4$
|
CH
‖ *cis*
CH
|
CH
‖ *trans*
CH
|
·CH
|
(CH$_2$)$_7$
|
COOR
(11)

</td></tr>
<tr><td>

CH$_3$
|
(CH$_2$)$_4$
|
CH—OOH
|
CH
‖ *trans*
CH
|
CH
‖ *cis*
CH
|
(CH$_2$)$_7$
|
COOR

(13)

</td><td>

CH$_3$
|
(CH$_2$)$_4$
|
CH
‖ *cis*
CH
|
CH—OOH
|
CH
‖ *cis*
CH
|
(CH$_2$)$_7$
|
COOR

(12)

</td><td>

CH$_3$
|
(CH$_2$)$_4$
|
CH
‖ *cis*
CH
|
CH
‖ *trans*
CH
|
CH—OOH
|
(CH$_2$)$_7$
|
COOR

(14)

</td></tr>
</table>

The shifting of the double bond also enables the new bond to take up the more stable *trans* configuration, and a *cis*:*trans* configuration is predicted for the structure of the conjugated radicals. Structures **9–11** can then give rise to the hydroperoxides **12–14** by oxidation and hydrogen abstraction as described above for oleates.

A number of alternative mechanisms have been proposed for linoleate oxidations. These often suggest initial attack at the double bond, complex formation, or 1:4 or 1:5 type addition.[13,14] For example, Khan[15] has favored the formation of an activated complex between oxygen and the linoleates, as shown in Figure 2.2, to account for the early stages of autoxidation, particularly hydroperoxide formation. This theory predicts only two

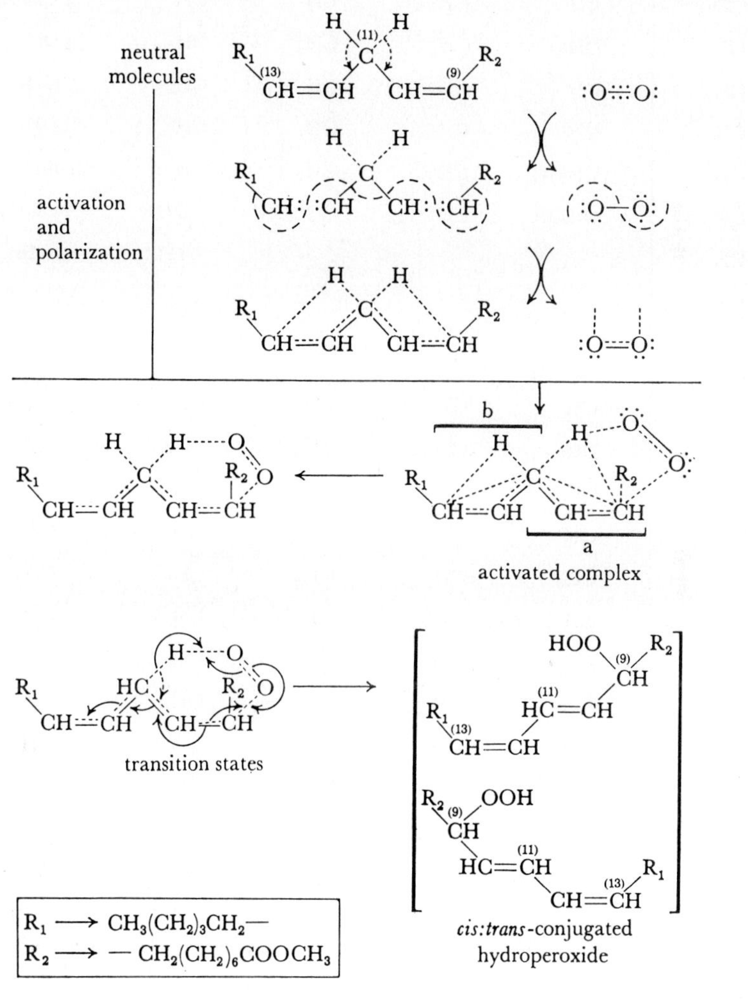

Figure 2.2. Possible mechanism for attack of oxygen on linoleates.[15]

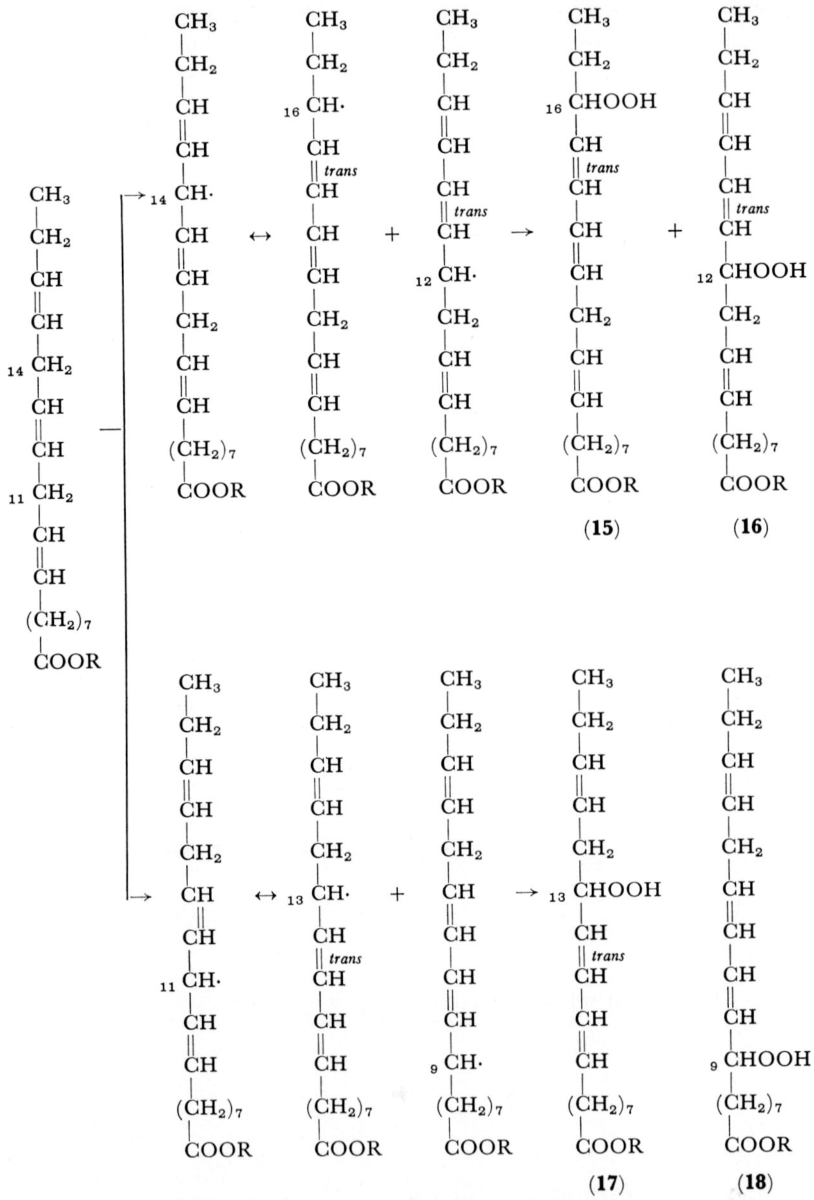

Figure showing columns of chemical structures:

$$
\begin{array}{c}
\text{CH}_3 \\
\text{CH}_2 \\
\text{CH} \\
\text{CH (double bond)} \\
{}_{14}\text{CH}\cdot \\
\text{CH} \\
\text{CH (double bond)} \\
{}_{14}\text{CH}_2 \\
\text{CH} \\
\text{CH (double bond)} \\
{}_{11}\text{CH}_2 \\
\text{CH} \\
\text{CH (double bond)} \\
(\text{CH}_2)_7 \\
\text{COOR}
\end{array}
$$

Top row of structures (left to right):

Structure A:
$$
\text{CH}_3-\text{CH}_2-\text{CH}=\text{CH}-{}_{14}\text{CH}\cdot-\text{CH}-\text{CH}-\text{CH}_2-\text{CH}-\text{CH}-(\text{CH}_2)_7-\text{COOR}
$$

$\leftrightarrow$

Structure B (with $trans$ double bond):
$$
{}_{16}\text{CH}\cdot-\text{CH}=\text{CH (}trans\text{)}-\text{CH}-\text{CH}-\text{CH}_2-\text{CH}-\text{CH}-(\text{CH}_2)_7-\text{COOR}
$$

$+$

Structure C (with $trans$ double bond):
$$
\text{CH}_3-\text{CH}_2-\text{CH}=\text{CH}-\text{CH}-\text{CH (}trans\text{)}-{}_{12}\text{CH}\cdot-\text{CH}_2-\text{CH}-\text{CH}-(\text{CH}_2)_7-\text{COOR}
$$

$\rightarrow$

Structure (15), with $trans$ double bonds:
$$
{}_{16}\text{CHOOH}-\text{CH}-\text{CH (}trans\text{)}-\text{CH}-\text{CH}-\text{CH}_2-\text{CH}-\text{CH}-(\text{CH}_2)_7-\text{COOR}
$$

$+$

Structure (16), with $trans$ double bonds:
$$
\text{CH}_3-\text{CH}_2-\text{CH}=\text{CH}-\text{CH}-\text{CH (}trans\text{)}-{}_{12}\text{CHOOH}-\text{CH}_2-\text{CH}=\text{CH}-(\text{CH}_2)_7-\text{COOR}
$$

Bottom row of structures (left to right):

Structure D:
$$
\text{CH}_3-\text{CH}_2-\text{CH}=\text{CH}-\text{CH}_2-\text{CH}=\text{CH}-{}_{11}\text{CH}\cdot-\text{CH}=\text{CH}-(\text{CH}_2)_7-\text{COOR}
$$

$\leftrightarrow$

Structure E (with $trans$ double bond):
$$
{}_{13}\text{CH}\cdot-\text{CH}=\text{CH (}trans\text{)}-\text{CH}=\text{CH}-(\text{CH}_2)_7-\text{COOR}
$$

$+$

Structure F:
$$
\text{CH}_3-\text{CH}_2-\text{CH}=\text{CH}-\text{CH}_2-\text{CH}=\text{CH}-\text{CH}=\text{CH}-{}_{9}\text{CH}\cdot-(\text{CH}_2)_7-\text{COOR}
$$

$\rightarrow$

Structure (17), with $trans$ double bond:
$$
{}_{13}\text{CHOOH}-\text{CH}-\text{CH (}trans\text{)}-\text{CH}=\text{CH}-(\text{CH}_2)_7-\text{COOR}
$$

Structure (18), with $trans$ double bonds:
$$
\text{CH}_3-\text{CH}_2-\text{CH}=\text{CH}-\text{CH}_2-\text{CH}=\text{CH}-\text{CH}=\text{CH}-{}_{9}\text{CHOOH}-(\text{CH}_2)_7-\text{COOR}
$$

Figure 2.3.   Autoxidation of linoleates.

hydroperoxide isomers (**13** and **14**) with no substitution at C_{11}. The failure to isolate the C_{11} hydroperoxide[15] is evidence in support of Khan's theory. Those who favor a Farmer-type mechanism have suggested that failure to isolate the C_{11} hydroperoxide is not unexpected, since the intermediate radical (9), if present, would exist predominantly as (10) and (11). However, Khan[15] has cited the studies of Harrison and Wheeler,[16] where radical attack on linoleates gives products resulting from reaction of a C_9, C_{11}, C_{13} radical. In addition, Moffitt and Coulson[17] have predicted that a resonating radical of type (9) would be approximately equally distributed between C_9, C_{11}, and C_{13}.

The autoxidation of linolenates is formally similar to the linoleates, but the presence of two highly activated methylene groups at C_{11} and C_{14} increases the number of possible hydroperoxide isomers. The presence of hydroperoxides in the early stages of linolenate oxidation has been established by Privett et al.[18] The isomers have been identified by Frankel et al.[19] as 15–18 with the hydroperoxide group at C_9, C_{12}, C_{13}, and C_{16}, and the corresponding double bond shifts. These isomers can be accounted for by Khan's theory or, as suggested by Frankel,[19] by postulating the greater stability of the conjugated radicals derived in a Farmer-type mechanism, as shown in Figure 2.3.

The synthetic procedure recently described by Bailey and Barlow[20] for the preparation of a pure linoleate hydroperoxides offers the possibility of more detailed and clear-cut studies on oxidation reactions. These authors prepared the p-phenylphenacyl derivative of linoleate hydroperoxide.

GENERAL REFERENCES

1. Bragdon, C. R., ed., *Film Formation, Film Properties and Film Deterioration*, Interscience Publishers, Inc., New York, 1958.
2. Wexler, H., *Chem. Rev.*, **64**, 591 (1964).

REFERENCES

1. Skellon, J. H., and D. M. Wharry, *Chem. Ind. (London)*, **1963**, 929.
2. Silbert, L. S., *J. Am. Oil Chemists' Soc.*, **39**, 480 (1962).
3. Criegee, R., *Ann.*, **84**, 522 (1936).
4. Farmer, E. H., and D. A. Sutton, *J. Chem. Soc. (London)*, **1946**, 10.
5. Swern, D., J. E. Coleman, H. B. Knight, C. Ricciuti, C. O. Willits, and C. R. Eddy, *J. Am. Chem. Soc.*, **75**, 3135 (1953).
6. Ross, J., A. I. Gebhart, and J. F. Gerecht, *J. Am. Chem. Soc.*, **71**, 282 (1949).
7. Boelhouwer, C., F. A. de Roos, and H. I. Waterman, *Chem. Ind. (London)*, **1953**, 1287.
8. Horikx, M. M., *J. Appl. Chem. (London)*, **14**, 50 (1964).
9. Cowan, J. C., *Ind. Eng. Chem.*, **41**, 294 (1949).
10. Chipault, J. R., E. C. Nickell, and W. O. Lundberg, *Offic. Dig. Federation Soc. Paint Technol.*, **23**, 740 (1951).
11. Kartha, A. S., *J. Sci. Ind. Res. (India)*, **17B**, 135 (1958).
12. Gunstone, F. D., and T. P. Hilditch, *J. Chem. Soc. (London)*, **1945**, 836; Gunstone, F. D., and T. P. Hilditch, *J. Chem. Soc. (London)*, **1946**, 1022; Hilditch, T. P., *J. Oil Colour Chemists' Assoc.*, **30**, 1 (1947).

13. Treibs, W., *Fette, Seifen, Anstrichmittel*, **54**, 3 (1952).
14. Khan, N. A., *J. Am. Oil Chemists' Soc.*, **30**, 273 (1953); Khan, N. A., *Pakistan J. Sci.*, **11**, No. 6, 290 (1959).
15. Khan, N. A., *Can. J. Chem.*, **37**, 1029 (1959).
16. Harrison, S. A., and D. H. Wheeler, *J. Am. Chem. Soc.*, **76**, 2379 (1954).
17. Moffitt, W. E., and C. A. Coulson, *Trans. Faraday Soc.*, **44**, 81 (1948).
18. Privett, O. S., C. Nickell, W. E. Tolberg, R. F. Paschke, D. H. Wheeler, and W. O. Lundberg, *J. Am. Oil Chemists' Soc.*, **31**, 23 (1954).
19. Frankel, E. N., C. D. Evans, D. G. McConnell, E. Selke, and H. J. Dutton, *J. Org. Chem.*, **26**, 4663 (1961).
20. Bailey, W. J., and G. L. Barlow, *J. Paint Technol.*, **42**, 287 (1970).
21. Rawls, H. R., and P. J. Van Santen, *J. Am. Oil Chemists' Soc.*, **47**, 121, (1970).

Conjugated Fatty Acids and Esters. Oils containing conjugated fatty acids dry very much faster than the isomeric nonconjugated oils. "Frosting," which is characteristic of wood oil or dehydrated castor oil, results from the preferential rapid surface dry, followed by wrinkling. In considering the mechanism by which conjugated systems dry, it is essential to realize that oxygen uptake and drying time are quite distinct. It has been clearly shown that a conjugated oil dries with very much less oxygen absorption than a nonconjugated oil, and this is illustrated by the viscosity increase of the methyl esters shown in Figure 2.4.[1]

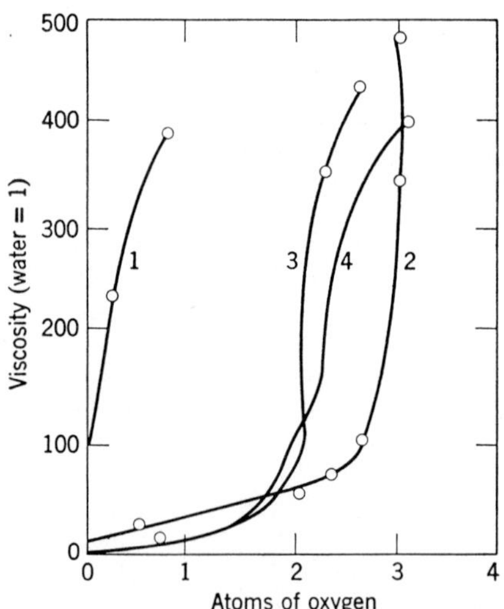

Figure 2.4 Viscosity increase of methyl esters of fatty acids when oxidized. *1*, Eleostearic acid methyl ester; *2*, linoleic acid methyl ester; *3*, linolenic acid methyl ester; and *4*, hexenic acid methyl ester.[1]

The major product resulting from the oxidation of methyl-α-eleostearate has been shown by O'Neill[2] to be a peroxide. The properties of this product suggested that it was a mixture, probably of the 1:4 cyclic peroxides **19** and **20**.

$$CH_3-(CH_2)_3-CH=CH-CH=CH-CH=CH-(CH_2)_7-COOCH_3$$

$$CH_3-(CH_2)_3-CH-CH=CH-CH-CH=CH-(CH_2)_7-COOCH_3$$
$$O\text{———}O$$

(19)

$$CH_3-(CH_2)_3-CH=CH-CH-CH=CH-CH-(CH_2)_7-COOCH_3$$
$$O\text{———}O$$

(20)

It should be noted that no conjugated unsaturation is present in structures **19** and **20**. Some 1:2, or even 1:6, addition of oxygen to the conjugated triene is also possible.

As with nonconjugated systems, the presence of a *cis* or *trans* configuration influences the rate of oxidation of conjugated double bonds. It has been claimed that the *cis* isomers are more readily oxidized than the *trans*,[3,4] although the evidence for this statement appears to be doubtful. For example, the relative rates of oxidation of α- and β-eleostearic acids, shown in Table 2.3,[3] are claimed to indicate the greater reactivity of the *cis* conjugated system. (Linolenic acid is included for comparative purposes.)

However, when the results of Table 2.3 are considered in conjunction with the mechanism of oxidation, it can be seen that the first products of

TABLE 2.3

OXIDATION OF TRIENOIC ACIDS AT 40°C[3]

Compound	Moles O$_2$/Mole Acid/100 min	Configuration
α-Eleostearic acid	2.68	*cis : trans : trans* conjugated
β-Eleostearic acid	1.02	*trans : trans : trans* conjugated
Linolenic acid	0.52	*cis : cis : cis* unconjugated

oxidation will be nonconjugated dienes, probably with a *cis:cis*, and *trans: cis* configuration as shown (the steric requirements for the formation of a cyclic peroxide result in the newly formed bond taking up a *cis* configuration).

$$CH_3—(CH_2)_3—\overset{trans}{CH{=}CH}—\overset{trans}{CH{=}CH}—\overset{cis}{CH{=}CH}—(CH_2)_7—COOH$$

α-eleostearic acid

$$\downarrow O_2$$

$$\begin{array}{c} O\text{————————}O \\ | \qquad\qquad | \end{array}$$

$$CH_3—(CH_2)_3—\underset{trans}{CH}—\overset{cis}{CH{=}CH}—CH—\overset{cis}{CH{=}CH}—(CH_2)_7—COOH$$

$$CH_3—(CH_2)_3—\overset{trans}{CH{=}CH}—\overset{trans}{CH{=}CH}—\overset{trans}{CH{=}CH}—(CH_2)_7—COOH$$

β-eleostearic acid

$$\downarrow O_2$$

$$CH_3—(CH_2)_3—\overset{trans}{CH{=}CH}—CH—\overset{cis}{CH{=}CH}—CH—(CH_2)_7—COOH$$

$$\begin{array}{c} | \qquad\qquad | \\ O\text{————————}O \end{array}$$

Therefore, the results can be interpreted as showing that a *cis:cis*, 1:4 unsaturated diene peroxide oxidizes more readily than a *trans:cis*, 1:4. This alternative explanation becomes of greater significance when attempting to explain the behavior of dehydrated castor oil, which is a mixture of nonconjugated and conjugated *cis* and *trans* isomers. This is discussed further on page 62.

In general, it is found that conjugated oils oxidize faster than the nonconjugated, and film formation takes place with much lower oxygen absorption. The oxygen adds to give predominantly a 1:4 peroxide.

REFERENCES

1. Treibs, W., *Fette, Seifen, Anstrichmittel*, **54**, 3 (1952).
2. O'Neill, L. A., *Chem. Ind. (London)*, **1954**, 384.
3. Wexler, H., *Chem. Rev.*, **64**, 591 (1964).
4. Myers, J. E., J. P. Kass, and G. O. Burr, *J. Am. Oil Chemists' Soc.*, **18**, 107 (1941).

FILM FORMATION AND DEGRADATION

The change of an oil film from a viscous sticky liquid to a dry hard film involves the decomposition of the hydroperoxides, peroxides, and other oxygenated products. Associated with the decomposition reactions is the formation of a cross-linked structure shown schematically in Figure 2.5.

The dried film still contains unsaturated carbon linkages, and the processes of oxidation and polymerization can continue over long periods. This

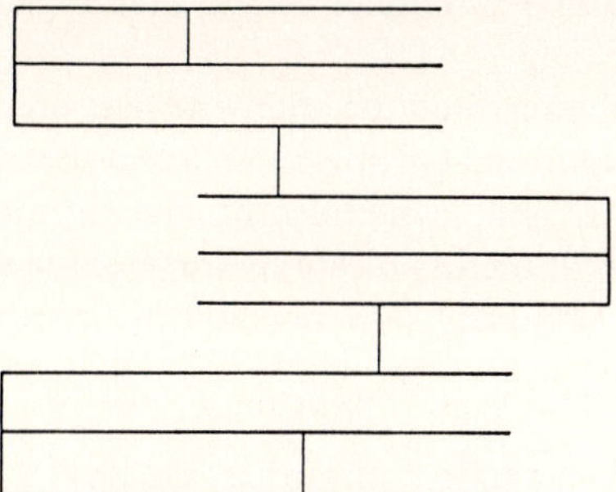

Figure 2.5.   Schematic representation of a cross-linked oil film.

eventually results in the degradation of the film by the loss of volatile chemical entities. As the degree of cross-linking becomes excessive, physical properties will also deteriorate. It should be emphasized that the various stages in autoxidation—namely, oxygen uptake, polymerization, and degradation—overlap to varying degrees.

Analysis of dried oil films indicates that various chemical cross-links are formed including peroxides, ethers, and carbon-carbon bonded structures. There is also an increase in the number of ester groups present, and volatile decomposition products that account for the familiar odors of air-drying oil-based coatings are liberated.

In view of the different initial products formed in the autoxidation of conjugated and nonconjugated structures, it is not surprising that different subsequent steps have also been proposed for film formation.

Nonconjugated Systems. The hydroperoxides formed from nonconjugated olefins usually decompose by the initial dissociation of the oxygen-oxygen bond of the hydroperoxy group. Equations 2.1 to 2.7 represent a generalized form of this decomposition and the possible subsequent reactions.

$$
\begin{aligned}
\text{ROOH} &\longrightarrow \text{RO·} \quad &+ \text{·OH} & \quad &(2.1)\\
2\,\text{ROOH} &\longrightarrow \text{RO·} \quad &+ \text{ROO·} + \text{H}_2\text{O} & \quad &(2.2)\\
\text{RO·} \quad + \text{R'H} &\longrightarrow \text{ROH} \quad &+ \text{R'·} & \quad &(2.3)\\
\text{RH} \quad + \text{·OH} &\longrightarrow \text{R·} \quad &+ \text{H}_2\text{O} & \quad &(2.4)\\
\text{R·} \quad + \text{R·} &\longrightarrow \text{R—R} & & \quad &(2.5)\\
\text{RO·} \quad + \text{R·} &\longrightarrow \text{R—O—R} & & \quad &(2.6)\\
\text{RO·} \quad + \text{RO·} &\longrightarrow \text{R—O—O—R} & & \quad &(2.7)
\end{aligned}
$$

R—H and R'—H = unoxidized fatty acid molecules. This scheme superficially explains the formation of alcohols (Eq. 2.3), ethers (Eq. 2.6), peroxides (Eq. 2.7), carbon bonded structures (Eq. 2.5), and water (Eqs. 2.2 and 2.4). Reaction 2.2 involves second-order kinetics.[1] The relative importance of the various reactions is influenced by the conditions, including the presence of catalysts. Higher temperatures probably favor carbon-carbon cross-links.[2]

The polarity of the medium, or of any solvent present during the oxidation of oils or oil-containing polymers, can affect the rate of peroxide/hydroperoxide breakdown and, consequently, the drying rate of the system. Khan[3] has shown that methyl oleate hydroperoxide decomposes rapidly in methyl oleate, but relatively slow breakdown occurred in the presence of

TABLE 2.4

DECOMPOSITION PRODUCTS OF FATTY ACIDS[4]

Carboxylic Acids	Ketones	Aldehydes	Volatile Gases
Formic acid	Methyl ethyl ketone	Acetaldehyde	Carbon monoxide
Acetic acid	Methyl vinyl ketone	Propionaldehyde	Carbon dioxide
Valeric acid	Methyl n-amyl ketone	Acrolein	Hydrogen
Caproic	Methyl n-hexyl ketone	Crotonaldehyde	
Heptanoic acid	Methyl n-heptyl ketone	Pentanal	
Caprylic acid	Methyl n-octyl ketone	2-Pentenal	
Pimelic acid	Di-n-butyl ketone	Hexanal	
Suberic acid	Diamyl ketone	3-Hexenedial	
Pelargonic acid	Dinonyl ketone	Heptenal	
Azelaic acid		2-Octenal	
Capric acid		Nonanal	
Sebacic acid		2-Nonenal	
Undecanoic acid		2, 4-Decadienal	
		2-Undecenal	

ethanol. This finding is of significance in the selection and formulation of solvents for oil-based polymers, and indicates the necessity to avoid high boiling polar solvents, which can remain in the film and interfere with hydroperoxide decomposition. It has been proposed that the solvent forms a stable intermediate involving hydrogen bonding with the hydroperoxide.

$$ROOH + R'OH \rightleftharpoons ROOH$$
$$HOR'$$

The decomposition products resulting from the oxidative-polymerization reactions arise from two main sources. First, the intermediate radicals shown in Eqs. 2.1 to 2.7 may react by alternative paths; and second, by-products may arise from the decomposition of the non-hydroperoxide oxygenated products. The chemical nature of the decomposition products varies widely and includes saturated and unsaturated acids, ketones, and aldehydes, as shown in Table 2.4.[4] O'Neill has shown that the main volatile products consist of aldehydes, derived from breakdown of the fatty acid chain, followed by formic acid, carbon dioxide and water. In the non-volatile fraction oxygenated products are present.[8] Detailed mechanisms are not well established for the formation of all of these compounds.

The formation of a conjugated double bond system that follows the initial autoxidation can lead to further oxidative attack by 1:4 addition either at the hydroperoxide or at subsequent stages, as suggested below.

$$
\begin{array}{c}
\overset{cis}{-CH=CH} - \overset{trans}{CH=CH} - CH - \\
| \\
OOH
\end{array}
$$

$$\downarrow O_2$$

$$
\begin{array}{c}
O-\!\!\!-\!\!\!-\!\!\!-\!\!\!-\!\!\!-O \\
| \quad\quad\quad | \\
-CH-CH=CH-CH-CH- \\
| \\
OOH
\end{array}
$$

Privett and Nickell[5] have confirmed that oxidation of conjugated linoleate hydroperoxides does occur; in addition, these workers have shown that such reactions may occur at all stages of the autoxidation. These reactions will contribute to film formation and to the onset of degradation; they possibly occur at a slower rate than with wood oil or similar compounds because of the *cis*:*trans* configuration of the bonds (see page 47), viscosity effects, and the presence of α-substituents or cross-links.

Conjugated Systems. The cross-linked films derived from conjugated and nonconjugated olefins differ in structure. A particularly notable feature of

$$
\begin{array}{c}
-CH-CH=CH-CH-CH=CH- \\
| \quad\quad\quad\quad\quad\quad | \\
O-\!\!\!-\!\!\!-\!\!\!-\!\!\!-O
\end{array}
$$

$$
\begin{array}{c}
OO\cdot \\
| \\
-CH-CH=CH-CH-CH=CH- \\
\\
\downarrow\quad -CH=CH-CH=CH-CH=CH- \\
\\
OO\cdot \\
| \\
-CH-CH=CH-CH-CH=CH- \\
-CH-CH=CH-CH-CH=CH- \\
| \\
-CH-CH=CH-CH-CH=CH-
\end{array}
$$

the conjugated oil is the presence of much more carbon-carbon bonding in the cross-links. Peroxide structures are present in the film as shown by reaction with hydrogen iodide but, since this treatment does not yield significant quantities of monomeric products, the peroxide links are not involved in polymer formation.[6] These facts are accommodated by a theory which suggests that the free radicals, produced by dissociation of the monomeric peroxide, attack the conjugated diene or triene monomer in a vinyl-type polymerization.[7] This will only yield low molecular weight products, probably dimers and trimers, because of the likely termination by the oxygen present.

As with nonconjugated oils, considerable chain scission occurs, yielding mainly peroxides and aldehydes. The mechanisms for these breakdown reactions are still not established but, in principle, they presumably follow the nonconjugated systems.

REFERENCES

1. Bawn, C. E. H., *J. Oil Colour Chemists' Assoc.*, **36**, 443 (1953).
2. Williamson, L., *J. Appl. Chem. (London)*, **3**, 301 (1953).
3. Khan, N. A., *Pakistan J. Sci.*, **12**, 95 (1960).
4. Wexler, H., *Chem. Rev.*, **64**, 591 (1964).
5. Privett, O. S., and C. Nickell, *J. Am. Oil Chemists' Soc.*, **33**, 156 (1956).
6. O'Neill, L. A., *Chem. Ind. (London)*, **1954**, 384.
7. Faulkner, R. N., *J. Appl. Chem. (London)*, **8**, 448 (1958).
8. O'Neill, L. A., X, *F. A. T. I. P. E. C. Congr.*, 225, (1970).

AUTOXIDATION CATALYSTS—DRIERS

Autoxidation, particularly of nonconjugated methylene-interrupted fatty acids and esters, does not proceed rapidly enough when considered from the viewpoint of commercially usable surface coatings; even those oils high in linolenic content may take days to dry. It is necessary, therefore, to speed up hydroperoxide formation and decomposition so that oil films become dry in up to twelve hours. For this purpose, catalysts or driers are used, for example, the hydrocarbon soluble salts—often called soaps, (octoates, naphthenates, linoleates)—of selected metals such as cobalt, manganese, iron, lead, calcium, and zinc, or aluminum complexes. The use of driers also alters the amount of oxygen required for a film to dry.

The metal driers in common use may be divided into two broad classes indicative of the mechanisms by which the drier operates. First, there are the so-called primary driers, such as cobalt or manganese salts, which in small amounts (0.005 to 0.2% metal based on oil) catalyze the drying process. In the second class are the auxiliary or promoter driers, which do not catalyze autoxidation when used alone, but have an activating effect on primary driers. These cause a number of effects including "through" dry, reduction in bloom, and increased drier stability. Typical examples of this second class are calcium, barium, and zinc salts. Lead functions mainly as a secondary drier but in some situations can act also as a primary drier. Often quite high percentages of the secondary drier are used (up to 0.6% metal based on oil).

The detailed mechanisms by which driers function are still being debated, but certain general principles are now established for primary driers. The position with secondary driers is less clear, but some possible mechanisms to explain their action are suggested below.

The primary driers appear to function as true catalysts, and are exemplified by the transition metal salts. Driers of this type can exist in two valency states, the higher valency being the less stable. To function as a drier, however, the oxidation from the lower to the higher valency must be readily brought about by the fatty acid hydroperoxide. Therefore, we have a situation in which the hydroperoxide and metal salts constitute a redox system, and possible reactions that can occur in the case of cobalt are shown in Eqs. 2.8 and 2.9.[1,2]

$$ROOH + Co^{2+} \longrightarrow Co^{3+} + RO\cdot + OH^- \tag{2.8}$$
$$ROOH + Co^{3+} \longrightarrow Co^{2+} + ROO\cdot + H^+ \tag{2.9}$$

It can be seen that the cobaltous ion is oxidized by the hydroperoxide (Eq. 2.8), whereas the cobaltic ion is reduced (Eq. 2.9), clearly showing the ability of a hydroperoxide to act as both oxidant and reductant. In some systems, however, it has been suggested that reaction 2.9 is not significant.[3]

The above equations explain only the hydroperoxide decomposition stage of autoxidation, but driers are also claimed to influence oxygen uptake and hydroperoxide formation. A possible mechanism is the direct interaction of cobaltic ions with the fatty acid (as shown in Eq. 2.10) to form an alkyl radical which, by subsequent addition of oxygen and hydrogen abstraction, can give a hydroperoxide (Eq. 2.11).[2,3]

$$RH + Co^{3+} \longrightarrow R\cdot + Co^{2+} + H^+ \tag{2.10}$$
$$R\cdot + O_2 + RH \longrightarrow ROOH + R\cdot \tag{2.11}$$

A similar proposal has been made[4] in which the cobaltic ion directly attacks the double bond. In fact, Uri[5,6] has suggested that even in highly purified oils, traces of metal ions are responsible for the onset of autoxidation.

The driers may also act as oxygen carriers, or may convert the oxygen to a more active form by the formation of an association type complex, as illustrated by Eq. 2.12.

$$Co^{3+} + O_2 \longrightarrow Co^{3+}\!-\!O\!-\!O\cdot \tag{2.12}$$

$$\nearrow R\!-\!CH_2\!-\!CH{=}CH\!-\!R'$$
$$Co^{3+}\!-\!OOH + R\!-\!\overset{\cdot}{C}H\!-\!CH{=}CH\!-\!R'$$
$$\downarrow$$
$$Co^{3+} + \overset{\cdot}{O}OH + R\!-\!\overset{\cdot}{C}H\!-\!CH{=}CH\!-\!R'$$

The catalytic activity of cobalt and manganese in the reactions shown above varies, one of the major differences being in the length of the induction period. Berry and Mueller[7] have compared cobalt and manganese driers in the oxidation of linseed oil. They have shown that manganese has a longer induction period but, once the reaction has started, it gives a faster oxidation rate than cobalt. These results suggest that cobalt is a more effective catalyst for hydroperoxide formation, whereas manganese favors the breakdown of oxygenated products. These results partially explain why mixtures of driers often give faster drying formulations than is possible with the individual component driers.

The enhanced drier activity imparted to primary driers by the auxiliary or secondary driers can be partly accounted for in similar terms to the manganese-cobalt example above. For example, it has been shown that lead driers speed up the rate of oxygen absorption by linoleates, but do not have an appreciable effect on hydroperoxide decomposition.[8] Hence, in manganese-lead or cobalt-lead drier combinations, the increased activity results from lead catalyzing the oxygen-uptake reaction, and the cobalt or manganese promoting the hydroperoxide breakdown or polymerization reaction. Other effects of the promoter driers are more difficult to explain, but it should be noted that the amounts of these compounds used (up to 0.6% metal based on oil) are far above the normal catalytic levels. It would seem that a possible explanation for the action of promoter driers is their preferential (compared to primary driers) formation of salts with the acid groups present in the fatty acid containing polymer (for example, an alkyd), or with acidic centers derived by oxidative breakdown of the oil. Support for this theory is given by the viscosity increase that takes place when calcium or zinc driers, in amounts equivalent to the acid value, are added to alkyd resins. This increased viscosity is particularly noticeable at high solids concentration (90 to 95%) where gelation can occur. This indicates that some of the initial drying of the film is due to salt-type cross-links. A further possible contribution to the initial tack-free dry stage could come from co-ordination-type complexes of hydroxyl groups of the polymer with, for example, calcium ions.

This suggestion is in line with the proposal that auxiliary driers "arrange" the molecules in the film.[9]

The suggestion that auxiliary driers form salts with polymer carboxyl groups also helps explain the stabilizing action of calcium on other driers; the calcium, being a stronger base, would preferentially neutralize acidic groups and maintain the primary drier in a free state.

The polarity of the solvent,[6] and the amount of water vapor above the film[10] play important roles in the rate of oxidation of drier-catalyzed oils and fatty acids. The effects of moisture are shown in Table 2.5.[10]

TABLE 2.5

EFFECT OF HUMIDITY ON METAL CATALYST COMBINATIONS[10]

Metals	Humidity Effect
Pb-Mn, Pb-Mn-Co, Co-Pb	Little effect above 20°C
Co, Co-Mn, Co-Mn-Zn	Large effect at 20 to 30°C Little effect above 40°C
Pb, Pb-Zn, Mn-Zn, Co-Zn	Drying rate depends on humidity

These results are of special practical significance; they show that by suitable selection of drier combinations it is possible to minimize the variation of drying rate with humidity. The precise theoretical reasons underlying the moisture effects are not established. However, it should be noted that it is necessary to have the dissociated metal ion, or a derived complex, to catalyze autoxidation; the moisture could be influencing the degree of dissociation of the drier salt and/or the possible formation of coordination complexes[11] with the metal ion.

The common anions used to prepare hydrocarbon-soluble drier salts (naphthenates, linoleates, tallates, octoates) can affect the properties of the surface coating and the efficiency of the metal ion as a drier. For example, in certain formulations the octoates reduce the drying time to approximately one half that obtained with naphthenates of tallates.[12] In other cases, the naphthenates are at least equivalent to the other anions. The likely effect of the anion on drier efficiency is difficult to predict[15] and the magnitude of the anion's effect is generally very much less than the other variables in the system (type of resin, cation etc.). Most drier solutions contain some free acid, and this can affect the ease with which the pigment is wetted, and the stability of the dispersions. In this regard naphthenates offer advantages over other driers.[13]

Once a film has dried, the presence of the drier metals can have a deleterious effect on the durability of the film, since continued attack by oxygen will slowly degrade the oil molecules. A suggested method of minimizing this effect is to use small quantities of an active metal drier consisting of isotopes of short half-life periods which decompose to less effective elements or isotopes. In this connection Co^{58} has been claimed to offer advantages over the usual cobalt naphthenate.[14]

REFERENCES

1. Bawn, C. E. H., *J. Oil Colour Chemists' Assoc.*, **40**, 1027 (1957).
2. Bawn, C. E. H., *Discussions Faraday Soc.*, **14**, 181 (1952).
3. Heaton, F. W., and N. Uri, *J. Lipid Res.*, **2**, 152 (1961).
4. Bawn, C. E. H., and J. A. Sharp, *J. Chem. Soc. (London)*, **1957**, 1854.
5. Uri, N., *Chem. Ind. (London)*, **1956**, 515.
6. Uri, N., *Nature*, **177**, 1177 (1956).
7. Berry, D. A., and C. R. Mueller, Am. Chem. Soc., Div. Paint, Plastics, Printing Ink Chem., Chicago Meeting, **18**, 44 (1958).
8. Williamson, L., *Proc., News Oil Colour Chemists' Assoc., (Australian Sections)*, **1**, No. 5, 12 (1964).
9. Bowles, R. F., *J. Oil Colour Chemists' Assoc.*, **33**, 97 (1950).
10. Kaufmann, H. P., and E. Gulinsky, *Deut. Farben-Z*, **11**, 90 (1957); via Wexler, H., *Chem. Rev.*, **64**, 591 (1964).
11. Bawn, C. E. H., *J. Oil Colour Chemists' Assoc.*, **40**, 1047 (1957).
12. Gardner, C., *Am. Paint J.*, **41**, 77 (1957).
13. Doadrio, A., and R. Montequi, *Grasas Aceites (Seville, Spain)*, **8**, 76 (1957).
14. Pallauf, F., Abshagen and Co., Akt.-Ges., Ger. Pat. 801,525 (1951), via *Chem. Abstr.*, **45**, 2234 (1951).
15. *Paint Driers and Additives.* Federation Series on Coatings Technology, Unit 11, Federation of Societies for Paint Technology, Philadelphia 1969.
    Philadelphis Society for Paint Technology, Technical Subcommittee on Drier Evaluation, *J. Paint Technol.*, **41**, 623 (1969).

COORDINATION COMPLEXES AS DRIERS

Over recent years, it has become common practice to add selected compounds to surface coatings to increase the efficiency of the metallic driers. These compounds have in common the ability to form coordination complexes with the transition metal ions, and the enhanced drier activity results from a lowering of the energy required for either oxygen coordination and activation (page 52) or the cobaltous to cobaltic oxidation.

Typical of the complexing agents is 1:10 phenanthroline; the effect of this agent on the drying time of a soya bean oil alkyd catalyzed by a series of driers is shown in Figure 2.6.[1]

Studies on the detailed structure of the 1:10 phenanthroline metal complexes[2] suggest that the cobalt species has a coordination number of 4, whereas manganese gives an equilibrium mixture of two species with an average coordination number of 1.5. These complexes may be formed prior to adding the drier to the polymer solution, or in the presence of the resinous components.[3]

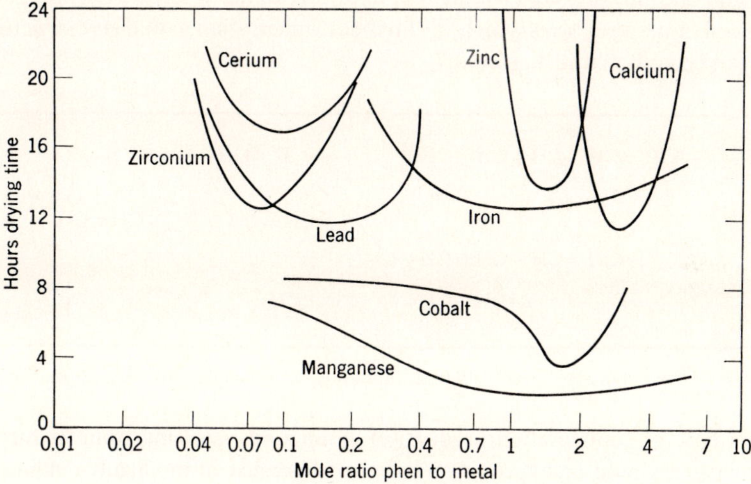

Figure 2.6. Influence of 1:10 phenanthroline on the drying time of an alkyd.[1]

The amine-drier complexes offer a number of other advantages in coating compositions, the most important of which is the prevention of the loss of drying potential on aging.[3]

The detailed theory on the function of amine activators has been discussed by a number of workers; it is suggested that the strongly electron donating ligand, 1:10 phenanthroline, splits the electrons of the $3d$ orbitals so that at least one orbital is energetically more attractive to oxygen.[4] Alternatively, the metal complex may be capable of facilitating the decomposition of hydroperoxides and peroxides.[4]

REFERENCES

1. Stearns, M. E., *Offic. Dig. Federation Soc. Paint Technol.*, **26**, 817 (1954).
2. Canty, W. H., G. K. Wheeler, and R. R. Myers, *Ind. Eng. Chem.*, **52**, 67 (1960).
3. Zettlemoyer, A. C., and D. M. Nace, Am. Chem. Soc. Div. Paint, Varnish, Plastics Chem. Atlantic Meeting, 1949, 145.
4. Myers, R. R., *Offic. Dig. Federation Soc. Paint Technol.*, **34**, 575 (1962).

YELLOWING AND DISCOLORATION OF OIL FILMS

The dried films of surface coatings that contain fatty acids are prone to yellowing in the absence of direct sunlight, a situation commonly encountered in interior paints and enamels. Yellowing appears to be associated with the unsaturated centers of the fatty acid molecules and is the result of autoxidation products, and/or their subsequent reaction with atmospheric ammonia.

The decomposition of conjugated hydroperoxides can lead to conjugated diene-ketone structures which, by further condensation, could give structures of the type shown in Figure 2.7.

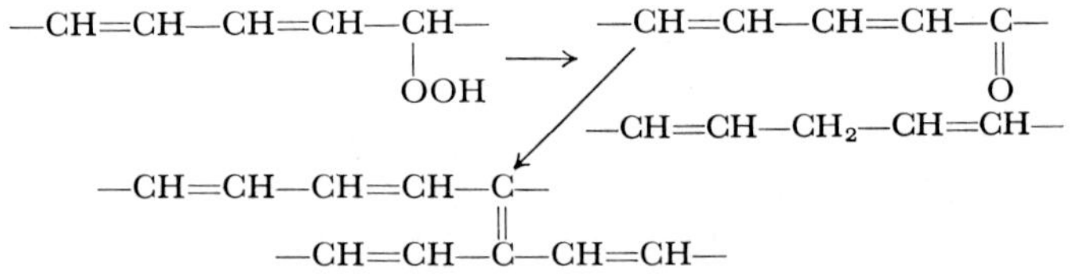

$$\mathrm{-CH{=}CH{-}CH{=}CH{-}CH{-}} \atop \mathrm{OOH} \longrightarrow \mathrm{-CH{=}CH{-}CH{=}CH{-}\overset{\text{O}}{\overset{\|}{C}}-}$$

$$\mathrm{-CH{=}CH{-}CH_2{-}CH{=}CH-}$$

$$\mathrm{-CH{=}CH{-}CH{=}CH{-}\overset{\|}{C}-} \atop \mathrm{-CH{=}CH{-}\overset{\|}{C}{-}CH{=}CH-}$$

Figure 2.7. Formation of colored compounds by hydroperoxide decomposition.

This type of condensation product, or compounds resulting from its further oxidation, would be expected to be colored because of the highly conjugated system. This mechanism is in conformity with the greater tendency of linseed oil (high in linolenic acid) to yellow. The inhibiting action of aldehydes that contain active methylene groups[1] is also consistent with this mechanism, since these could compete in the condensation reaction for the diene-ketone. Thus it has been shown that trialdehyde oils, the triglyceride of azelaaldehydic acid,[5] and methyl azelaaldehyde[6] improve the color stability of certain linseed oil compositions.

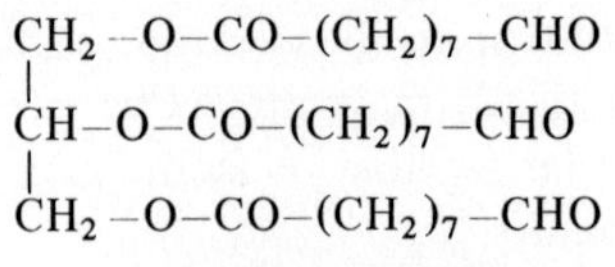

$$CH_2-O-CO-(CH_2)_7-CHO$$
$$CH-O-CO-(CH_2)_7-CHO$$
$$CH_2-O-CO-(CH_2)_7-CHO$$

"Trialdehyde Oil"
(Azelaaldehydic acid triglyceride).

The known inhibiting action of zinc oxide on yellowing is also explicable in terms of the reaction proposed, since it may inhibit conjugated-ketone formation by reaction with the slightly acidic hydroperoxide.[2] This results in a cross-linking reaction and is a contributing cause to the development of brittle films with compositions containing zinc oxide.

$$2 \ —CH=CH—CH=CH—CH— \quad + ZnO$$

$$\begin{array}{c} | \\ OOH \end{array}$$

$$\downarrow$$

$$—CH=CH—CH=CH—CH—$$
$$\begin{array}{c} | \\ O \\ | \\ O \\ | \\ Zn \quad + H_2O \\ | \\ O \\ | \\ O \\ | \end{array}$$
$$—CH=CH—CH=CH—CH—$$

Alternatively, the yellowing of a drying oil film has been related to reaction of 1:4 diketones,[3] or carbonyl compounds capable of enolization[4] formed by autoxidation of the oil, with atmospheric nitrogenous bases (particularly ammonia) to give substituted pyrroles. Model compounds of this series are known to readily undergo oxidation or condensation reactions resulting in colored compounds, and it has, therefore, been suggested that yellowing of oil films is a related process.

$$—C—CH_2—CH_2—C— \rightleftharpoons —C=CH—CH=CH—$$

with O (doubly bonded) on the diketone carbons and OH on the enol form, reacting with NH_3 to give the pyrrole ring:

$$\begin{array}{c} CH——CH \\ \| \quad\quad \| \\ —CH \quad CH— \\ \backslash \quad / \\ NH \end{array}$$

which by oxidation and condensation gives yellow-colored compounds.

Related to this type of reaction is the problem of discoloration with certain water-solubilized alkyds and modified oils, where the use of ammonia as the solubilizing base enhances color formation in the dried film.

REFERENCES

1. Privett, O. S., M. L. Blank, J. B. Covell, and W. O. Lundberg, *J. Am. Oil Chemists' Soc.*, **38**, 22 (1961).
2. Elm, A., *Offic. Dig. Federation Soc. Paint Technol.*, **29**, 351 (1957).
3. O'Neill, L. A., *Paint Technol.*, **27**, 44 (1963); O'Neill, L. A., S. M. Rybicka, and T. Robey, *Chem. Ind.* (*London*), **1962**, 1796.
4. Franks, F., and B. Roberts, *J. Appl. Chem.* (*London*), **13**, 302 (1963).
5. McManis, G. E., L. E. Gast, and J. C. Cowan, *J. Paint Technol.*, **38**, 740, (1966).
6. McManis, G. E., E. W. Swain, L. E. Gast and J. C. Cowan, *J. Paint Technol.*, **43**, (554), 53 (1971).

MODIFIED OILS

The raw or refined oils find only limited use as such in surface coatings, partly because of their poor drying characteristics. Various methods of modifying the oils have, therefore, been developed to improve the drying potential, and to enhance the possibility of other chemical reactions. These treatments are of importance not only to the oils, but also to the wide range of oil-containing polymer systems.

Oils, or the derived fatty acids, are treated or modified for the following reasons.

1. To alter and/or control the subsequent film-forming reactions.

2. To convert nondrying to semidrying or drying oil structures.

3. To alter and improve the structure for subsequent chemical modification by converting nonconjugated to conjugated structures, and/or changing the geometric isomers present.

4. To reduce the reactivity so that subsequent chemical reactions may be controlled, that is, the "frosting" and gelation of highly conjugated oils.

Some of the more common methods of carrying out these objectives are considered below.

Isomerized Oils. The properties of the naturally occurring oils are related to both the degree and type of unsaturation present, and to the particular geometrical isomerism about the double bonds. It has been shown that in mono-ethenoid systems, a *cis* double bond undergoes catalytic oxidation more readily than a *trans*, and that the drying rate of a conjugated oil is much faster than the nonconjugated. Similarly, with maleic anhydride (page 70), styrene, or other vinyl monomers (Chapter 4), and in thermal polymerization, the reactions depend on the type of unsaturation in the oil. The isomerization of oils or fatty acids is, therefore, generally aimed at increasing the amount of conjugated unsaturation and/or interconverting *cis* and *trans* isomers.

cis-trans Isomerism. In the fatty acid series, the simplest *cis*-olefin is oleic acid, and the corresponding *trans* isomer is elaidic acid. Conversion of oleic to elaidic acid is commonly referred to as elaidinization, and this term is often used in a more general sense to indicate conversion of a *cis* to a *trans* form.

The *trans* form of the nonconjugated fatty acids is usually the more stable, and consequently the isomerization of the naturally occurring *cis* isomers is relatively easy. The catalysts commonly employed are oxygen,[1] iodine,[2] nitrogen dioxide,[1] sulfur,[3] or selenium.[4] These probably form some type of unstable olefine complex which on decomposition gives the *trans* isomer.

Photosensitized isomerization is a more recent technique especially with the nonconjugated olefins. Testa[5] has been able to convert methyl oleate to methyl elaidate using benzophenone as photosensitizer. Monoenoic and dienoic fatty acid methyl esters have been isomerized to the *trans* form by treatment with thiols and diphenylphosphine in the presence of azobisisobutyronitrile.[16] The equilibrium mixture contained 75-80% *trans* double bonds and there was no migration of the double bond.

The conversion of a *trans* configuration to the *cis* is obviously more difficult, and requires the imparting of energy. This has been done by UV irradiation.[6] A possible application of this technique is in improving the drying potential of nonconjugated *trans* olefins resulting from dehydration, or other chemical modifications, of oils.

Nonconjugated to Conjugated Isomerization. The vegetable oils commonly isomerized to conjugated systems are generally those with substantial proportions of linoleic and linolenic acid—mainly soya bean, safflower, and linseed oils. These oils are isomerized by a number of alkaline catalysts including potassium hydroxide,[7] potassium *tert*-butoxide,[8] and sodium amide in liquid ammonia.[9] The suggested mechanism for this reaction is the removal of a proton from the active methylene group at C_{11} and/or C_{14}, and the formation of an anion with a number of possible resonating structures as shown in Figure 2.8.

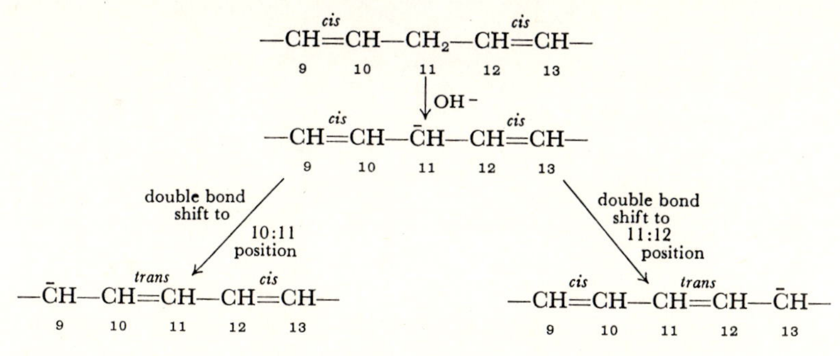

Figure 2.8. Possible mechanism for isomerization of nonconjugated linoleic and linolenic acids/esters with alkali.

It can be seen that some of the double bonds remain in the original position (for example, with linoleic acid either the C_{9-10} or the C_{12-13} bonds) and retain the original *cis* configuration. However, the double bond that shifts does so via the resonating intermediate structure, and so can take up the more stable *trans* configuration when it reforms. Therefore, the isomerization gives rise to a change in configuration, as well as the position of the unsaturation.

Alternative methods of conjugating oils probably involve free radical intermediates of analogous structures to the anions shown above. Typical catalysts for this type of mechanism are quinones, notably anthraquinone and 2-chloroanthraquinone.[10]

Conversion of a *cis*:*trans* conjugated isomer to the all *trans* configuration is often desirable, especially where reactions of the Diels-Alder type are to be undertaken. Catalysts for this change include iodine,[11] and a number of relatively unstable iodine derivatives.

Dehydration. Efforts to convert non- or semidrying oils to more unsaturated structures have generally aimed at the removal of a molecule (HX) from the fatty acid chain as shown in Figure 2.9. Attempts to introduce the substituent X, generally a halogen or hydroxyl, into positions α or β to the double bond of existing fatty acids (Types 1 and 2 in Figure 2.9), have not

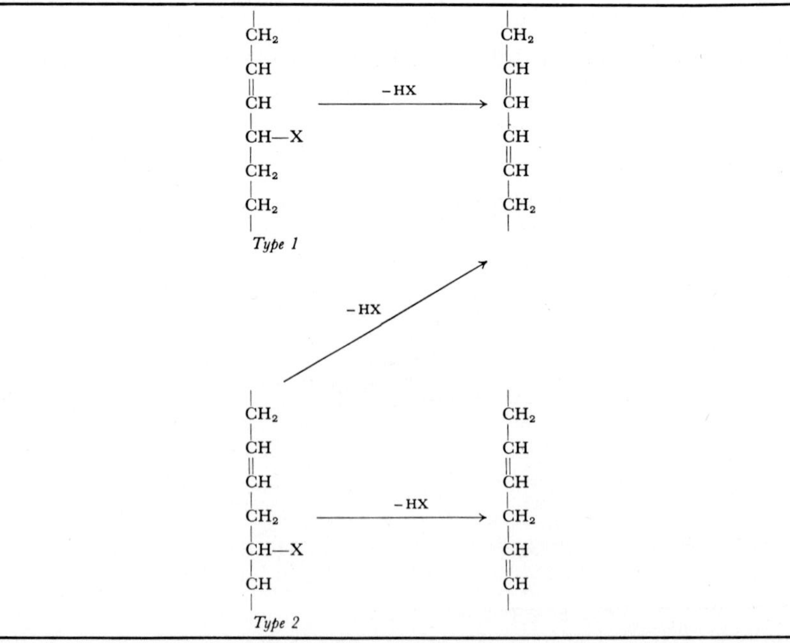

Figure 2.9.   Possible methods of converting oils to more highly unsaturated structures.

resulted in processes of major importance. However, ricinoleic acid, which as its glyceride occurs as castor oil, is 12-hydroxy oleic acid (Type 2 in Figure 2.9 where $X = OH$), and is a naturally occurring acid suitable for dehydration. Recently a more highly unsaturated hydroxy acid, 9-hydroxy, 10:12-octadecadienoic acid has been isolated from the oil from Cape Marigold, and offers the possibility of a commercial process for producing a conjugated triene.[12]

Dehydration of ricinoleic acid would be expected to yield predominantly the resonance-stabilized conjugated 9:11 isomer. However, in actual practice, the dehydrated oils contain only 25 to 30% conjugation; the major product is a mixture of the 9:12 isomers. Even so, the composition of dehydrated castor oil (D.C.O.) is attractive for the formulation of surface coatings; it is intermediate between the nonconjugated linoleic based oils (soya, safflower, etc.) and the highly unsaturated and conjugated tung or wood oil.

In general, the catalysts used to dehydrate castor oil are acidic, and the probable course of the reaction is the esterification of the C_{12} alcohol group followed by a splitting out of the acid as indicated in Figure 2.10 for phthalic anhydride.

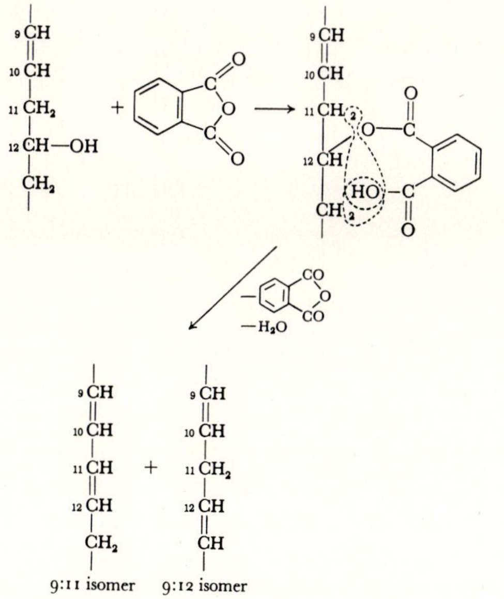

Figure 2.10. Possible mechanism for the dehydration of castor oil by phthalic anhydride.

The unexpectedly high proportion of 9:12 isomer is probably related to the structure of the intermediate cyclic complex prior to dehydration. In fact, significant amounts of the newly formed double bond have the *cis* structure [13] which would support the above suggestion.

In considering the properties of D.C.O. it is necessary to take into account not only the relative amounts of the conjugated and nonconjugated dienes, but also the geometrical configuration of these isomers. The somewhat drastic conditions used in dehydration cause changes in both the configuration and the position of the original *cis* 9–10 double bond. As a consequence, some 10:12-diene structure and some di-olefins with a *trans*-9–10 unsaturation are present.

It is well known that D.C.O., and some isomerized oils, dry to give films with a "sweat" or "after-tack," and there are two predominant causes of this phenomenon. First, the formation of an estolide, by partial hydrolysis and then re-esterification, or by alcoholysis, can in some cases limit dehydration and cause the nondrying structure [14] to be retained.

$$-CH{=}CH-CH_2-CH-CH-$$

$$\begin{array}{c} | \\ O \\ | \\ CO \\ | \\ (CH_2)_7 \\ | \\ CH \\ \| \\ CH \\ | \\ CH_2 \\ | \\ CHOH \\ | \\ (CH_2)_5 \\ | \\ CH_3 \end{array}$$

(Castor oil fatty acid estolide)

Second, and probably more important, it has been suggested that some of the position or geometric isomers present are not readily oxidizable and, hence, these isomers cause "after-tack." A number of papers [6] have suggested that the offending isomers are those containing *trans* unsaturation, and while, by analogy with elaidic/oleic systems, this may be true for nonconjugated systems, it is difficult to reconcile with other evidence in the conjugated series. A study of molecular models of *cis*:*cis*, *cis*:*trans*, and *trans*:*trans* conjugated octadecadienoic acids shows that oxygen can readily form a 1:4 cyclic peroxide by addition to the *trans*:*trans* isomer, but that ring formation would be difficult with the *cis*:*trans* and virtually impossible with the *cis*:*cis* compounds. These results are identical with the findings on Diels-Alder type addition[15] which has steric requirements similar to oxygen attack. It seems reasonable to conclude that the conjugated isomer responsible for "after-tack" is that with the *cis*:*cis* configuration; further work on isomerizing this isomer to the *trans*:*trans* would be worthwhile. One piece of evidence quoted to support the greater reactivity of *cis* conjugated compounds to oxidation is the greater oxygen uptake of α-eleostearic acid (*cis*:*trans*:*trans*) compared to the β-isomer (*trans*:*trans*:*trans*) (see page 47). This conclusion does not appear to be justified when the mechanism of the reaction is considered as pointed out on page 48.

REFERENCES

1. Khan, N. A., *Pakistan J. Sci. Ind. Res.*, **2**, 45 (1959).
2. Tolberg, W. E., and D. H. Wheeler, *J. Am. Oil Chemists' Soc.*, **35**, 385 (1958).
3. Griffiths, H. N., and T. P. Hilditch, *J. Chem. Soc. (London)*, **1932**, 2315.
4. Bertram, S. H., *Chem. Weekblad*, **33**, 3 (1936).
5. Testa, A. C., *J. Org. Chem.*, **29**, 2461 (1964).
6. Cowan, J. C., *Ind. Eng. Chem.*, **41**, 294 (1949).
7. Kass, J. P., and G. O. Burr, *J. Am. Chem. Soc.*, **61**, 3292 (1939).
8. White, H. B., and F. W. Quackenbush, *J. Am. Oil Chemists' Soc.*, **36**, 653 (1959).
9. Abu-Nasr, A. M., and R. T. Holman, *J. Am. Oil Chemists' Soc.*, **32**, 414 (1955).
10. Rushman, D. F., and E. M. G. Simpson, *Trans. Faraday Soc.*, **51**, 237 (1955).
11. Nichols, P. L., S. F. Herb, and R. W. Riemenschneider, *J. Am. Chem. Soc.*, **73**, 247 (1951).
12. Mills, M. R., *J. Oil Colour Chemists' Assoc.*, **47**, 187 (1964).
13. Body, D. R., and F. B. Shorland, *Chem. Ind. (London)*, **1961**, 1665; *J. Am. Oil Chemists' Soc.* **42**, 5 (1965).
14. Bolley, D. S., *J. Am. Oil Chemists' Soc.*, **36**, 518 (1959).
15. Paschke, R. F., and D. H. Wheeler, *J. Am. Oil Chemists' Soc.*, **32**, 469 (1955).
16. Sgoutas, D. S., and F. A. Kummerow., *Lipids*, **4**, No. 4, 283, (1969).

POLYMERIZED AND MALEINIZED OILS

The relatively long time required for an oil film to dry has been a limiting factor in the use of oils in surface coatings. One of the approaches used to shorten the drying time has been to carry out some of the oxidation or polymerization reactions prior to film application. In other words the molecular weight of the initial film former is higher, and fewer cross-links are required to reach the dry stage. Other reasons for polymerizing or pretreating oils include the prevention of frosting or rapid surface dry, and the formation of dimerized and modified acids for use in other formulations.

When air or oxygen is passed through an oil which is heated at up to 130°C, oxidation occurs to give a product termed "blown oil". The reaction may be accelerated by conventional driers, and the degree of blowing is related to the proposed end use.

Cross-linking of oil molecules is also achieved by thermal or peroxide induced polymerization, and these methods give products with more of the desirable carbon-carbon cross-links (as distinct from some oxygen bonded systems in blown oils). These oils are usually bodied to a predetermined standard viscosity, hence, they are often termed "stand oils".

The rate at which an oil polymerizes thermally is related to the type of unsaturation; conjugated olefins react very much faster than the non-conjugated, and *trans* conjugated isomers react faster than the *cis*. This is shown in Table 2.6.[1] The rates were measured by comparing the time to form 60% polymer.

TABLE 2.6

COMPARATIVE RATES OF POLYMERIZATION OF
C_{18} ACID ESTERS [1]

C_{18} Acid Ester	Relative Rate of Polymerization
9:12-*cis*:*trans*	0.74
9:12-*cis*:*cis*	1.0
9:12-*trans*:*trans*	1.2
9:12:15-*cis*:*cis*:*cis*	2.4
9:11-*cis*:*trans*	5.8
9:11-*trans*:*trans*	26.0
9:11:13-*cis*:*trans*:*trans*	170
9:11:13-*trans*:*trans*:*trans*	340

The mechanism by which thermal polymerization occurs is generally regarded as being of the Diels-Alder type addition of a conjugated diene to a double bond. The rates indicated in Table 2.6 are in conformity with this mechanism since, for example, the all *trans* β-eleostearic acid reacts faster than the α-eleostearic acid (one *cis* bond). It has been shown by a study of molecular models that the *trans* conjugation is much preferred for cyclic addition. Hence, the β-form has two possible points of attack and the α- only one, as shown below. Kinetic studies also support the view that increased probability of reaction, and not a reduction in activation energy, is responsible for the greater activity of the β-isomer.[2]

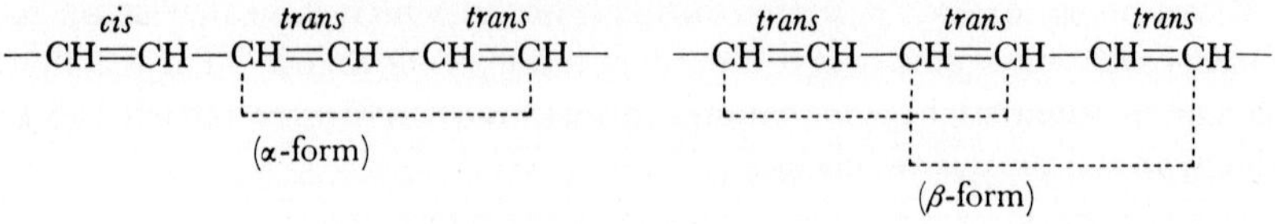

| *cis* | *trans* | *trans* | | *trans* | *trans* | *trans* |

—CH=CH—CH=CH—CH=CH—   —CH=CH—CH=CH—CH=CH—

(α-form)   (β-form)

Since the dimers, trimers, etc. formed from the eleostearates do not have diene conjugation,[1,3] it has been proposed that the central double bond of a second molecule is the dienophile.[1] The dimer from a β-eleostearate could then have structures of the type **21** to **24**, depending on whether addition is to the 9:11 or 11:13 conjugated double bonds, and if the arrangement is of the head-tail or head-head type.

$$CH_3—(CH_2)_3—CH{=}CH—\overset{11}{C}H—CH{=}CH—\overset{9}{C}H—(CH_2)_7—COOR$$
$$CH—————CH$$
$$CH_3—(CH_2)_3—CH{=}CH \qquad\qquad CH{=}CH—(CH_2)_7—COOR$$

(**21**)

$$CH_3—(CH_2)_3—CH{=}CH—\overset{11}{C}H—CH{=}CH—\overset{9}{C}H—(CH_2)_7—COOR$$
$$CH—————CH$$
$$ROOC—(CH_2)_7—CH{=}CH \qquad\qquad CH{=}CH—(CH_2)_3—CH_3$$

(**22**)

$$CH_3—(CH_2)_3—\overset{13}{C}H—CH{=}CH—\overset{11}{C}H—CH{=}CH—(CH_2)_7—COOR$$
$$CH—————CH$$
$$CH_3—(CH_2)_3—CH{=}CH \qquad\qquad CH{=}CH—(CH_2)_7—COOR$$

(**23**)

$$CH_3—(CH_2)_3—\overset{13}{C}H—CH{=}CH—\overset{11}{C}H—CH{=}CH—(CH_2)_7—COOR$$
$$CH—————CH$$
$$ROOC—(CH_2)_7—CH{=}CH \qquad\qquad CH{=}CH—(CH_2)_3—CH_3$$

(**24**)

The thermal polymerization of tung oil is carried out to reduce the tendency of films to "frost" or dry rapidly on the surface. Tung oil varnishes are prone to wrinkling in poorly ventilated ovens because the gaseous reaction products induce a rapid surface polymerization; this can be minimized by a prior thermal treatment which is called "gas proofing." However, the complete removal of the "frosting" tendency of tung oil is difficult; gelation often occurs during the thermal treatment before the "frost-free" stage is reached.[3] Gas proofing is often carried out in the presence of another resinous material, (for example, a phenolic resin) to give better control over the reaction.

Conjugated dienes derived from dehydrated castor oil or isomerized linoleic based oils also undergo thermal polymerization. Paschke et al.[4] have shown that with the 10:12-*trans*:*trans* linoleate the reaction mixture consists of products derived from addition to the 10 and 12 double bonds of the second molecule. Therefore, both bonds have a function as a dienophile.

The relatively slow thermal polymerization of nonconjugated oils and acids is generally attributed to the necessity to first isomerize and conjugate some of the unsaturation; this is the rate-controlling step. The conjugated diene can then add to any of the double bonds available to give structures, from linoleates, of the general type[5] shown below.

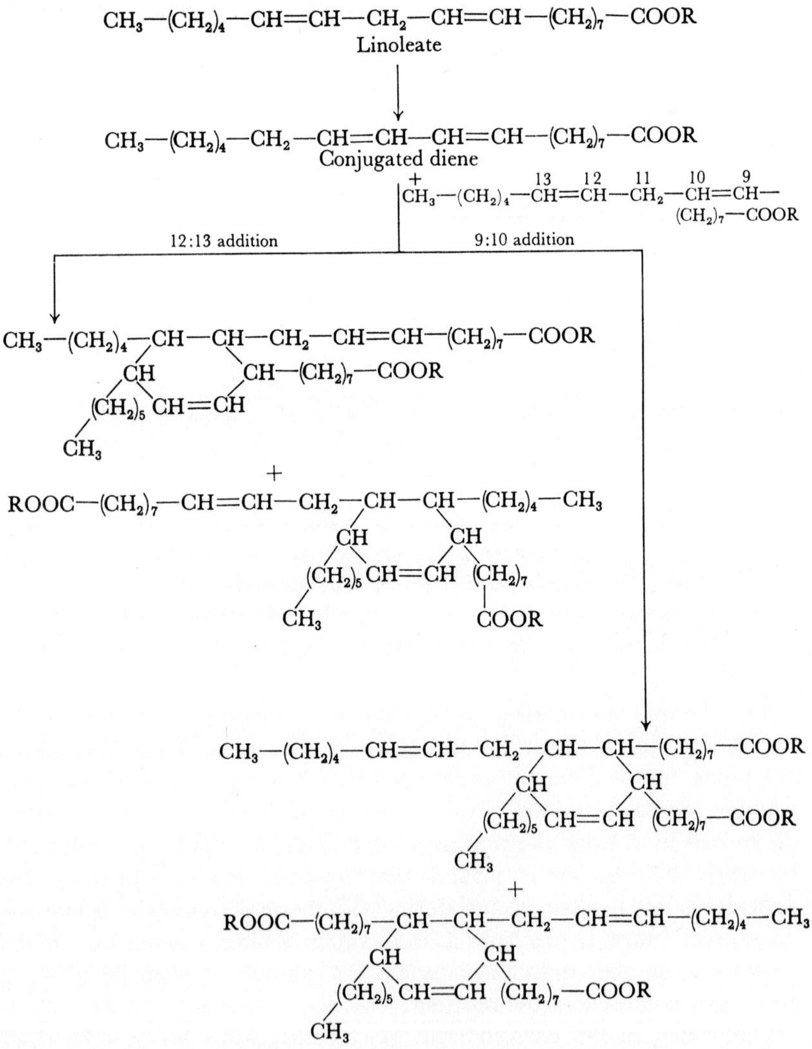

$$CH_3-(CH_2)_4-CH=CH-CH_2-CH=CH-(CH_2)_7-COOR$$

Linoleate

$$CH_3-(CH_2)_4-CH_2-CH=CH-CH=CH-(CH_2)_7-COOR$$

Conjugated diene

+

$$\overset{13}{CH_3}-(CH_2)_4-\overset{13}{CH}=\overset{12}{CH}-\overset{11}{CH_2}-\overset{10}{CH}=\overset{9}{CH}-(CH_2)_7-COOR$$

12:13 addition   9:10 addition

$$CH_3-(CH_2)_4-CH-CH-CH_2-CH=CH-(CH_2)_7-COOR$$

with pendant groups:

$$\begin{array}{c}CH \\ (CH_2)_5\ CH=CH \\ CH_3\end{array} \qquad CH-(CH_2)_7-COOR$$

+

$$ROOC-(CH_2)_7-CH=CH-CH_2-CH-CH-(CH_2)_4-CH_3$$

with pendant groups:

$$\begin{array}{c}CH \\ (CH_2)_5\ CH=CH \\ CH_3\end{array} \qquad \begin{array}{c}CH \\ (CH_2)_7 \\ COOR\end{array}$$

$$CH_3-(CH_2)_4-CH=CH-CH_2-CH-CH-(CH_2)_7-COOR$$

with pendant groups:

$$\begin{array}{c}CH \\ (CH_2)_5\ CH=CH \\ CH_3\end{array} \qquad \begin{array}{c}CH \\ (CH_2)_7-COOR\end{array}$$

+

$$ROOC-(CH_2)_7-CH-CH-CH_2-CH=CH-(CH_2)_4-CH_3$$

with pendant groups:

$$\begin{array}{c}CH \\ (CH_2)_5\ CH=CH \\ CH_3\end{array} \qquad \begin{array}{c}CH \\ (CH_2)_7-COOR\end{array}$$

Some intramolecular cyclization is also likely to occur.[6] The mechanism by which the original nonconjugated system thermally isomerizes to the conjugated isomer is still uncertain, but a likely explanation is that suggested by Rushman and Simpson[7] where one molecule adds a hydrogen atom, abstracted from an active methylene of another molecule, as shown.

$$\text{—CH=CH—CH}_2\text{—CH=CH—} \quad\quad \text{—CH—CH—CH}_2\text{—CH=CH—}$$

$$+ \quad\quad\quad\quad\quad\quad\quad \overset{|}{\text{H}} \quad + $$

$$\text{—CH=CH—CH}_2\text{—CH=CH—} \longrightarrow \quad \text{—CH=CH—CH—CH=CH—}$$

$$\text{—CH—CH=CH—CH=CH—} \quad\quad \text{—CH=CH—CH=CH—CH—}$$

$$\quad\overset{|}{\quad} \text{—CH—CH—CH}_2\text{—CH=CH—} \quad\quad \overset{|}{\quad} \text{—CH—CH—CH}_2\text{—CH=CH—}$$

$$\quad\quad \overset{|}{\text{H}} \quad\quad\quad\quad\quad\quad \overset{|}{\text{H}}$$

$$\text{—CH}_2\text{—CH=CH—CH=CH—} \quad\quad \text{—CH=CH—CH=CH—CH}_2\text{—}$$

$$+ \quad\quad\quad\quad\quad\quad\quad\quad +$$

$$\text{—CH=CH—CH}_2\text{—CH=CH—} \quad\quad \text{—CH=CH—CH}_2\text{—CH=CH—}$$

The diene radical can then isomerize and, by reversal of the original hydrogen abstraction process, lead to either a 9:11 or 10:12 conjugated isomer.

It has been pointed out previously that the formation of a conjugated 9:11 or 10:12 system by isomerization of one bond leads to a *cis*:*trans*, or *trans*:*cis*, relationship, and it is preferable to add isomerization catalysts to form the *trans*:*trans* isomer which reacts most readily in the Diels-Alder reactions.

Dimerization of fatty acids or their derivatives can also be carried out by a dehydrogenation procedure using a free radical initiator (generally di-*tert*-butyl peroxide). This process is particularly applicable to nonconjugated and mono-olefins, and has also been used with fully saturated fatty acids. The dimer acids differ from those derived from thermal polymerization in that they have an open chain structure and contain unsaturation equivalent to the original fatty acid.

Treatment of methyl stearate at 130°C with di-*tert*-butyl peroxide results in dimerization, via the carbon atom α- to the ester group,[8] and the formation of *tert*-butanol.

$$2\ \text{CH}_3\text{—(CH}_2)_{16}\text{—COOCH}_3 + \text{(CH}_3)_3\text{C—O—O—C(CH}_3)_3$$

$$\downarrow$$

$$\begin{array}{l} \text{CH}_3\text{—(CH}_2)_{15}\text{—CH—COOCH}_3 \\ \quad\quad\quad\quad\quad\overset{|}{} \quad\quad\quad\quad + 2\ \text{(CH}_3)_3\text{COH} \\ \text{CH}_3\text{—(CH}_2)_{15}\text{—CH—COOCH}_3 \end{array}$$

Similarly, treatment of methyl oleate gives a dimerized product. However, in this case, the peroxide attack is predominantly (90 to 95%) at the methylene groups α- to the unsaturated portion of the molecule, since these are much more susceptible to radical reactions than the methylene α- to the carboxy group. A mixture of acids involving coupling at C_8, C_9, C_{10}, and C_{11}, is obtained;[9] this is to be expected in view of the resonating nature of the radicals from such attack.

$$CH_3—(CH_2)_6—\overset{10}{C}H{=}\overset{9}{C}H—CH_2—(CH_2)_6—COOCH_3$$

tert-butyl peroxide

attack at C_{11} — attack at C_8

$$CH_3—(CH_2)_5—\overset{\cdot}{C}H—CH{=}CH—(CH_2)_7—COOCH_3$$

$$CH_3—(CH_2)_5—CH{=}CH—\overset{\cdot}{C}H—(CH_2)_7—COOCH_3$$

$$CH_3—(CH_2)_6—\overset{10}{C}H{=}\overset{9}{C}H—\overset{\cdot}{C}H—(CH_2)_6—COOCH_3$$

$$CH_3—(CH_2)_6—\overset{\cdot}{C}H—CH{=}CH—(CH_2)_6—COOCH_3$$

Linoleates and linolenates give dimer acids or esters with a greater possible number of structural isomers because of the increased number of resonance forms of the intermediate radicals.[10] The reactions are analogous to those described in the autoxidation studies and gives rise to some conjugated double bonds.

Clay catalyzes the dimerization of linoleic acid and results in the formation of a product with a monocylic structure (main product) and some bicyclic and tricyclic products. The monocyclic structure possibly arises from a Diels-Alder type addition reaction and the bicyclic product from a free-radical coupling followed by intramolecular ring closure. The dimer formed from linoleic acid by the clay catalyst has a higher ratio of monocyclic to bicyclic dimer than is formed by thermal polymerization.[20]

The polymer-forming reactions induced in oils by heat or peroxides are of interest not only in the formation of bodied or stand oils but also in the preparation of dimer fatty acids for use in alkyds and polyamides. The function of the dimer acids in these formulations is to provide a dibasic acid with an aliphatic chain separating the carboxyl groups, thereby conferring flexibility on the polymer molecule.

Reaction of Oils with Maleic Anhydride and Other Dienophiles. The maleinization of oils or fatty acids is a reaction of significance in the preparation of water-soluble resins, since it results in an increase in acid groups for salt formation and solubilization (see page 311). The reaction also occurs during some polyesterifications, often to a minor, but nevertheless significant extent and can result in premature gelation.

The reaction of maleic anhydride with conjugated dienes, such as the 9:11 or 10:12 fatty acids, proceeds readily at temperatures as low as 80°C

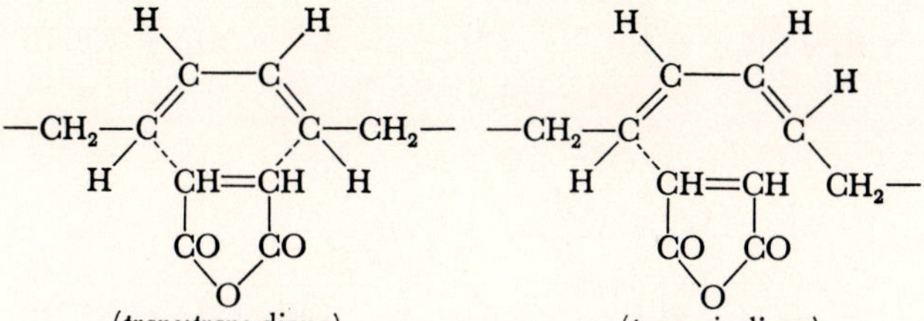

$$\textit{(trans:trans-diene)} \qquad \textit{(trans:cis-diene)}$$

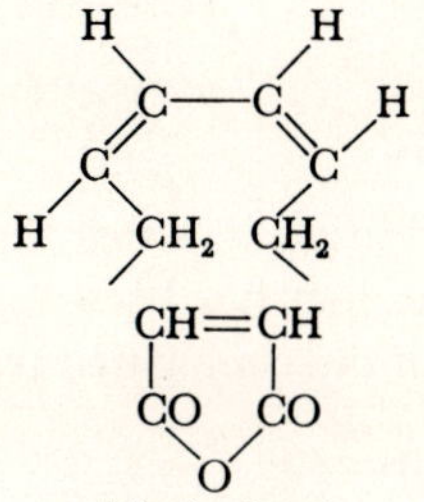

$$\textit{(cis:cis-diene)}$$

provided the double bonds of the oil are in a *trans*:*trans* configuration.[11] With a *cis*:*trans* diene, reaction is more difficult; a *cis*:*cis* compound does not react. This difference in reactivity of the various geometrical isomers can be appreciated from the illustration above.

The product of reaction between 9:11-*trans*:*trans* linoleic acid and maleic anhydride is shown below.[11]

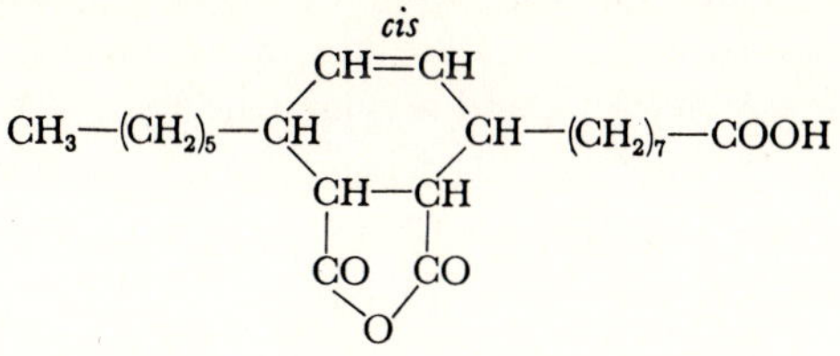

$$CH_3-(CH_2)_5-CH \cdots CH-(CH_2)_7-COOH$$

When conjugated trienes (such as those available from wood oil) are maleinized, there is the possibility of position isomers being formed, particularly with the all *trans* β-eleostearic acid. In this case, the maleic

anhydride can give the adducts shown below depending on whether the
9:11 or 11:13 bonds are involved. Both isomers are formed.[12]

$$CH_3-(CH_2)_3-\underset{\underset{CO\quad CO}{\underset{\diagdown O \diagup}{|\qquad|}}}{\overset{\overset{cis}{CH=CH}}{\underset{CH-CH}{\diagup\qquad\diagdown}}}\overset{trans}{CH-CH=CH-(CH_2)_7-COOH}$$

$$CH_3-(CH_2)_3-\overset{trans}{CH=CH}-\underset{\underset{CO\quad CO}{\underset{\diagdown O \diagup}{|\qquad|}}}{\overset{\overset{cis}{CH=CH}}{\underset{CH-CH}{\diagup\qquad\diagdown}}}CH-(CH_2)_7-COOH$$

With α-eleostearic acid, addition is across the *trans*:*trans*-11:13 bonds; in
fact, the structure of the adduct has been used to establish the 9-*cis*, 11-
trans, 13-*trans* relationship of the conjugated triene.[13]

$$CH_3-(CH_2)_3-\underset{\underset{CO\quad CO}{\underset{\diagdown O \diagup}{|\qquad|}}}{\overset{\overset{cis}{CH=CH}}{\underset{CH-CH}{\diagup\qquad\diagdown}}}\overset{cis}{CH-CH=CH-(CH_2)_7-COOH}$$

As is to be expected, the β-eleostearic acid reacts at a faster rate than the
α-isomer,[13] and with dienophiles other than maleic anhydride the rate
varies with the substituent groups. Maleic anhydride, chloromaleic anhydride, and methyl maleic anhydride, have reaction rates with β-eleostearic
acid[13] in the ratio of 19.1:14.6:1.0. Maleate or fumarate esters may be
used as alternative sources of dienophilic compounds.[14] Other compounds
that have been added to conjugated fatty acids include acrylic acid,[11]
methyl vinyl ketone,[15] acrolein, and acrylonitrile.[11]

$$-CH_2-CH=CH-CH=CH-$$
$$+\ CH_2=CH$$
$$|$$
$$COOH$$
$$\downarrow$$

$$-CH_2-\underset{\underset{X\quad Y}{|\qquad|}}{\overset{\overset{CH=CH}{\diagup\qquad\diagdown}}{\underset{CH-CH}{\diagdown\qquad\diagup}}}CH-$$

$$(X = H\ or\ COOH)$$
$$(Y = COOH\ or\ H)$$

The reaction of the nonconjugated 1:4 diene structure with maleic
anhydride (and similar compounds) is more difficult than with a conjugated

diene. Two types of product can result depending on the conditions used. Isomerization of the 1:4 diene to a *cis*:*trans*-1:3 system, followed by elaidinization, results in a *trans*:*trans* conjugated diene[16] which undergoes a typical Diels-Alder type addition. The methods used to develop the *trans*:*trans* conjugated isomer have been discussed previously (see page 60). The alternative method by which nonconjugated and even mono-olefins can add maleic anhydride involves the active methylene groups, and results in the formation of substituted succinic acids. These reactions require more drastic conditions (temperature approximately 200°C)[17] than a simple Diels-Alder addition but are common in resin production. The product of reaction from oleic acid-based oils and maleic anhydride is a mixture in which the maleic residue is attached at carbons 8, 9, 10, and 11. These isomers can result from intermediate radicals or carbonium ions in an analogous manner to the hydroperoxides formed by oxidation.

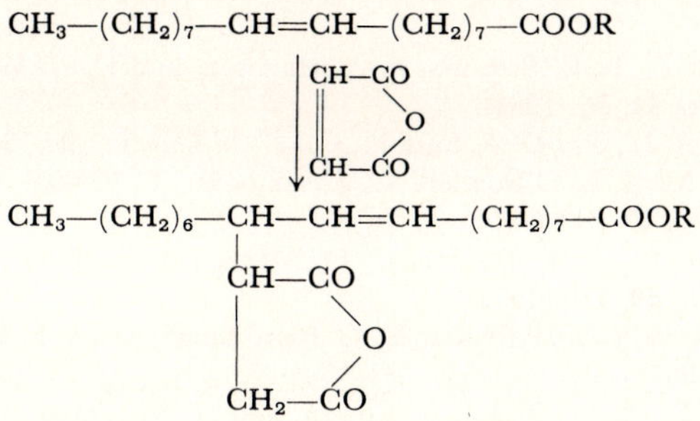

$$CH_3-(CH_2)_7-CH{=}CH-(CH_2)_7-COOR$$

$$CH_3-(CH_2)_6-CH-CH{=}CH-(CH_2)_7-COOR$$

With linoleic- and linolenic-based oils, the product is a more complicated mixture because of the increased number of position isomers (compare with autoxidation) and the possibility of more than one maleic anhydride residue adding to each fatty acid molecule.[18,19]

For example Rheineck and Khoe[21] have recently shown that the first adduct formed from methyl linoleate and maleic anhydride is a succinyl derivative presumably at the active methylene group at C_{11}. A double bond shift then occurs to yield a conjugated isomer with which a second maleic anhydride reacts via a Diels-Alder addition to give a product with a cyclohexane structure. One of the possible isomers is shown below

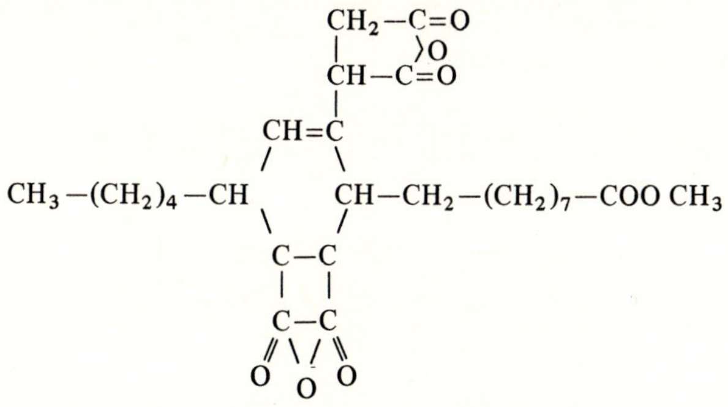

In practice it is common to use non-conjugated oils to control the reactivity of conjugated oils during maleinigation. For example a blend of up to 50% dehydrated castor oil in linseed oil has been used.[22]

REFERENCES

1. Wheeler, D. H., *Offic. Dig. Federation Soc. Paint Technol.*, **23**, 661 (1951).
2. Sims, R. P. A., *J. Am. Oil Chemists' Soc.*, **32**, 94 (1955).
3. Rheineck, A. E., and S. C. G. Peng, *Offic. Dig. Federation Soc. Paint Technol.*, **36**, 879 (1964).
4. Paschke, R. F., L. E. Peterson, and D. H. Wheeler, *J. Am. Oil Chemists' Soc.*, **41**, 723 (1964).
5. Firestone, D., *J. Am. Oil Chemists' Soc.*, **40**, 247 (1963).
6. Paschke, R. F., and D. H. Wheeler, *J. Am. Oil Chemists' Soc.*, **31**, 208 (1954).
7. Rushman, D. F., and E. M. G. Simpson, *Trans. Faraday Soc.*, **51**, 230 (1955); Rushman, D. F., *J. Oil Colour Chemists' Assoc.*, **40**, 814 (1957).
8. Harrison, S. A., L. E. Peterson, and D. H. Wheeler, *J. Am. Oil Chemists' Soc.*, **42**, 2 (1965).
9. Paschke, R. F., L. E. Peterson, S. A. Harrison, and D. H. Wheeler, *J. Am. Oil Chemists' Soc.*, **41**, 56 (1964).
10. Clingman, A. L., and D. A. Sutton, *J. Am. Oil Chemists' Soc.*, **30**, 53 (1953).
11. Teeter, H. M., J. L. O'Donnell, W. J. Schneider, L. E. Gast, and M. J. Danzig, *J. Org. Chem.*, **22**, 512 (1957).
12. Bickford, W. G., E. F. Du Pre, C. H. Mack, and R. T. O'Connor, *J. Am. Oil Chemists' Soc.*, **30**, 376 (1953).
13. Bickford, W. G., J. S. Hoffmann, D. C. Heinzelman, and S. P. Fore, *J. Org. Chem.*, **22**, 1080 (1957).
14. Miller, W. R., E. W. Bell, J. C. Cowan, and H. M. Teeter, *J. Am. Oil Chemists' Soc.*, **38**, 235 (1961).
15. Placek, L. L., and W. G. Bickford, *J. Am. Oil Chemists' Soc.*, **36**, 463 (1959).
16. Teeter, H. M., E. W. Bell, J. L. O'Donnell, M. J. Danzig, and J. C. Cowan, *J. Am. Oil Chemists' Soc.*, **35**, 238 (1958).
17. Bickford, W. G., G. S. Fisher, L. Kyame, and C. E. Swift, *J. Am. Oil Chemists' Soc.*, **25**, 254 (1948).
18. Plimmer, H., *J. Oil Colour Chemists' Assoc.*, **32**, 99 (1949).
19. Teeter, H. M., M. J. Geerts, and J. C. Cowan, *J. Am. Oil Chemists' Soc.*, **25**, 158 (1948).
20. Wheeler, D. H., A. Milun, and F. Linn., *J. Am. Oil Chemists Soc.*, **47**, 242, (1970).
21. Rheineck, A. E., and T. H. Khoe, *Fette, Siefen, Anstrich.*, **71** (8), 644 (1969).
22. Ghanen, N. A., and A. M. M. Nasser, *Farbe u. Lack*, **75** (5), 419 (1969).

CHAPTER 3

VARNISHES AND ALKYD RESINS

Varnishes and alkyd resins are formed by modifying vegetable oils with natural or synthetic resins. In alkyd resins the oil is combined chemically with the components of a polyester resin, whereas varnishes are often a physical solution of the natural or synthetic resin in the oil; some varnishes are formed by chemical reaction between the oil and resin components. Compared with unmodified oils, alkyds and varnishes are characterized by improved gloss and hardness, better resistance properties, and reduced drying times. These properties are the result of the higher initial average molecular weight, the formation of a film of higher molecular weight and complexity, and the production of less decomposition products on aging. Figure 3.1 illustrates this latter point for a range of vehicles subjected to accelerated weathering.[1]

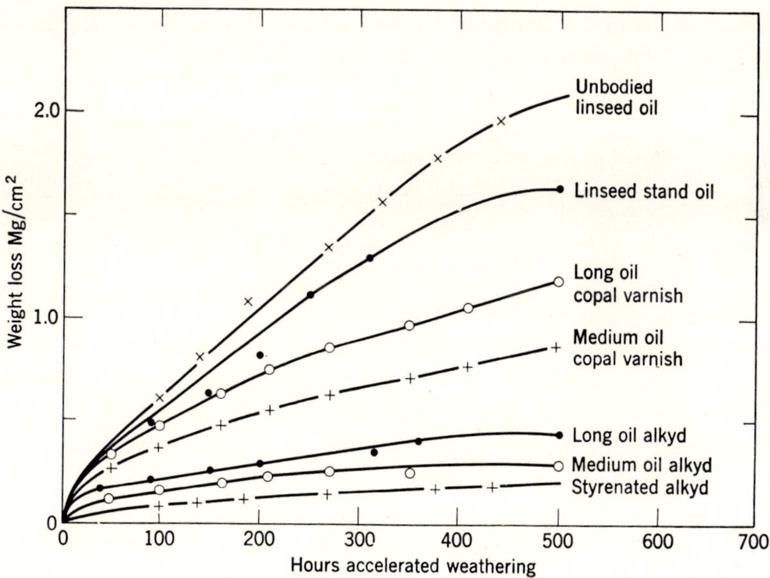

Figure 3.1. The relationship between polymer composition and artificial weathering resistance.[1]

75

VARNISHES

Many varnishes are simply a solution of a natural or synthetic resin or gum in oil. The dispersion or solubilization of the resin in the oil is aided by heating. In some cases, the oil may be partially bodied by thermal polymerization to improve or modify the film properties and to give acceptable solution viscosities.

Typical resins used include the low molecular weight (approximately 1200) coumarone-indene copolymers, prepared by cationic polymerization of the monomers.

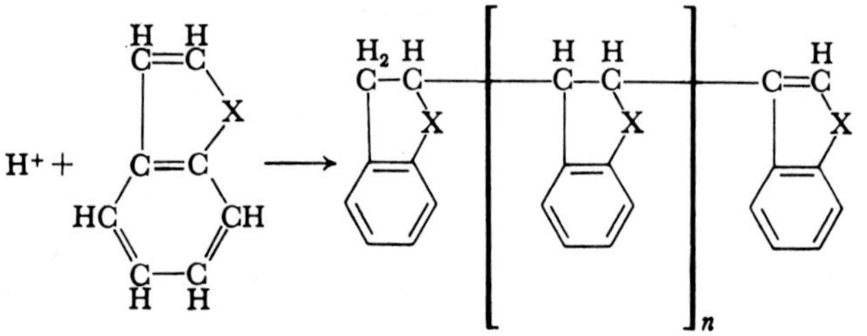

$$(X = O \text{ or } CH_2)$$

Rosin or its glyceryl ester (known as ester gum) is also commonly used, and with this type of resin there is the possibility of some ester interchange with the oil. The main constituent of rosin is abietic acid.

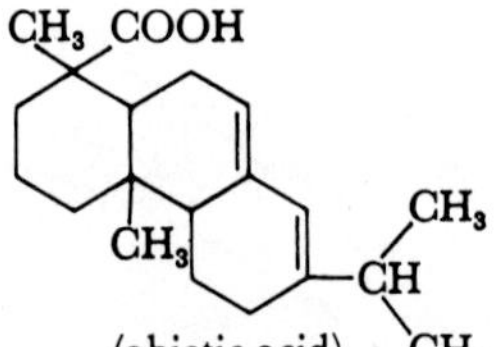

(abietic acid)

The use of the pentaerythritol ester of abietic acid (PE ester gum) gives improved properties as a result of the increased molecular weight and complexity. Also, the esterification of pentaerythritol with rosin is more readily accomplished than that of glycerol, since it has only primary hydroxyl groups and has less steric hindrance.[2]

Rosin, or its esters, may be improved for varnish use by a maleinization process which increases the potential carboxyl functionality of the molecule from one to three. It appears likely that the rosin isomerizes to a conjugated diene—with both bonds in the same ring—before the maleic anhydride adds to the molecule.

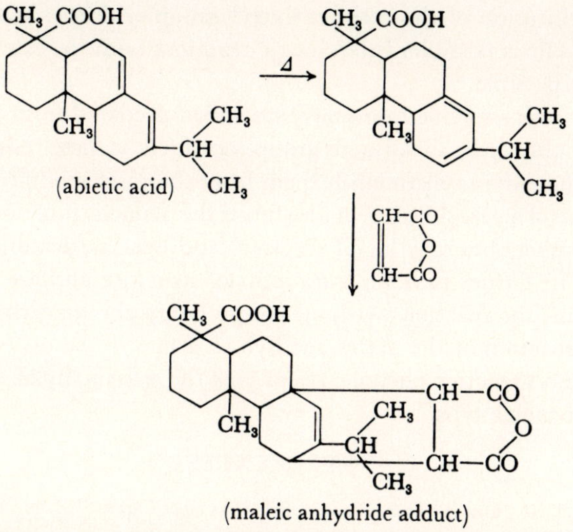

CH_3 COOH

(abietic acid) CH_3

$\xrightarrow{\Delta}$

CH_3 COOH

CH–CO
O
CH–CO

CH_3 COOH

(maleic anhydride adduct)

Quantitative studies of the amount of maleic anhydride that can be reacted with a given rosin have shown that the uptake of maleic anhydride is greater than predicted by the abietic acid content. Smith and Wise[4] have concluded that other rosin acids, including those present in hydrogenated rosin, react to form substituted succinic anhydride structures.

Esterification of the rosin/maleic anhydride adducts is used to form high molecular weight polymers which are useful in a number of applications.

Rosin will also react with phenolic resins (see page 260) to give modified phenolic resins that are soluble in vegetable oils. The reaction is probably as shown below,[3] and leads to chroman ring formation.

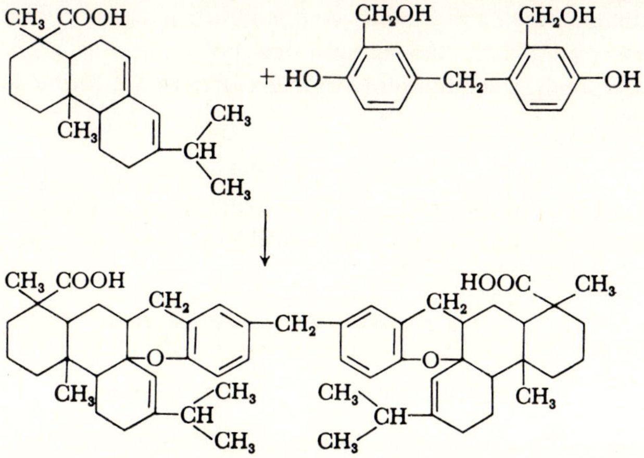

CH_3 COOH

$+$ HO–CH_2OH–CH_2–CH_2OH–OH

CH_3 COOH HOOC CH_3

Some esterification of the rosin carboxyl group with the hydroxy groups of the phenolic is possible. Interchange reactions with the ester links in the oil are also possible.

Phenolic resins soluble in oil have also been prepared from phenols with a large oleophilic *para* substituent group (octyl, etc.); these resins are known as 100% phenolics to distinguish them from the rosin modified resins. The introduction of a *para* substituent also limits the phenolic functionality to two. The 100% phenolics may be oil reactive (and heat hardenable) or non-oil reactive. The former have an *ortho*methylol structure and are of the resole type of resin; the reactions with oil are probably chroman ring formation, and some reaction at the active methylene group of the oil (see also page 261). The non-reactive phenolic resins lack the *ortho*methylol structure and are of the novolac type.

REFERENCES

1. Bullett, T. R., and A. T. S. Rudram, *J. Oil Colour Chemists' Assoc.*, **42**, 778 (1959).
2. Hercules Powder Company, *Alkyd Report No. 1*.
3. Hultzsch, K., *Chemie der Phenolharze*, Springer-Verlag, Berlin, 1950, p. 78, p. 156.
4. Smith, C. D., and J. K. Wise, *J. Paint Technol.*, **41**, 338 (1969).

ALKYD RESINS

Alkyd resins represent a class of polymers that are used extensively in surface-coating formulations because of their low cost and versatility.[1-3] The word *alkyd* was originally coined by Kienle[4] from the component monomers—*al*cohol and *acid*—and in the broad sense refers to a polyester. However, in the coatings industry, polyesters with unsaturation in the backbone, as distinct from a side chain, are given special distinction and termed "unsaturated polyesters." The term alkyd is often used in a restrictive manner to describe a fatty acid modified polyester.[5] This differentiation between alkyds and unsaturated polyesters can be justified on chemical grounds, since the film-forming reactions are distinctly different (Chapter 5).

REFERENCES

1. Rooney, J. F., *Offic. Dig. Federation Soc. Paint Technol.*, **36**, 32 (1964).
2. Martens, C. R., *Alkyd Resins*, Reinhold Publishing Corp., New York, 1961.
3. *Chem. Eng. News*, **42**, 100 (1964).
4. Kienle, R. H., and C. S. Ferguson, *Ind. Eng. Chem.*, **21**, 349 (1929).
5. Spitzer, W. C., *Offic. Dig. Federation Soc. Paint Technol.*, **36**, 16 (1964).

THEORY OF POLYESTERIFICATION

In order to form a polymer molecule by esterification reactions, it is necessary for both the alcohol and acid monomers to be at least bifunctional.[1]

For example, equimolar amounts of ethylene glycol and adipic acid can yield a polymer of the following structure:

$$n\ HO—CH_2—CH_2—OH + n\ HOOC—(CH_2)_4—COOH$$

$$\downarrow -(2n-1)H_2O$$

$$HO—CH_2—CH_2—O—\left[CO—(CH_2)_4—COOCH_2CH_2O\right]_{n-1}—CO—(CH_2)_4—COOH$$

In the above system, both the glycol and dibasic acid are able to achieve their maximum functionality of two. However, if we consider the esterification of 1 mole of adipic acid with 2 moles of ethylene glycol then, on the average, the glycol has an actual functionality of one, and the idealized product of reaction is as shown.

$$2\ HO—CH_2—CH_2—OH + HOOC—(CH_2)_4—COOH$$

$$\downarrow -2H_2O$$

$$HO—CH_2—CH_2—OCO—(CH_2)_4—COO—CH_2—CH_2—OH$$

Therefore, the actual functionality is related to the mole ratio of the reactants, and the actual and potential functionality are not necessarily the same.

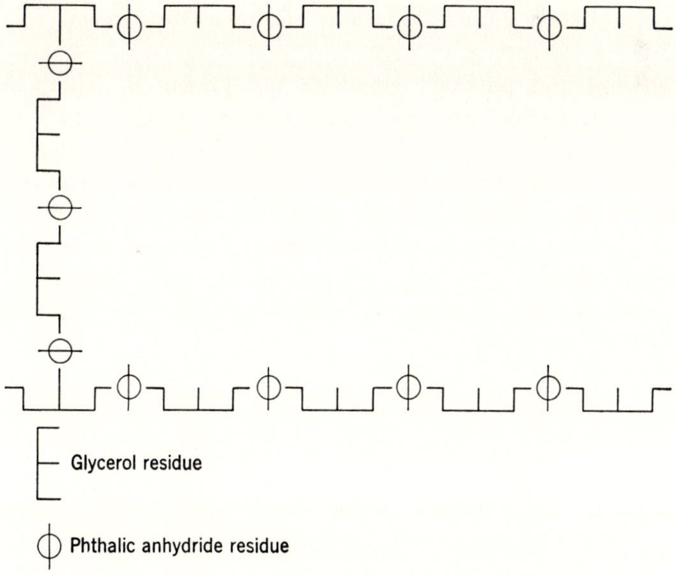

The use of more than the stoichiometric requirement of one component in a polyesterification assumes even greater importance in more highly functional systems capable of cross-linking and gelation, and which are used in many alkyd-type formulations. In the condensation of glycerol with phthalic anhydride, the structure of the polyester that is formed depends on the mole ratio of the reactants. With equivalent concentrations of hydroxyl and carboxyl groups (that is, two molecules of glycerol and three of phthalic anhydride), gelation is possible, since a cross-linked structure can form and each component can reach its maximum functionality.

As the condensation of glycerol and phthalic anhydride proceeds, the functionality of the molecules formed increases, whereas in a linear polyester, where each component is bifunctional, the polymer molecules are also bifunctional.

(functionality of four—two carboxyl and two hydroxyl groups)

If the glycerol and phthalic anhydride are present in equimolecular amounts, that is, 1 mole of glycerol (three hydroxyl groups) and 1 mole of phthalic anhydride (two carboxyl groups), then the actual functionality of the glycerol is limited to two, that is, the number of carboxyl groups available. To illustrate this point, we shall assume that all the primary hydroxyl groups of glycerol react before any secondary hydroxyls react. A linear polymer will result.

Therefore, the use of more glycerol than the stoichiometric amount required by the carboxyl groups present reduces the functionality. In the example given above, the actual functionality becomes two. Most alkyd formulations are such that the glycerol has an actual functionality of slightly more than two. The actual functionality of a polyol can be calculated from the following equation.[2]

$$F_{OH(actual)} = \frac{F_{OH(potential)}}{1 + n}$$

where

$$F_{OH(potential)} = \text{maximum functionality if all groups can react}$$

and

$$n = \text{decimal proportion of the polyol hydroxyl in excess of the carboxyl groups available}$$

For example, where glycerol is present in 50% excess (see above)

$$F_{OH} = \frac{3}{1 + 0.5} = 2$$

The term "excess polyol" is often used to indicate the presence of non-stoichiometric quantities of alcohol and acid groups. Unfortunately, there is not general conformity in the manner in which this term is calculated; the following formulas are used.

$$\text{Excess polyol} = \frac{\text{Weight of polyol in excess of stoichiometric amount}}{\text{Total weight of polymer}} \times 100$$

or

$$\text{Excess polyol} = \frac{\text{Percent of polyol in excess of stoichiometric amount}}{\text{Total percent of polyol}} \times 100$$

The use of monobasic acids is an additional means of controlling the functionality of a system by reducing the acid functionality. For example, an equimolar mixture of fatty acid, phthalic anhydride, and glycerol ideally can give the linear polymer shown in a simplified form below:

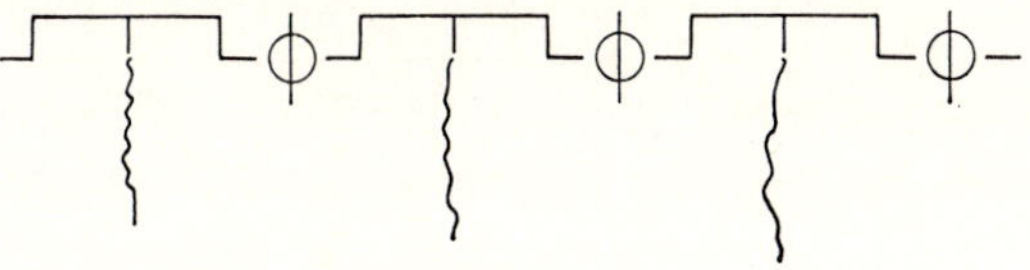

where $\quad \sim\!\sim\!\sim$ = fatty acid residue

Here, the acid functionality is given by

$$F_{acid} = \frac{[\text{mole acid}_1 \times \text{functionality}] + [\text{mole acid}_2 \times \text{functionality}]}{\text{Total mole}}$$

$$= \frac{[1 \times 2] + [1 \times 1]}{1 + 1}$$

$$= 1.5$$

The overall average functionality of the system

$$F_{\text{av}} = \frac{\text{Total equivalents}}{\text{Total mole}}$$

and in the case above

$$F_{\text{av}} = \frac{[3 \times 1] + [1 \times 1] + [1 \times 2]}{1 + 1 + 1}$$
$$= 2$$

An alternative method of looking at this system is to consider the fatty acid and glycerol as the equivalent of a glycol.

The mathematical treatment of functionality becomes more difficult when excess polyol is present; it is necessary to calculate and then apply the actual functionality. However, suitable procedures have been developed, mainly as a result of the original efforts of Flory[3] and Carothers,[4] and these are useful in calculating the degree of reaction at which gelation will occur. Most alkyds are required to have a low acid number, and gel point calculations can be used to indicate a formulation that will have a low acid number before gelation occurs. Using the terminology of Carothers,[4] the gel point equation is derived as follows.

Let F_{av} = average degree of functionality (that is, number of functional groups per monomer molecule)

and N_0 = number of monomer molecules initially present

then the number of functional groups initially present

$$= N_0 F_{\text{av}}$$

If N = the number of molecules after reaction has occurred, then the number of functional groups lost = $2(N_0 - N)$

Since the extent of reaction (p) = the fraction of functional groups lost

$$p = \frac{2(N_0 - N)}{N_0 F_{\text{av}}}$$

Now the average degree of polymerization (x) = N_0/N

$$\therefore p = \frac{2}{F_{\text{av}}} - \frac{2}{x F_{\text{av}}}$$

Since at the gel point x is very large, the extent of reaction on gelation (p_g) is given by

$$p_g = \frac{2}{F_{\text{av}}}$$

This equation has been developed assuming that equivalent amounts of hydroxyl and carboxyl groups are present. This situation rarely occurs in commercial alkyd formulations, and the actual functionality of the polyol must be used in calculating the entity F_{av}. As an example, consider the alkyd composition shown in Table 3.1.

TABLE 3.1

FUNCTIONALITY IN AN ALKYD FORMULATION

Composition	Equivalents	Maximum Functionality	Moles
Fatty acid	1.0	1	1.0
Phthalic anhydride	2.22	2	1.11
Glycerol	4.08	3	1.36
	7.30		3.47

Then, disregarding any excess of polyol (that is, Carothers' equation),

$$F_{av} = \frac{\text{Total equivalents}}{\text{Total moles}}$$

$$= \frac{7.30}{3.47} = 2.104$$

This suggests the composition will gel, but this is incorrect as will be shown below. The glycerol is in excess and only 3.22 equivalents (that is, the amount of acid) can react. Carothers' equation can be applied if we assign zero functionality to the excess polyol,[2] then

$$F_{av} = \frac{\text{equivalents}}{\text{mole}} = \frac{3.22 + 3.22}{3.47}$$

$$= 1.86$$

and

$$p = \frac{2}{F_{av}} = \frac{2}{1.86} = 1.07$$

$$= 107\% \text{ reaction}$$

that is, the composition will not gel.

Conversely, the equation can be used to calculate the modification necessary to a formulation so that gelation will occur at a given acid value.

In the example given above, if the required acid value at gel was 0 (that is, $p = 1$), then

$$p = \frac{2}{F_{\mathrm{av}}}$$

$$1 = \frac{2}{F_{\mathrm{av}}}$$

$$\therefore F_{\mathrm{av}} = 2$$

Therefore, the number of molecules must be one half the number of equivalents and, since the total equivalents must equal twice the acid equivalents, we have

$$\text{Total equivalent} = 6.44$$

$$\text{Therefore, total moles} = \frac{\text{total equivalents}}{2} = \frac{6.44}{2}$$

$$= 3.22$$

$$\text{Mole of glycerol} = 3.22 - (\text{mole acid})$$

$$= 3.22 - 2.11 = 1.11$$

that is, it is necessary to reduce the glycerol from 1.36 to 1.11 mole.

Carothers' definition of gel point, and the modifications of this concept mentioned in a number of references,[5-7] assume, in effect, that an infinite number-average molecular weight corresponds to gelation. On the other hand, Flory,[3] and others,[8,9] have developed their theories of gelation around the concept that the weight-average molecular weight is infinitely large at gel point. On this basis, the derived formula for calculating gel point becomes (see Refs. 3, 8, and 9, for details of derivation):

$$p_A \text{ (gel)} = \sqrt{\frac{\epsilon}{2(1 - \lambda)}}$$

where

$$\epsilon = \frac{\text{equivalents of hydroxyl}}{\text{equivalents of acid}}$$

$$\lambda = \frac{\text{equivalents of carboxyl from monobasic acid}}{\text{total equivalents of carboxyl from monobasic and dibasic acid}}$$

This equation is theoretically more accurate than those derived from Carothers' definition of gel point.[9] Nevertheless, in practice, the theoretical and actual gel points often do not correspond despite numerous modifications and correction factors which have been applied to both the Carothers and Flory equations.[5-12] The equations are of limited use in designing an alkyd resin from theoretical considerations alone.[5] Their value is in providing a basis for systematically modifying a formulation that has gelled prematurely, or that does not have acceptable acid value/viscosity characteristics. Possible reasons for the lack of agreement between theoretical and practical gel points are the following:

1. Intramolecular esterification reactions that result in ring closure[13] and are, therefore, chain terminating, are not allowed for.

$$CH_2-OH$$
$$CH-OH \ + \ \text{(phthalic anhydride)}$$
$$CH_2-OH$$

$$\downarrow$$

$$CH_2-OH$$
$$CH-O-CO-$$
$$CH_2-O-CO-$$

2. Alternative reactions which can increase the functionality without loss of carboxyl groups are neglected. Two common examples are etherification of the polyol, and the reaction of unsaturated dibasic acids, such as maleic anhydride with the fatty acid or polyol.[2]

$$2 \ \begin{matrix} CH_2-CH-CH_2 \\ | \quad\ | \quad\ | \\ OH \quad OH \quad OH \end{matrix} \xrightarrow[-H_2O]{H^+} \begin{matrix} CH_2-CH-CH_2-O-CH_2-CH-CH_2 \\ | \quad\ | \qquad\qquad\qquad | \quad\ | \\ OH \quad OH \qquad\qquad\qquad OH \quad OH \end{matrix}$$

$$\sim\!\!\sim\!\!\sim COOH \ + \ \begin{matrix} CH-CO \\ \| \qquad\quad O \\ CH-CO \end{matrix}$$

$$\downarrow$$

$$\sim\!\!\sim\!\!\sim COOH$$
$$\begin{matrix} CH-CO \\ \| \qquad\quad O \\ CH-CO \end{matrix}$$

(where $\sim\!\!\sim\!\!\sim COOH$ = fatty acid)

(or isomers)

$$\begin{matrix} CH-CO \\ \| \qquad\quad O \\ CH-CO \end{matrix} + \begin{matrix} CH_2OH \\ | \\ CHOH \\ | \\ CH_2OH \end{matrix} \longrightarrow \begin{matrix} CH_2OH \\ | \\ CHOH \\ | \\ CH_2 \\ | \\ O \\ | \\ CH-CH_2 \\ | \qquad | \\ CO \quad CO \\ \ \ \diagdown\!O\!\diagup \end{matrix}$$

3. The theory assumes all groups are equally reactive and available for reaction. This is now known to be incorrect and, hence, the calculation of the functionality becomes unreliable.[14]

4. The calculated gel points could correspond with the *initial* formation of branched or three-dimensional polymer, whereas the practical gel points represent *significant* amounts of three-dimensional polymer.[12]

REFERENCES

1. Bevan, E. A., *J. Oil Colour Chemists' Assoc.*, **26**, 155 (1943).
2. Spitzer, W. C., *Offic. Dig. Federation Soc. Paint Technol.*, **36**, 16 (1964).
3. Flory, P. J., *J. Am. Chem. Soc.*, **58**, 1877 (1936).
4. Carothers, W. H., *Trans. Faraday Soc.*, **32**, 39 (1936).
5. Patton, T. C., *Alkyd Resin Technology*, Interscience Publishers, Inc., New York, 1962.
6. Johnston, C. W., *Offic. Dig. Federation Soc. Paint Technol.*, **32**, 1327 (1960).
7. Vaughan, C. L. P., and F. E. Schmitt, *Offic. Dig. Federation Soc. Paint Technol.*, **30**, 1131 (1958).
8. Jonason, M., *J. Appl. Polymer Sci.*, **4**, 129 (1960).
9. Christensen, G., *Offic. Dig. Federation Soc. Paint Technol.*, **36**, 28 (1964).
10. Lilley, H. S., *J. Appl. Polymer Sci.*, **5**, S16 (1961).
11. Pinner, S. H., *J. Polymer Sci.*, **21**, 153 (1956).
12. Bobalek, E. G., E. R. Moore, S. S. Levy, and C. C. Lee, *J. Appl. Polymer Sci.*, **8**, 625 (1964).
13. Kienle, R. H., P. A. Van der Meulen, and F. E. Petke, *J. Am. Chem. Soc.*, **61**, 2258 (1939).
14. Solomon, D. H., and J. J. Hopwood, *J. Appl. Polymer Sci.*, **10**, 981 (1966).

METHODS OF PREPARING AND PROCESSING ALKYDS

Fatty acid modified polyesters are prepared by a number of processes, and each has commercial or technical considerations to recommend it. The four most common procedures are:

1. The fatty acid process.
2. The monoglyceride or alcoholysis process.
3. The acidolysis process.
4. The fatty acid/oil process.

The fatty acid process in its simplest form involves the simultaneous condensation of the polyol, dibasic acid, and fatty acid, usually at temperatures of 200 to 240°C. The molecular weight or the degree of condensation can be followed by acid value and viscosity measurements. A modification of this process, developed by Kraft[1] and sometimes referred to as the "high polymer technique," involves the incremental addition of the fatty acid so that chain-terminating reactions are limited.

Most fatty acids are naturally available as their triglyceride and, since these oils are usually cheaper than the corresponding fatty acids, they are widely used in the commercial production of alkyd resins. However, when a glyceride oil (except hydroxylated oils), polyol, and dibasic acid are heated together, a heterogeneous mixture results because of the preferential condensation of the polyol and dibasic acid. This mixture has no value as a coating vehicle. Alcoholysis or acidolysis of the oil, prior to the condensation stage, overcomes the problem of compatibility. Alcoholysis is by far the most common process and, since glycerol was used in the majority of earlier alkyd formulations, this technique of preparing alkyd resins is known as the monoglyceride process. Usually, two moles of glycerol are reacted with one mole of oil at approximately 240°C in the presence of a basic catalyst. The product can be represented in an idealized manner by the following equation:

$$
\begin{array}{ccc}
CH_2OH & CH_2OCOR & CH_2OCOR \\
| & | & | \\
2\ CHOH \ + & CHOCOR \ \longrightarrow\ 3 & CHOH \\
| & | & | \\
CH_2OH & CH_2OCOR & CH_2OH \\
\end{array}
$$

(R = fatty acid residue)

Other polyols are often used in the alcoholysis of the oil, but the product is sometimes described as a monoglyceride. After the monoglyceride formation, the dibasic acid is added, and the mixture is processed as for the fatty acid process.

Acidolysis is an alternative to alcoholysis, and is shown in the following equation:

$$
\begin{array}{l}
CH_2{-}O{-}COR \\
| \\
CH{-}O{-}COR \ + \\
| \\
CH_2{-}O{-}COR
\end{array}
\quad
\begin{array}{c}
COOH \\
\bigcirc \\
COOH
\end{array}
$$

$$
\downarrow
$$

$$
\begin{array}{l}
\\
CH_2{-}O{-}CO{-}\bigcirc{-}COOH \ + R{-}COOH \\
| \\
CH{-}O{-}COR \\
| \\
CH_2{-}O{-}COR
\end{array}
$$

This process usually requires higher temperatures than alcoholysis, but is used where there are special problems associated with the solubility or reactivity of the dibasic acid. For example, iso- and terephthalic acids are relatively insoluble in the common solvents when compared with o-phthalic acid or anhydride, and this sometimes presents problems in the conventional monoglyceride process. Acidolysis of these acids overcomes this difficulty.[2] After the acidolysis, the polyol and any additional components are added, and the condensation reaction is carried out as before.

The fatty acid/oil method of preparing alkyds involves the direct reaction of a fatty acid/vegetable oil/polyol/dibasic acid mixture. The ratio of fatty acid to oil must be such that a homogeneous reaction mixture results. A variation of the method is to add the oil after the alkyd preparation, but care is necessary to avoid incompatibility. This process has cost advantages over a fatty acid process, and it enables higher viscosity alkyds to be prepared, particularly where the overall fatty acid content is high (for example, above 60% oil length).[3]

Each of the above processes is in commercial use, but the majority of alkyds are prepared by the fatty acid or monoglyceride processes.

The use of fatty acids instead of vegetable oils offers a number of technical advantages[4,5] and disadvantages. The advantages include the following ones.

1. Greater freedom of formulation, since any polyol or polyol blend can be used.

2. Mixtures of fatty acids—not available as the glycerides—may be used, and/or the fatty acids may be purified, and natural antioxidants removed before processing.

3. No catalyst is necessary. This reduces problems associated with oxidation and color development.

4. Processing is more reproducible since formation of a monoglyceride is not yet fully controlled (see page 113).

5. Molecular weight distribution can be controlled by the use of, for example, the "high polymer technique" (see page 107).

The disadvantages of using fatty acids are as follows.

1. They are more corrosive and are susceptible to discoloration on storage.

2. They often have higher melting points than oils and require preheating equipment to facilitate handling them as liquids.

3. They are more expensive than oils but, in assessing this cost, it is necessary to consider the overall formula cost. Where glycerol is used, then the fatty acid formulation is undoubtedly more expensive than that based on oil. With other polyols the difference may be slight since, for example, pentaerythritol is cheaper than glycerol.

Alkyds of the same composition but prepared by different methods do not always have the same properties either as a resin solution or as a dried coating. This aspect is considered further on page 103.

Oils that contain free hydroxyl groups in the fatty acid chain (castor and hydrogenated castor oil) do not usually require a monoglyceride stage; they can be condensed directly with the polyol and dibasic acid. Dehydrated castor oil alkyds can be prepared either by preforming the dehydrated castor oil and then using a monoglyceride process, or by an in situ method where the oil is dehydrated during the early part of the condensation process. This is usually carried out by heating at 310°C (approximately) until the viscosity drops to a minimum, and then continuing the reaction at normal esterification conditions (220 to 250°C). The initial drop in viscosity results from the removal of the hydroxyl group from the oil molecule, with the consequent loss of the hydrogen bonds responsible for the high viscosity of castor oil.

Two methods are used to manufacture alkyds. They are referred to as the fusion or solventless process, and the solvent or solution process. The fusion process is the older method and utilizes equipment similar to that used for varnish making. The components of the alkyd are heated in kettles, sometimes fitted with crude condensing equipment. The volatile products of reaction and some of the resin components are either discharged to the atmosphere or collected in fume chambers. The solvent process uses equipment capable of condensing the volatile vapors, separating the water, and returning the organic distillate to the kettle. In both the fusion and solvent methods, the reaction mixture is stirred; an inert gas or solvent vapor atmosphere is maintained to minimize oxidation of unsaturated organic components.

The fusion process is still widely used, since the equipment costs are comparatively low, and suitable kettles are often already in existence where varnishes are being made. The fusion process is often acceptable for alkyds containing a high percentage of oil (oil length 60% and over). These have less phthalic anhydride and polyol than short oil length resins and the problems associated with loss by volatilization are less. However, the solvent process is preferred in formulations where the alkyd specification and performance are critical. It offers much better control of temperature and composition, since the resin components are not lost. Also, the solvent (usually xylene) continually cleans the resin from the kettle sides and enables a more uniform alkyd to be prepared that is free from gel particles. The solvent also reduces the viscosity, and this results in more effective agitation, which contributes to faster reaction and easier water removal. The presence of solvent also alters the gel point, and enables higher molecular weight and better air drying vehicles to be prepared.[6]

REFERENCES

1. Kraft, W. M., G. T. Roberts, E. G. Janusz, and J. Weisfeld, *Am. Paint J.*, **41**, 96 (1957).
2. Opp, C. J., *Offic. Dig. Federation Soc. Paint Technol.*, **32**, 1477 (1960); Carlston, E. F., *J. Am. Oil Chemists' Soc.*, **36**, 28 (1959).
3. Martin, S. R. W., in H. W. Keenan, collator, *Paint Technology Manual, Part 3, Convertible Coatings*, Chapman and Hall, London, 1962.
4. *Fatty Acids in Alkyd Manufacture*, Tech. Bull. No. 2, by Victor Wolf Ltd., 1962.
5. Patton, T. C., *Alkyd Resin Technology*, Interscience Publishers, Inc., New York, 1962.
6. Bobalek, E. G., E. R. Moore, S. S. Levy, and C. C. Lee, *J. Appl. Polymer Sci.*, **8**, 625 (1964).

THE EFFECTS OF CHEMICAL COMPOSITION ON THE PREPARATION AND FILM-FORMING REACTIONS OF ALKYD RESINS

Commercial alkyds are prepared from balanced formulas chosen so that the desired molecular weight or degree of esterification is reached at the required acid and hydroxyl numbers, which are usually a predetermined number of units before the gel point. Therefore, in specifying an alkyd, it is necessary to nominate the mole ratio of polyol/dibasic acid/fatty acid/ modifiers, the molecular weight or acid value, and the method of preparation since this can influence the molecular weight distribution. While the use of mole ratios is a desirable method[1,2] for specifying alkyds, it is far less widely used than the term *oil length*, which is a carry-over from the varnish formulations, and refers to the percentage of oil in a resin. Alkyds are referred to as short (less than 45%), medium (45 to 55%), or long (more than 55%) oil (length) but these divisions are arbitrary ones. Alkyds also vary in the type of oil or fatty acid on which they are based, and this governs the properties and potential end uses of the resin as shown in Table 3.2.

It is not intended to discuss the influence of formulation changes on properties, since this is well documented in other textbooks and in the literature of the raw material suppliers. However, formulations where the chemistry is unusual, or deviates from that expected in a simple esterification reaction, will be considered.

Where the mole ratio of fatty acid (C_{18} acid) : glycerol : phthalic anhydride = 1:1:1, the oil length is 61% (approximately), and the system is bifunctional. At higher oil lengths, the polymer size will be limited unless other components are added, or the processing is varied. At lower oil lengths, the functionality increases, and gelation will occur unless formulation

TABLE 3.2

EFFECT OF OIL LENGTH AND TYPE OF OIL ON THE PROPERTIES AND USES OF ALKYDS

Oil Type	Oil Length (%)	Typical Oil	Properties
Oxidizing type	60 or more	Linseed, safflower, soya bean, tall oil fatty acids, wood oil in blends with other oils, dehydrated castor oil.	Soluble in aliphatic solvents. Compatible with oils and medium oil length alkyds, good drying characteristics. Films are flexible with reasonable gloss and durability.
Oxidizing type	45–55	Linseed, safflower, soya bean, tall oil fatty acids, wood oil in blends with other oils.	Soluble in aliphatic or aliphatic-aromatic solvent mixtures. Good drying characteristics, durability, and gloss.
Oxidizing type	45 or less	Linseed, safflower, soya bean, tall oil fatty acids, wood oil in blends with other oils and dehydrated castor oil.	Soluble in aromatic hydrocarbons. Low tolerance for aliphatic solvents. Usually cured at elevated temperatures either by heating with manganese driers or with urea- or melamine-formaldehyde resins.
Non-oxidizing type	40–60	Coconut oil, castor oil, hydrogenated castor oil.	Soluble in aliphatic-aromatic solvent blends. Usually used as a plasticizer for thermoplastic polymers such as nitrocellulose.
Non-oxidizing type	40 or less	Coconut oil, castor oil, hydrogenated castor oil.	Soluble in aromatic solvents. Used as a reactive plasticizer which chemically combines with other resin entities, e.g. melamine formaldehyde resin.

modifications are made. The simplest approach is to increase the polyol concentration (that is, lower the actual functionality). Typical excess glycerol figures, based on the glycerol used in the esterification, are given in Table 3.3.

TABLE 3.3

TYPICAL EXCESS GLYCEROL FIGURES
IN ALKYDS OF VARIOUS OIL LENGTHS

Oil Length (%)	Excess Glycerol (%)
60	0–10
55–60	10–17
40–55	17–30
30–40	30–35

With other trifunctional polyols, similar figures to those listed in Table 3.3 will apply but, with the tetrafunctional pentaerythritol, other methods are necessary below approximately 50% oil length.[3] These methods are also applicable to other polyols. A low molecular weight monobasic acid, such as benzoic acid, can be used to reduce the functionality of pentaerythritol; for example, equimolar amounts of pentaerythritol and benzoic acid can be regarded as equivalent in functionality to a triol. Alternatively, mixtures of pentaerythritol with other polyols of functionality three or less, may be used: the most common polyol used to reduce the functionality of pentaerythritol is ethylene glycol. Equimolar blends of pentaerythritol and ethylene glycol offer the following advantages over glycerol:

1. The blend is usually cheaper.
2. The blend contains primary hydroxyl groups only, and gives faster esterification rates.
3. Different molecular structures, and molecular distributions in the final polymer, are obtained.

The main disadvantage in using ethylene glycol is in controlling or preventing its loss from the resin kettle, since it is relatively volatile. Yet another method of reducing the functionality of polyols involves the use of formaldehyde (paraform); this gives a cyclic formal (**1**) which, in the case of pentaerythritol, reduces the functionality to two.

The linear formal is apparently not formed, since the viscosity of the alkyd was not higher than the unmodified pentaerythritol alkyd.[4,5] The paraform

$$HO-CH_2 \diagup \negmedspace\negmedspace \diagdown CH_2-OH$$
$$\underset{HO-CH_2 \diagdown \negmedspace\negmedspace \diagup CH_2-OH}{C} + HCHO \longrightarrow \underset{O-CH_2 \diagdown \negmedspace\negmedspace \diagup CH_2-OH}{CH_2 \diagdown \negmedspace\negmedspace \diagup \overset{O-CH_2 \diagup \negmedspace\negmedspace \diagdown CH_2-OH}{C}}$$

$$(\mathbf{1})$$

(cyclic formal of
pentaerythritol)

also reduces phthalic anhydride losses by sublimation; it is not effective in this regard with isophthalic acid.[4] Formaldehyde-modified alkyds have faster drying times[4] than the corresponding unmodified resins, and this is possibly related to the introduction of an additional oxidizable methylene group (that is, that of the formal), and to the increased rate of hydroperoxide breakdown to be expected in the absence of hydroxyl groups (see page 50). Polyols undergo a number of reactions besides esterification during the preparation of an alkyd. The extent to which these reactions occur will depend on the polyol structure, on the reaction conditions, and on the

---

$$\underset{\overset{|}{OH} \quad \overset{|}{OH} \quad \overset{|}{OH}}{CH_2-CH-CH_2}$$

$$\downarrow -H_2O$$

$$\underset{\overset{|}{OH}}{CH_2-CH_2-CHO} \longleftarrow \underset{O \qquad \overset{|}{OH}}{CH_2-CH-CH_2} \longrightarrow \underset{\overset{|}{OH} \quad \overset{\|}{O}}{CH_2-C-CH_3}$$

(propanol-1-al-3) (glycidol) (propanol-1-one-2)

$$\downarrow -H_2O \qquad\qquad \downarrow \overset{CH_2OH}{\underset{CH_2OH}{+\,CHOH}}$$

$$CH_2{=}CH-CHO$$

(acrolein)

$$\underset{CH_2OH}{+\,CHOH} \qquad \underset{\overset{|}{CH_2OH} \quad \overset{|}{CH_2OH}}{\underset{CHOH \quad CHOH}{CH_2-O-CH_2}}$$

(diglycerol)

$$\underset{\overset{|}{CH_2-O} \diagup}{\overset{CH_2OH}{\underset{\overset{|}{}}{\overset{|}{CH}-O} \diagdown} \negmedspace CH-CH{=}CH_2}$$

(1:2 allylidene glycerol)

Figure 3.2. Thermal decomposition of glycerol.[6-8]

presence of trace amounts of catalysts. Under the action of heat, polyols dehydrate to form a variety of products; this reaction is more pronounced with the vicinal, or 1:2 glycol structure. For example, glycerol is claimed to yield the products shown in Figure 3.2.[6,8]

The tendency for a polyol to polymerize through ether formation increases with the temperature and in the presence of alkaline catalysts,[9] as shown in Table 3.4.[9] The catalysts used to promote monoglyceride formation are effective in polyether formation.

TABLE 3.4

EFFECT OF TEMPERATURE AND CATALYSTS ON POLYETHER FORMATION DURING GLYCEROLYSIS OF LINSEED OIL (mole ratio = 2/1)[9]

Catalyst (0.1% on Oil)	Temperature	Polyglycerol Estimated as Diglycerol (Percent of Total Glycerol Recovered)
Na_3PO_4	240	3.5
Na_3PO_4	280	27.3
NaOH	240	8.0
NaOH	280	25.6
CaO	240	1.3
CaO	280	7.7

Acid-catalyzed etherification may take place during the condensation stage of alkyd manufacture, and this is particularly likely where a relatively strong acid, such as isophthalic acid or maleic acid, is present.[10]

$$2\ \begin{array}{c} CH_2OH \\ | \\ CHOH \\ | \\ CH_2OH \end{array} \xrightarrow{H^+} \begin{array}{c} CH_2OH \\ | \\ CHOH \\ | \\ CH_2 \end{array} \begin{array}{c} CH_2OH \\ | \\ CHOH \\ | \\ O{-}CH_2 \end{array}$$

This reaction accounts for the functionality of isophthalic acid apparently being greater than two. Evidence supporting the etherification theory includes the formation of water of reaction in excess of that required for esterification, and the presence of ether bands in the infrared spectrum of the polyols separated from the alkyd.[10] This evidence has been challenged by Geoghegan and Bambrick[32] who found only small quantities of ethers in pentaerythritol alkyds. They favor the theory that the reported differences in gel-point between *ortho*phthalic and *iso*phthalic alkyds are due to cyclic (intramolecular) esterification of some of the *ortho*phthalic acid. Supporting evidence for intramolecular reactions in *ortho*phthalic alkyds is given by Crawford and Sutton[33] who analyzed a glycerol-phthalic anhydride–stearic acid alkyd, and concluded that it was possible that considerable quantities of the intramolecular products shown below exist in alkyd resins.

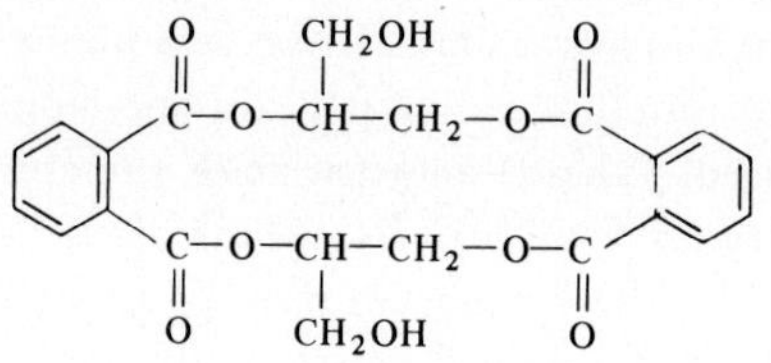

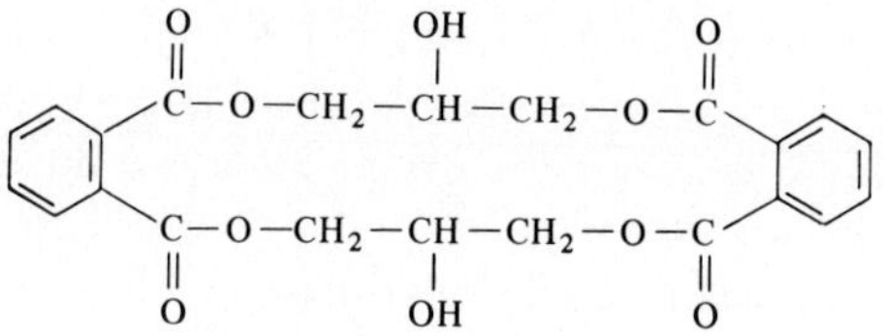

Technical grades of pentaerythritol often contain up to 15% of a poly-pentaerythritol, and this gives rise to greater rates of viscosity increase in alkyd manufacture, because of the higher functionality. Some technical grades of pentaerythritol contain sodium and/or calcium formates which act as esterification catalysts,[11] and further accelerate the viscosity increase.

Polyols can add across carbon-carbon double bonds to give either a hydroxy acid or, by cyclization, a lactone (see page 132); this is yet another reaction competing with esterification.

The fatty and dibasic acids present in an alkyd reaction mixture can also undergo reactions in addition to esterification. Unsaturated fatty acids may polymerize, by the mechanism discussed in Chapter 2, to dimer acids with a functionality of two. Temperature is important in controlling this reaction; for example, in a 60% linseed oil alkyd, only 6% of polymeric acids are formed after 30 hours at 240°C, whereas only 7 hours at 280°C gave 25% polymeric acids.[9] Addition of unsaturated compounds across, or adjacent to, the fatty acid unsaturation can also occur (page 68); for example, the reaction of maleic anhydride with a fatty acid gives a trifunctional acid (or ester). This reaction, together with the addition of polyols across the maleic double bond and the etherification of the polyol, explains why small amounts of maleic anhydride give a correspondingly large increase in the viscosity of the alkyd. The functionality of maleic anhydride is greater than two, and the actual functionality will depend on the relative extents of the various reactions, that is,

    (1) Polyol etherification
    (2) Polyol addition to the maleic double bond
    (3) Maleic addition to the fatty acid

The mechanism by which alkyds are converted from a liquid to a dry film depends on the alkyd structure and on the drying conditions. Where the alkyd is essentially a plasticizer, as in nitrocellulose lacquers, no chemical reactions are involved in film formation. Hence, in a plasticizing alkyd, fatty acids which are fully saturated or which contain only one double bond per molecule are usually used. Where the fatty acids present in the alkyd are derived from the semidrying or drying oils, the alkyd can undergo autoxidation at room temperature; the oxygen attacks the unsaturated area of the fatty acid molecule. The general mechanism of film formation is similar to that described for the glyceride oils, but with a number of important differences. First, the molecular weight of the alkyd is higher than the molecular weight of an oil; consequently, the number of cross-links needed to give a dry film is less, and the drying time is reduced. Second, since relatively few cross-links are required to dry an alkyd film, the less unsaturated semidrying oils can be used to give alkyds with good drying potential. Third, the presence of hydroxyl and carboxyl groups can influence hydroperoxide decomposition. The hydroxyl group inhibits breakdown of the hydroperoxide by complex formation,[12] whereas the acids facilitate the decomposition. These polar groups contribute to the good adhesion to substrates shown by most alkyds. The carboxyl and hydroxyl groups also respond readily to calcium-type driers; they form salts or coordination compounds that give a rapid initial dry and possibly accelerate subsequent cross-linking (see below). This is clearly shown by gelation occurring when

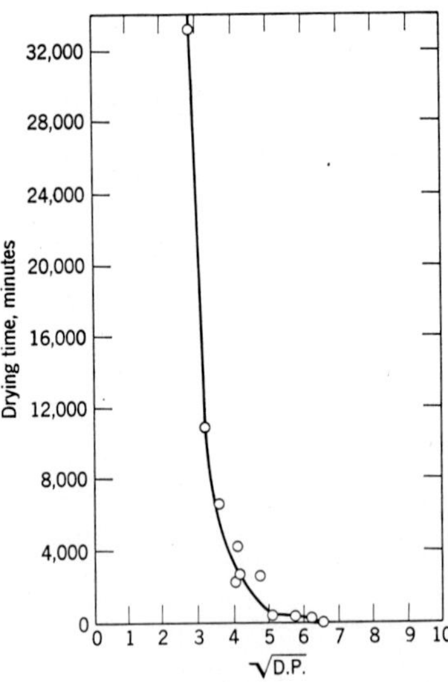

Figure 3.3.   Influence of degree of polymerization on drying times of a linoleic-based alkyd. (D.P. = degree of polymerization.)[19]

calcium naphthenate is added to certain alkyds when the solids concentration is high (95% or over). If necessary, the free hydroxyl content of an alkyd can be reduced by reaction with an isocyanate[13] or diketene (see page 138). In choosing an alkyd formulation for use in air-drying coatings, it is necessary to balance two opposing requirements; the need for maximum cross-linking potential which suggests a high fatty acid content or long oil length, and a high polyester content for maximum durability, since subsequent film breakdown is predominantly a result of continued oxidation of the fatty acid residues. In practice, this means that alkyds with oil lengths of approximately 50% are commonly used in air-drying enamels.[14,15] A further compromise is necessary in deciding the degree of polymerization or molecular weight of the alkyd. Whereas high molecular weight resins dry more readily, they usually exhibit poor can stability. A figure of five acid value units before the gel is usually suggested for air-drying alkyds.[14,15] However, this is a somewhat superficial specification for an alkyd, as shown by the enlightening studies of Bobalek, Levy, Moore, and co-workers.[16-19] For an alkyd to possess rapid drying properties, the presence of microgel particles of high molecular weight, or containing aggregates of molecules, is necessary. Therefore, the marked change in drying rate as the gel point is approached (shown in Figure 3.3[19]) is related to the formation of microgel structures, and is not solely related to the increase in molecular weight. This is clearly shown by the blending of a high and low molecular weight alkyd; as little as 20% of the high molecular weight alkyd reduced the drying time by 85% of the original (Figure 3.4[19]). This information could offer a method of

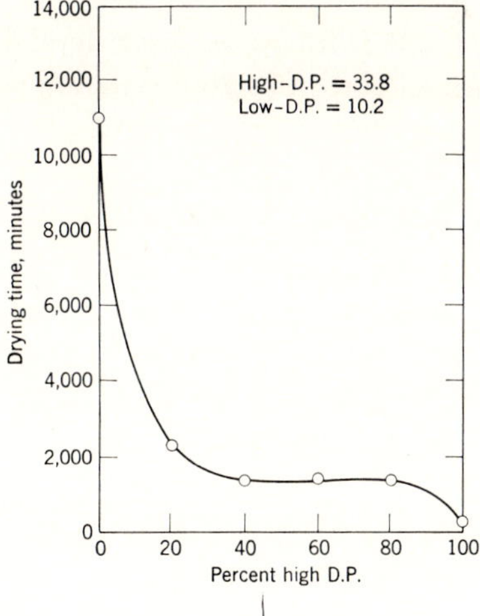

Figure 3.4. Influence of microgel fraction on drying times of linoleic-based alkyds.[19]

doping batches of resin to give constant drying times. The presence of the microgel or nucleating species also influences the drying of the film by creating a reaction environment where the oxygen uptake, initiation, and propagation reactions are accelerated, and where the termination reactions are inhibited. As final proof of the importance of the microgel, the precipitation of this fraction from an alkyd substantially reduces the ability of the alkyd to dry.[19]

The autoxidation of a fatty acid residue is accelerated by increasing temperatures and, hence, alkyds based on unsaturated fatty acids can be cured rapidly at elevated temperatures (120°C). Some stoving enamels are based solely on an alkyd, and the usual drier is simply a manganese salt. This drier has a marked temperature coefficient and a negligible inhibition period at high temperatures.[20] The cross-links formed under these conditions are more likely to be of the carbon-carbon type; consequently, these films tend to be more durable than those produced by room temperature dry. The faster rate of reaction at the higher temperature means that fewer fatty acid residues are necessary in the alkyd; therefore, oil lengths of 40 to 50% are used for stoving enamels. The higher percentage of polyester component in these enamels (compared to air-drying resins) also contributes to the improved durability. Under the stoving conditions, little further polyesterification occurs, and this is not a primary film-forming reaction. This can be readily demonstrated by comparison with non-drying oil alkyds that do not dry under these conditions.

The advent of urea, melamine, and phenol formaldehyde resins which are soluble in organic solvents, has led to the development of the present type of alkyd stoving enamels. Some thermosetting acrylics also use these resins. Initially, small amounts (5%) of these formaldehyde-based polymers were added to the stoving alkyd to control wrinkling tendencies, but it was soon noted that improved hardness, exterior durability, alkali resistance, and rate of dry also resulted.[21] These observations led to the present-day finishes containing up to 30% of the formaldehyde condensation product. In these systems, the alkyd is plasticizing the other polymer system. Because

$$\boxed{\text{Polymer}}-CH_2OH + \boxed{\text{Alkyd}}-OH \xrightarrow{H^+}$$

$$\boxed{\text{Polymer}}-CH_2-O-\boxed{\text{Alkyd}} + H_2O$$

$$\boxed{\text{Polymer}}-CH_2OR + \boxed{\text{Alkyd}}-OH \xrightarrow{H^+}$$

$$\boxed{\text{Polymer}}-CH_2-O-\boxed{\text{Alkyd}} + ROH$$

$$\boxed{\text{Polymer}}-CH_2OR + \boxed{\text{Alkyd}}-COOH \longrightarrow$$

$$\boxed{\text{Polymer}}-CH_2-O-\overset{\displaystyle O}{\overset{\|}{C}}-\boxed{\text{Alkyd}} + ROH$$

Figure 3.5. Reactions between an alkyd and formaldehyde condensation polymers.

of color requirements, urea and/or melamine formaldehyde resins are used mainly in topcoats; phenolic resins are used in undercoats or primers because of their superior corrosion resistance properties. The chemical reactions responsible for the cure of the various formaldehyde polymers and alkyds are similar. If we initially consider an alkyd based on saturated fatty acids, or even a simple saturated polyester of the poly(ethylene adipate) type, then the carboxyl and hydroxyl groups are the only functional groups in the molecule. The formaldehyde polymers contain methylol and methylol-ether groups, and the cross-linking reactions involve both etherification or ether-interchange of these groups with the hydroxyl, and ester formation with the carboxyl group of the alkyd or polyester.[22–25] The carboxyl group also catalyzes the etherification and ether-interchange reactions (Figure 3.5). With drying-oil alkyds, additional cross-linking also takes place by concurrent autoxidative polymerization, at the hydroxyl groups formed on the fatty acid residues by oxidation, and by addition of the phenol-formaldehyde polymer to the double bond to form a chroman-type structure as shown below.

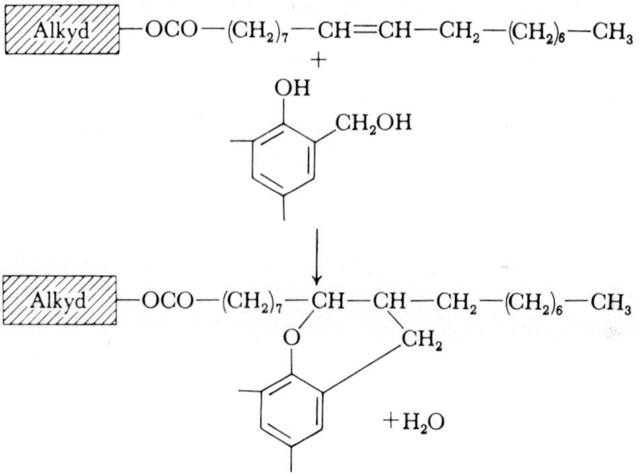

$$\boxed{\text{Alkyd}}-OCO-(CH_2)_7-CH=CH-CH_2-(CH_2)_6-CH_3$$

$$+$$

$$\text{phenol with } OH \text{ and } CH_2OH \text{ groups}$$

$$\downarrow$$

$$\boxed{\text{Alkyd}}-OCO-(CH_2)_7-CH-CH-CH_2-(CH_2)_6-CH_3$$

$$\text{chroman-type ring} \quad +H_2O$$

The enol forms of melamine- and urea-formaldehyde resins (shown below) react in a similar manner to the phenol formaldehyde.

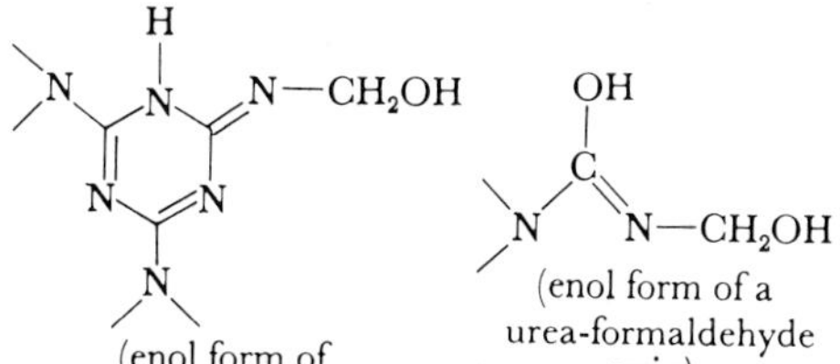

$$\text{(enol form of a melamine-formaldehyde resin)} \qquad \text{(enol form of a urea-formaldehyde resin)}$$

Consequently, at the same alkyd/melamine formaldehyde ratio, a drying-oil alkyd gives a harder film than a nondrying-oil alkyd. Conversely, less melamine resin is required by a drying oil alkyd to reach a specified hardness. This fact is utilized in the formulation of, for example, automotive topcoats, since the reduction in melamine resin gives a cheaper product. However, the presence of carbon unsaturation in the molecule gives rise to discoloration on repeated baking (a situation commonly met in two-tone color systems) and to a less durable system as a result of the increased rate of oxidative degradation. Therefore, it is usual to use the drying oil alkyd/ melamine system in dark shades and the non-drying formulation for pastel colors or for high quality dark colors.

The melamine resin can also intercondense with itself, and the extent to which this reaction occurs will depend on the type of resin, the catalyst system, and the temperature. Fast-curing resins are more prone to self-condensation, since the etherification of a methyl group—a comparatively large number of which are present—is difficult with high molecular weight polyols such as an alkyd resin. The fast cure is really a measure of the time to reach a given hardness which, in this case, results from self-condensation of the melamine resin rather than from co-condensation with the alkyd (see also page 246).

From the chemistry of the cross-linking reactions, it can be seen that the concentration of hydroxyl and carboxyl groups in alkyd resins designed for reaction with formaldehyde condensation polymers, must be controlled. Unlike the groups in air-drying formulations, these hydrophilic groups are lost during the film-forming reaction. Since one of the limiting specifications for these systems is a comparatively high solids content at application viscosity (approximately 50% solids), the molecular weights of the alkyds used are lower than for air-drying resins. Lower molecular weights can be counterbalanced by an increase in the concentration of hydroxyl and carboxyl groups. In fact, it is possible that quite different film properties will result from the use of low-molecular-weight, highly functional alkyds, compared to high-molecular-weight, less-functional resins. By applying the concept of block copolymers to this system, structures of the type shown below are possible and these should exhibit different film properties.

| Alkyd | MF | Alkyd | MF | Alkyd |

| Alkyd | MF | Alkyd |

(MF = melamine-formaldehyde condensate)

For a chemical group—such as a hydroxyl or carboxyl—to take part in a cross-linking reaction, it must be available to, and reactive with, the interacting group. As a cross-linking structure commences to form, some groups will not be available because of steric factors, or the restricted movement of a molecule, once one group has chemically combined with another molecule. In addition, many of the hydroxyl groups present in alkyds are not sufficiently reactive, or available, to react even with monomeric molecules. However, the available groups can be measured by simple acetylation procedures.[26]

The type of functional group will also be of importance; primary hydroxyls are more reactive than secondary. However, there is not much scope for controlling an alkyd from this point of view, since the primary hydroxyl groups are also the more reactive in ester-forming reactions. On the other hand, strong acids are more effective catalysts than weak acids, and an alkyd in which the free carboxyls are stronger than phthalate half esters would be of interest. Such an alkyd can be prepared by carrying out a half ester reaction with a suitable anhydride, for example, tetrachlorophthalic anhydride, at a low temperature (150°C), after the alkyd has been prepared.

This method is preferred to adding the tetrachlorophthalic anhydride to the initial alkyd charge, since it eliminates the possibility of the stronger acid forming a diester which would not assist in the curing reaction. Alternatively, the acid catalyst may be added as a separate component, and some of the more common acids used are butyl acid tetrachlorophthalate, butyl acid maleic, and sulfonic acids. These catalysts are often added when touch-up procedures for alkyd/nitrogen resin-based enamels demand a stoving temperature lower than the main curing temperature. Other resin components, particularly formaldehyde condensates, may also be present in the touch-up catalyst. Often, however, these lower temperature catalyzed cures give inferior properties to the normal stoving schedules, mainly as a result of increased self condensation of the formaldehyde-derived polymer and the consequent presence of unreacted alkyd.[27]

Alkyd resins react at a slow, but finite, rate with formaldehyde condensates at room temperature. Therefore, it is necessary to stabilize these formulations to give acceptable can stability. Some control is possible by using the minimum number of functional groups in each polymer species, but this approach alone will not yield stable systems. Both the type and the amount of solvent present will have an effect on stability. In one system studied, a major improvement resulted from a change in the xylene/butanol ratio from 4:1 (poor) to 3:2 (satisfactory), and a lowering of the total solids from 50% to 40%.[28] The use of butanol as a stabilizer is possibly related to the formation of bonded structures with the methylol and hydroxyl groups of the resin components. Chemical stabilizers are also used, but to be worthwhile they must be easily removed, or lost, from the formulation when it is heated to the curing temperature. Generally, these chemical methods involve the masking of the acid groups; the temporary removal of the catalyst results in increased stability. The formation of amine salts is a common method used either for the acidic groups of the alkyd, or for the added acid catalyst.[29] The amine salts are termed latent catalysts.

$$\text{Alkyd—COOH} + R_3N \rightleftharpoons \text{Alkyd—COO}^-R_3\overset{+}{N}H$$

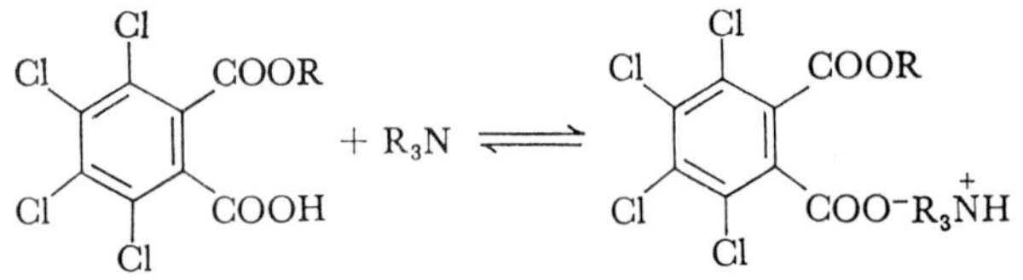

Low molecular weight tertiary amines, such as triethylamine, are commonly used to form the acid salt; primary and secondary amines are usually avoided, since they can result in a "depolymerization" of the formaldehyde resin and aminolysis of the alkyd. This applies particularly where free methylol groups, and not methylol ethers, are present—a situation found in most practical systems.

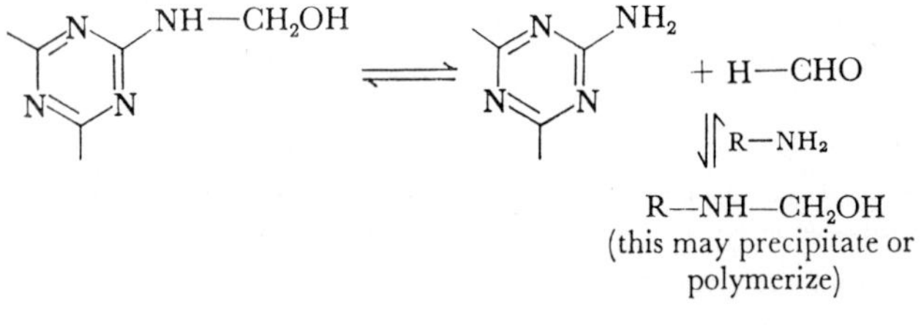

Alternatively, thermolabile esters can be used as a means of masking the carboxyl groups, and a typical example is the tetrahydropyranyl ester formed from dihydropyran.[30]

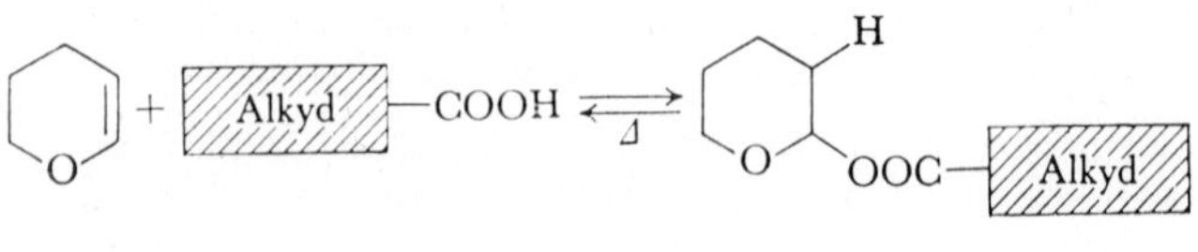

Alkyd/amine formaldehyde resin blends are sometimes used as coatings that dry at room temperature and, to achieve adequate cure, it is necessary to add acids which are much stronger than those used at elevated temperatures. Mineral acids (particularly hydrochloric acid), boron trifluoride, phosphoric acid, and acid phosphates[31] are among the acids used. The cured films can sometimes contain relatively large amounts of self-condensed formaldehyde polymer and, in cases where this adversely affects the film properties, it is worth considering the use of a preformed condensate with the alkyd. Such a system would give a film in which both of the original polymer entities are chemically incorporated.

$$\boxed{Alkyd}\!-\!OH + \boxed{Melamine\ resin}\!-\!CH_2OR$$
$$|$$
$$CH_2OH$$

$$\downarrow$$

$$\boxed{Alkyd}\!-\!O\!-\!CH_2\!-\!\boxed{Melamine\ resin}$$
$$|$$
$$CH_2OH$$

$$\downarrow\ \substack{H^+ \\ curing\ reaction}$$

$$\boxed{Alkyd}\!-\!O\!-\!CH_2\!-\!\boxed{Melamine\ resin}$$
$$|$$
$$CH_2 \qquad + H_2O + HCHO$$
$$|$$
$$\boxed{Alkyd\!-\!O}\!-\!CH_2\!-\!\boxed{Melamine\ resin}$$

REFERENCES

1. Kraft, W. M., *Offic. Dig. Federation Soc. Paint Technol.*, **29**, 780 (1957).
2. Sunderland, E., *Farg. Lack.*, **7**, 118; *Paint Technol.*, **26**, No. 1, 25 (1962).
3. Hercules Powder Company, Alkyd Report No. 4
4. Smith, P. E., *Can. Paint Varnish Mag.*, **1964**, 21.
5. Staddon, A. W. E., *Paint Manuf.*, **30**, 279 (1960).
6. Hauschild, R., and J. Petit, *Bull. Soc. Chim. France*, **1956**, 878.
7. Petit, J., *Offic. Dig. Federation Soc. Paint Technol.*, **20**, 977 (1948).
8. Tremain, A., *J. Oil Colour Chemists' Assoc.*, **40**, 737 (1957).
9. Brett, R. A., *J. Oil Colour Chemists' Assoc.*, **41**, 428 (1958).
10. Brown, R., H. Ashjian, ar.d W. Levine, *Offic. Dig. Federation Soc. Paint Technol.*, **33**, 539 (1961).
11. Gourley, S., and W. Dunlop, *J. Oil Colour Chemists' Assoc.*, **40**, 247 (1957).
12. Khan, N. A., *Pakistan J. Sci.*, **12**, 95 (1960).
13. Karl H. Hauck, and Fritz Hecker-Over, Ger. Pat., 1,045,652 (1958).
14. Berryman, D. W., *J. Oil Colour Chemists' Assoc.*, **42**, 393 (1959).
15. Patton, T. C., *Alkyd Resin Technology*, Interscience Publishers, Inc. New York, 1962.
16. Bobalek, E. G., C. C. Lee, and E. R. Moore, *Offic. Dig. Federation Soc. Paint Technol.*, **34**, 416 (1962).

17. Bobalek, E. G., and M. T. Chiang, *J. Appl. Polymer Sci.*, **8**, 1147 (1964).

18. Moore, E. R., S. S. Levy, and C. C. Lee, Papers presented to 139th Meeting of Am. Chem. Soc., 92 (1961).

19. Bobalek, E. G., E. R. Moore, S. S. Levy, and C. C. Lee, *J. Appl. Polymer Sci.*, **8**, 625 (1964).

20. Uyehara, H., *Offic. Dig. Federation Soc. Paint Technol.*, **27**, 794 (1955).

21. Hodgins, T. S., A. G. Hovey, S. Hewett, W. R. Barrett, and C. J. Meeske, *Ind. Eng. Chem.*, **33**, 769 (1941).

22. Seidler, R., and H. J. Graetz, *Fette, Seifen, Anstrichmittel*, **64**, 1135 (1962).

23. Wohnsiedler, H. P., Am. Chem. Soc., New York Meeting, 53 (1960); *Paint Varnish Prod.*, **50**, 178 (1960).

24. Paint Research Station (London), Research Memorandum No. 297.

25. Solomon, D. H., *Rev. Pure Appl. Chem.*, **13**, 171 (1963).

26. Solomon, D. H., and J. J. Hopwood, *J. Appl. Polymer Sci.*, **10**, 981 (1966).

27. Goppel, J. M., P. Bruin, and J. J. Zonsveld, VI, *F.A.T.I.P.E.C. Congr.*, **1962**, 31.

28. Zwicky, H., *Double Liaison*, **1962** (No. 79), 71.

29. Shelley, J. P., Rohm and Haas Co., Ger. Pat., 1,053,698, U.S. Pat., 2,871,209 (1959).

30. Bowman, R. E., and W. D. Fordham, *Chem. Ind. (London)*, **1951**, 742.

31. Gaynes, N. I., U.S. Pat. 2,982,745 (1961).

32. Geoghegan, J. T., and W. E. Bambrick, *J. Paint Technol.*, **42**, 490, (1970).

33. Crawford, L. M. R., and D. A. Sutton, *Chem. and Ind. (London)*, No. 38, 1232 (1970).

MOLECULAR WEIGHT DISTRIBUTION

Synthetic polymers consist of mixtures of macromolecules of different sizes or different molecular weights.[1] In the practical measurement of molecular weight distribution, selective solvency techniques are often used; separation may be on a molecular size, or polarity basis, depending on the particular system being studied. Here, the term molecular weight distribution will be used in the broad sense to cover differences in both polarity and molecular weight.

The majority of fractionation studies on alkyds have been carried out by using a solvent which will dissolve only the low molecular weight fractions, and leave as a residue the higher molecular weight material. Often the lower alcohols—such as methanol, ethanol, isopropanol, and butanol—have been used. These studies have clearly demonstrated the part played by the various fractions of an alkyd in film-forming reactions, particularly autoxidation. The low molecular weight fractions are generally non-film forming, whereas the residues are either gels, or fractions which will dry rapidly,[2-4] (Table 3.5).

The removal of the low molecular weight fraction also greatly reduces the tendency of the dried films to rivel.[3,4] However, the residual high molecular weight alkyd fraction is unstable in the absence of the lower

TABLE 3.5

EFFECT OF EXTRACTION ON THE AIR DRYING PROPERTIES OF AN ALKYD RESIN

Alkyd	Amount Extracted and Alcohol Used	Drying Time (Min)	
		Before Extraction	Residue After Extraction
70% Linseed oil glycerol phthalate	37% with isopropanol	120	105
60% Linseed oil glycerol phthalate	10% with methanol	120	90
60% Linseed oil glycerol phthalate	36% with isopropanol	120	30
60% Linseed oil glycerol phthalate	59% with butanol	120	20
40% Linseed oil glycerol phthalate	47% with methanol	120	45

fractions which effectively serve as plasticizers. Alternative materials that function as plasticizers, and yet do not induce riveling, may be added as shown in Table 3.6.[4] These results were obtained on a 70% linseed oil glycerol-phthalate alkyd, from which 37% had been extracted with isopropanol.

Therefore, extraction offers a method of improving the drying, and riveling, properties of alkyds. The extracted material can be recycled into following batches,[2] and the process is commercially and technically feasible. It is likely that the wetting and dispersing properties of the alkyd are primarily related to the low molecular weight polar fractions, and suitable surfactants would need to be added to compensate for the removal of these fractions. Other methods of fractionating alkyds use a miscible solvent-nonsolvent system, whereby incremental precipitation of the alkyd is possible. Acetone-water, which will tend to separate fractions on a molecular weight basis, and benzene-petroleum ether, which will favor separation by polarity, are two common solvent-nonsolvent systems used. Dioxane-water has been used to develop a turbidity titration method whereby integral distribution curves

TABLE 3.6

THE EFFECT OF ADDED PLASTICIZER ON THE DRYING TIME AND FILM APPEARANCE OF AN ISOPROPANOL INSOLUBLE ALKYD RESIDUE [4]

Added Plasticizer (10%)	Touch Dry Time 25 μ Film (Min)[a]	Appearance of 75 μ Film After 24 Hours
None	105	Very slightly riveled
Alkali refined linseed oil	120	Severely riveled
Castor oil	150	Severely riveled
Olive oil fatty acids	135	Severely riveled
Olive oil	150	Not riveled
Liquid paraffin	150	Not riveled
Dibutyl phthalate	135	Not riveled

[a] The resin that contained the nondrying plasticizers still gave tack-free films.

can be determined rapidly.[28] Separations of this type have substantiated the findings of the alcohol extraction methods. They have also indicated that a 55% oil length alkyd is essentially linear,[5] and the intermediate fractions of a long oil pentaerythritol alkyd had the best gloss retention.[6] Low temperature phase separation can also be used to fractionate alkyds.[7]

Gel Permeation Chromatography is a newer and more exact technique which has been used to measure the molecular weight distribution of alkyd resins.[29]

Since the molecular weight distribution of an alkyd has an important influence on the properties and reproducibility of the resin, successive batches of an alkyd formulation must have comparable distributions. Until recently, it has been widely stated that such is the case, mainly as a result of studies[8] on simple linear polyesters. For example, Flory[8] has shown that if two polymers of decamethylene adipate—one of low and the other of high average molecular weight—are mixed and heated under reflux, under conditions such that the number of molecules remains constant (that is, number-average molecular weight constant), interchange between the polymer species occurs. The distribution changes are illustrated in Figure 3.6.[9]

As the molecular weight distribution became narrower, the properties of the mixture changed and, in particular, the viscosity decreased. This work of Flory's has demonstrated that for normal processing of simple polyesters, the amount of polyol and polybasic acid, and the acid number are sufficient to characterize the polymer. These results of Flory have often been assumed by other workers to apply to polyesters in general (including alkyds). However, recent evidence[9–18] shows that the molecular weight distribution of an alkyd resin depends on the processing conditions,[30] and in particular on the

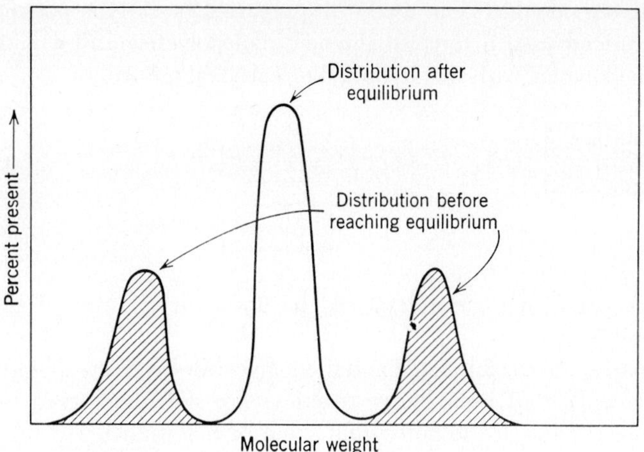

Figure 3.6. Schematic representation of changes in molecular weight distribution of simple polyesters.[9]

order in which the components are reacted, on the method (solvent or fusion) used, and on the process (fatty acid or monoglyceride). In the absence of competing side reactions, the dependence of molecular weight distribution on processing conditions clearly shows that interchange reactions (Eqs. 3.1–3.3) are slow, compared to polyesterification (Eq. 3.4).

$$ROH + R^1COOR^2 \rightleftharpoons R^1COOR + R^2OH \qquad \text{(alcoholysis)} \qquad (3.1)$$
$$R^3COOH + R^1COOR^2 \rightleftharpoons R^3COOR^2 + R^1COOH \qquad \text{(acidolysis)} \qquad (3.2)$$
$$R^1COOR^2 + R^3COOR^4 \rightleftharpoons R^1COOR^4 + R^3COOR^2 \qquad \text{(ester interchange)} \qquad (3.3)$$
$$ROH + R^3COOH \rightleftharpoons R^3COOR + H_2O \qquad (3.4)$$

In the more simple linear polyesters prepared from dibasic acids and a glycol, it has been shown that, where mixtures of acids are used, the polymer structure depends on the order and time of addition. For example, with maleic/isophthalic acid, it is possible to concentrate the maleate residues at the center of the chain (by early addition) or at the chain ends (late addition) (see page 139). Similar results are obtained in alkyds prepared by the fatty acid process; the "high polymer technique," which involves the incremental addition of the monobasic acid to the partly polymerized polyol/dibasic acid mixture, gives a molecular weight distribution which differs from that given in the normal process (all the ingredients are added simultaneously).[10] The "high polymer technique" alkyds are usually of higher viscosity, and have improved drying characteristics, possibly as a result of a high order of stability of the microgel fraction.[19] In the particular case of tall oil alkyds, the hardness is improved.[20] The concept that interchange reactions are comparatively slow has been used[21] to prepare a block-type polymer; the example quoted used a terephthalic acid/diol polyester and a preformed alkyd-type polymer with free hydroxyl or carboxyl groups.

This block copolymer gave improved hardness, drying time, toughness, and durability.[21]

A comparison of the fatty acid and monoglyceride processes should afford further evidence that interchange reactions are slow, compared to polyesterification. It should be noted that some earlier studies which claimed equivalent properties from alkyds prepared by the two processes, often used alkyds of less complex structures than current formulations. The property evaluations on which the comparisons were based, were not always critical (by present standards) or capable of detecting small, but significant, differences. More recent evidence suggests that differences of commercial significance do exist between alkyds of the same composition but prepared by the fatty acid or monoglyceride process. This is not surprising if the two processes are examined in more detail, since it is found that the reactions involved are quite different. In the fatty acid process, the polyol, dibasic acid, and fatty acid are reacted simultaneously. The fatty acids compete with the phthalic anhydride, and phthalate half ester, for the available hydroxyl groups. The relative rates of these reactions have been discussed by Goldsmith,[22] who claims that the primary hydroxyl groups of glycerol react more readily with phthalic carboxyls than with fatty acids; the reverse applies to secondary hydroxyls. With the monoglyceride process there is no fatty acid competing in the esterification process. For comparison purposes if the monoglyceride is assumed to be predominantly α-monoglyceride then, in the absence of interchange, structures **2** and **3** should result from the fatty acid and monoglyceride processes respectively. That is,

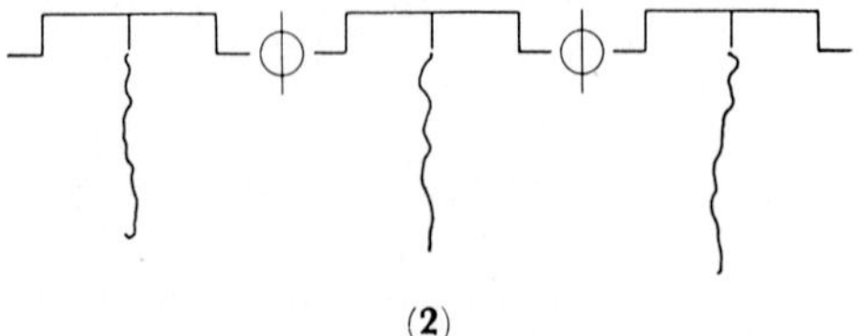

(**2**)

(idealized structure of an alkyd made from fatty acids)

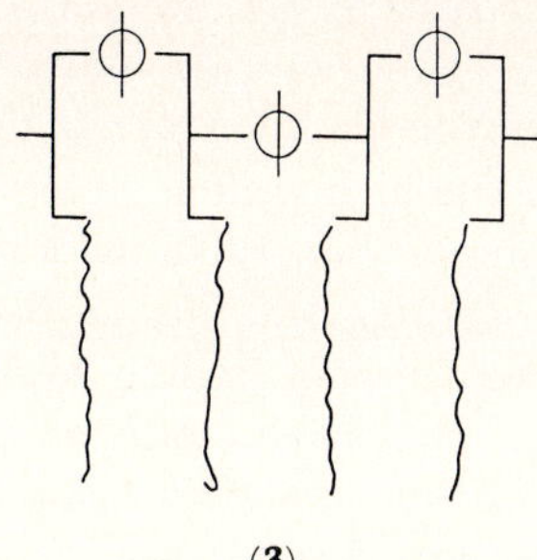

(3)

(idealized structure of an alkyd made from α-monoglyceride)

the fatty acid is in the β-position in fatty acid process, and in the α-position with the monoglyceride.[9,23]

Differences between alkyds prepared by the monoglyceride process have been noted by a number of authors,[13-18] and related to the composition at the monoglyceride stage. The alcoholysis reaction between glycerol and an oil can give rise to a number of products, some of which are shown in Figure 3.7.

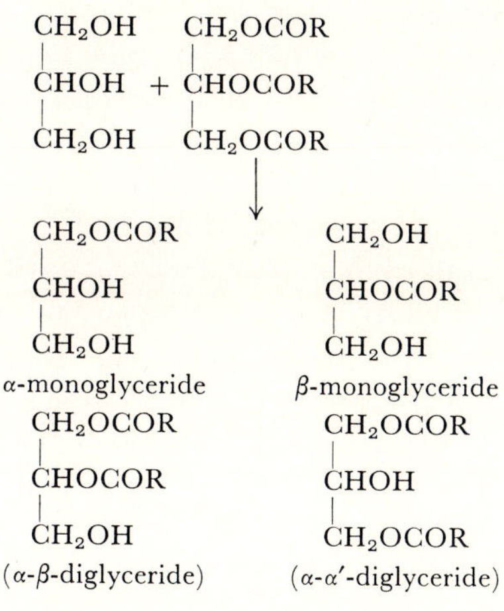

Figure 3.7.   Products resulting from the glycerolysis of an oil.

In addition to the above products, the different fatty acid residues can also be distributed over the various structures. Studies on a 33% coconut oil/glycerol phthalate alkyd (chosen because thermal polymerization of the oil is minimized) have shown that the gloss, and gloss retention on overbake of automotive enamels, prepared by the addition of pigment and a melamine resin,[16] is related to the α-monoglyceride content. High α-monoglyceride contents are preferred. The condensation rates of the alkyds also varied with the monoglyceride composition; high free glycerol contents (low α-monoglyceride) gave faster esterification rates than more highly alcoholyzed mixtures (high α-monoglyceride), and this has been related to the greater number of primary hydroxyl groups initially available for esterification.[16] The acid value/viscosity relationships differed in the alkyds, and this suggests a difference in molecular weight distribution; fractionation studies have confirmed this difference.[16]

Alkyds prepared from castor oil, and hydrogenated castor oil, are of interest in connection with molecular weight distribution theories for a number of reasons. First, the presence of a hydroxyl group on the fatty acid chain renders the oil miscible with the glycerol-phthalic anhydride mixture, and it is not essential to first interchange the oil with glycerol. Second, competitive reaction of the glycerol hydroxyls with those on the fatty acid chain could lead to a different molecular structure, depending on the degree of alcoholysis of the oil. Third, and most important, these alkyds are usually produced without a monoglyceride stage (that is, no α-monoglyceride); this is in apparent contradiction to findings on other oils where it has been suggested that the maximum degree of alcoholysis of the oil gives the best alkyd. Recent work[17] has shown that, depending on the method of preparation, castor oil alkyds differ in their molecular weight distribution, in the ratio of acid value to viscosity, and in the properties given in pigmented enamels where melamine formaldehyde resins are the cross-linking agent. The alkyds prepared direct from the oil gave films with better gloss and humidity resistance. With hydrogenated castor oil, the results were significantly different to those obtained with castor oil; as the condensation of the oil, glycerol, and phthalic anhydride proceeded, the resins became increasingly incompatible, and unchanged hydrogenated oil separated from the resin mixture. On the other hand, alkyds prepared by first reacting hydrogenated castor oil and glycerol to a monoglyceride, were compatible and gave satisfactory coatings. These results highlight a number of significant points in connection with alkyd studies. At the present state of knowledge, it is not possible to predict whether a high or low degree of alcoholysis of the oil will give the best alkyd for a given purpose. The previously held views, that the hydroxyl groups of hydrogenated castor oil are more reactive in esterification reactions than are the glycerol hydroxyl groups,[24] are not consistent with the separation of unchanged oil from the resin mixture.

Further evidence supporting the contention that the properties of an alkyd are related to the degree of glycerolysis of the oil comes from a study

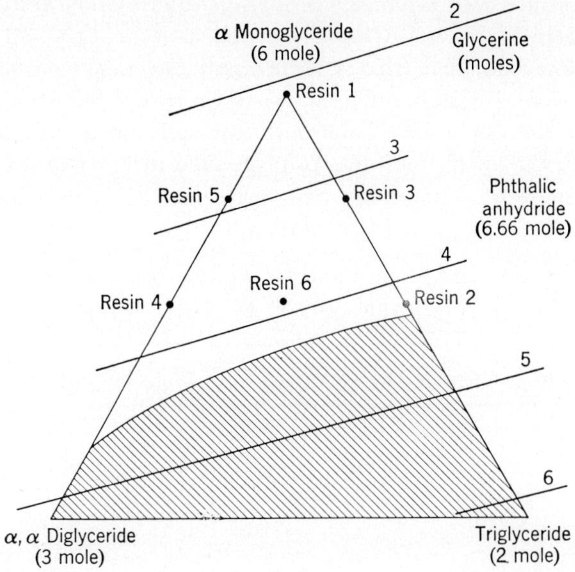

Figure 3.8. Glyceride compositions used in the preparation of 52% oil length alkyd.

of a drying-oil alkyd.[18] In addition, the results clearly show that there is
not one preferred distribution for a given alkyd composition; the required
distribution is dependent on the type of formulation in which the alkyd is
used. The alkyd studied corresponded to a 52% oil length and had a mole
ratio of fatty acid (linoleic acid): glycerol:phthalic anhydride of 1.0:1.36
1.11. Resins were prepared from the range of glyceride mixtures shown in
Figure 3.8. These glyceride compositions were prepared by mixing the

TABLE 3.7

VISCOSITY OF 52% OIL LENGTH ALKYDS AT SOLIDS CONTENT OF 55% IN
WHITE SPIRIT[18]

Resin Number (Refer to Figure 3.8)	Viscosity (Gardner Holdt Tubes)
1	F
2	$U + \frac{1}{2}$
3	$G + \frac{1}{4}$
4	R
5	$I + \frac{1}{4}$
6	$S - \frac{1}{2}$

previously synthesized α-monoglyceride, α,α'-diglyceride, and triglyceride of linoleic acid.

At the same acid value (that is, same number-average molecular weight), the alkyds were distinctly different, as shown by the following properties: viscosity at a given solids (Table 3.7), the solids at a constant viscosity (Table 3.8), the air-drying times of pigmented drier-catalyst films (Table

TABLE 3.8

SOLIDS OF 52% OIL LENGTH ALKYD AT VISCOSITY OF F (GARDNER HOLDT) IN WHITE SPIRIT [18]

Resin Number (Refer to Figure 3.8)	Solids Content
1	55.0
2	47.4
3	52.8
4	48.7
5	54.0
6	48.5

3.9), and the gloss of films formed by cross-linking with a melamine-formaldehyde resin (Table 3.10). The acid value-viscosity relationship and the measured molecular weight distribution of the resins also differed; resin 1 (prepared from α-monoglyceride) had the highest amount of high molecular weight fraction. This result is in apparent contradiction of the viscosity

TABLE 3.9

DRYING TIMES OF 52% OIL LENGTH ALKYD [18]

Resin Number (Refer to Figure 3.8)	Drying Time [a] with 1.5% Lead; 0.2% Cobalt as Catalyst (Minutes)
1	6
2	24
3	18
4	42
5	30
6	54

[a] Drying time was measured on a Beck-Koller type machine.

TABLE 3.10

GLOSS OF 52% OIL ALKYDS WHEN CROSS-LINKED WITH A MELAMINE-FORMALDEHYDE RESIN. Pigment, TiO_2, P/B = 100/100, Baked 30 Minutes at 120°C

Resin Number (Refer to Figure 3.8)	60° Gloss
1	36
2	90
3	90
4	84
5	90
6	85

figures (Table 3.7), but has been explained by the suggestion that this high molecular weight material is not present in true solution but as a microgel. This satisfactorily explains the low viscosity, but very much faster air drying, of resin 1.

The low gloss of films prepared by cross-linking resin 1 with a melamine-formaldehyde resin is probably related to the low available hydroxyl number of this resin (Table 3.11). These results suggest that the hydroxyl

TABLE 3.11

HYDROXYL VALUES OF 52% OIL LENGTH ALKYDS[18]

Resin Number (Refer to Figure 3.8)	Theoretical Hydroxyl Value	Hydroxyl Value by Acetylation
1	124.2	35.0
2	122.9	61.5
3	123.1	51.0
4	124.1	47.7
5	123.5	37.0
6	124	54.0

groups in the microgel fraction are those not available for chemical reaction.

Not all alkyds show the differences between theoretical and measured hydroxyl values shown in Table 3.11.

Examination of a number of alkyds[31] with formulations typical of those in common use showed that the measured hydroxyl value corresponded to the theoretical value when the system had a functionality of about two or the percentage of the carboxyl groups esterified was low (Table 3.12).

TABLE 3.12

HYDROXYL VALUES OF A SERIES OF COMMERCIALLY AVAILABLE ALKYD RESINS.

Alkyd composition, molar ratios	Percentage esterifi-cation[a]	Average function-ality[b]	Hydroxyl value	
			Theoretical	Observed
Safflower fatty acids/glycerol/ pentaerythritol/phthalic anhydride = 3.0/1.0/1.49/2.95	87.0	2.12	49.0	48.2
Linseed fatty acids/glycerol/ phthalic anhydride = 3.0/3.0/3.0	95.3	2.00	10.6	10.8
Safflower fatty acids[c] glycerol/phthalic anhydride = 3.0/4.56/4.50	97.2	2.31	64.2	32.5
Coconut fatty acids/glycerol/ phthalic anydride = 3.00/5.99/6.09	93.4	2.20	113.0	65.0
Coconut fatty acids/trimethylol propane/benzoic acid/ phthalic anhydride = 3.0/5.76/0.27/5.16	98.3	2.17	102.5	76.0

[a] Carboxyl basis.
[b] Average functionality is the ratio of total equivalents to total moles.
[c] Contains fatty acid derived from bodied safflower oil of viscosity 5 poise.

The results discussed above clearly indicate the need to be able to control, and measure, the degree of reaction between the oil and glycerol in the manufacture of alkyd resins. Failure to reproduce the same monoglyceride composition in successive batches of a resin will lead to variations in the molecular weight distribution; this variation will be of greatest importance where stringent specifications have to be met. The common method of following a monoglyceride reaction in alkyd resin manufacture is a solubility test in an alcohol. Methanol and/or ethanol are the most commonly used alcohols for this purpose. The solubility of a monoglyceride mixture in an alcohol is a guide to the compatibility of this mixture with either phthalic anhydride, or with the dibasic acid to be used in the condensation stage of resin manufacture. However, the fact that a monoglyceride mixture is soluble in a given volume of an alcohol does not indicate the amount of α-mono-glyceride present; it appears that a number of vastly different compositions

—which vary in the amounts of mono-, di-, and triglyceride, and free glycerol contents—will meet a given solubility specification. For example, Mort[13] has shown that glycerolysis products of linseed oil with 23% and 41% α-monoglyceride contents both pass the solubility test in methanol. Similar results have been noted by others.[16-18] For example it has been shown[32] that in the sodium hydroxide catalysed glycerolysis of linseed oil alcohol solubility is reached at about 35% monoglycerides and that the equilibrium composition contains about 45% monoglycerides. Titration of the mono-glyceride mixture by methanol is not an accurate measure of the composition;[25] for example, mixtures that contain 23.8% and 33.3% monoglyceride (α and β) give the same titration figure.[25] Mort[13] has suggested that the attainment of equilibrium in the glycerolysis of linseed oil can be gauged by measuring the electrical conductivity of the solution; this method could provide a cheap and simple technique for large scale control of glycerolysis reactions. Similarly, the electrical conductivity of oil/pentaerythritol mixtures can be used to indicate the point at which unreacted pentaerythritol is at a minimum.[26] Recent results[27] on the system glycerol/linseed oil, with a sodium hydroxide catalyst, show that extended times of heating of the mixture give small changes in conductivity for significant increases in α-monoglyceride contents. Therefore, in using electrical conductivity as a control, it is necessary to have limits on the time of reaction also. Conductivity is also a useful control for the esterification stage of alkyd manufacture.[26]

The catalyst used to promote the glycerolysis of an oil has an important bearing on the composition of the monoglyceride mixture. The results shown in Table 3.13 have been compiled from the findings of Shibayama;[25]

TABLE 3.13

INFLUENCE OF CATALYST ON THE ALCOHOLYSIS REACTION OF AN OIL[25]

Catalyst	Percent Monoglyceride	Percent Diglyceride	Percent Triglyceride	Percent Glycerol
Ca(OH)$_2$	47.4	26.1	9.7	16.8
PbO	27.1	27.1	25.2	20.6

they indicate that lead oxide is more effective in converting triglyceride (oil) to diglyceride, while calcium hydroxide gives higher yields of monoglyceride. These findings are consistent with the lead catalyst being present in the oil phase, and the calcium hydroxide present in the glycerol phase.

Therefore, for effective control of the properties and molecular weight distribution of an alkyd resin, the alcoholysis of the oil must be reproducible.

The use of dilution tests with alcohols for control of the alcoholysis reaction is inadequate. Electrical conductivity is a far more reliable method, especially when used in conjunction with a time limit on the reaction.

REFERENCES

1. Flory, P. J., *Principles of Polymer Chemistry*, Cornell University Press, New York, 1953.
2. Wright, H. J., and R. N. Du Puis, *Ind. Eng. Chem.*, **36**, 1004 (1944).
3. Brett, R. A., *J. Oil Colour Chemists' Assoc.*, **41**, 428 (1958).
4. Bullett, T. R., and A. T. S. Rudram, *J. Oil Colour Chemists' Assoc.*, **42**, 778 (1959).
5. Helme, J. P., J. Molines, G. Bosshard, and M. Rouzier, *VI, F.A.T.I.P.E.C. Congr.*, **1962**, 258.
6. Topham, W. G., *Chem. Ind. (London)*, **1963**, 401.
7. Seavell, A. J., *J. Oil Colour Chemists' Assoc.*, **42**, 319 (1959).
8. Flory, P. J., *J. Am. Chem. Soc.*, **64**, 2205 (1942).
9. Solomon, D. H., *J. Oil Colour Chemists' Assoc.*, **48**, 282 (1965).
10. Kraft, W. M., G. T. Roberts, E. G. Janusz, and J. Weisfeld, *Am. Paint J.*, **41**, 96 (1957).
11. Bult, R., *Verfinst. T.N.O., Circ.*, **89** (**1958**), 95.
12. Tawn, A. R. H., *J. Oil Colour Chemists' Assoc.*, **39**, 223 (1956).
13. Mort, F., *J. Oil Colour Chemists' Assoc.*, **39**, 253 (1956).
14. Runk, R. H., *Ind. Eng. Chem.*, **44**, 1124 (1952).
15. Wilson, R., and A. H. Robson, *Offic. Dig. Federation Soc. Paint Technol.*, **27**, 111 (1955).
16. Fletcher, J. R., L. Polgar, and D. H. Solomon, *J. Appl. Polymer Sci.*, **8**, 659 (1964).
17. Solomon, D. H., and J. J. Hopwood, *J. Appl. Polymer Sci.*, **10**, 993.
18. Solomon, D. H., and J. J. Hopwood, *J. Appl. Polymer Sci.*, in the press.
19. Bobalek, E. G., E. R. Moore, S. S. Levy, and C. C. Lee, *J. Appl. Polymer Sci.*, **8**, 625 (1964).
20. Chicago Paint and Varnish Production Club, *Offic. Dig. Federation Soc. Paint Technol.*, **31**, 1364 (1959).
21. Reiser, K., E. S. J. Fry, and M. Jonason, Berger, Jenson and Nicholson Ltd., Brit. Pat. 968,322 (1964) via *Chem. Abstr.*, **61**, 12159 (1964).
22. Goldsmith, H. A., *Ind. Eng. Chem.*, **40**, 1205 (1948).
23. Spitzer, W. C., *Offic. Dig. Federation Soc. Paint Technol.*, **36** (**475**), 16 (1964).
24. Fry, D. G., Paint Research Station (England), Research Memorandum No. 260 (1958).
25. Shibayama, K., *Mitsubishi Denki Laboratory Reports*, **1** (No. 4), 97 (1960).
26. McKee, R. S., and A. W. E. Staddon, *J. Oil Colour Chemists' Assoc.*, **44**, 497 (1961).
27. Solomon, D. H., and J. D. Swift, *J. Oil Colour Chemists' Assoc.*, **49**, 915 (1966).
28. Ivanfi, J., and E. Ady, *J. Paint Technol.*, **42**, 199 (1970).
29. Lesnini, D. G., *J. Paint Technol.*, **38**, 498 (1966).
30. Nagata, T., *J. Appl., Polymer Sci.*, **13**, 2277, 2601 (1969).
31. Solomon, D. H., and J. J. Hopwood, *J. Appl., Polymer Sci.*, **10**, 985 (1966).
32. Rheineck, A. E., R. Bergseth, and B. Sreenivasan, *J. Am. Chem. Soc.*, **46**, 447 (1969).

CHAPTER 4

VINYL AND ACRYLIC
MODIFIED OILS AND ALKYDS

The commercial availability of vinyl and acrylic monomers, and the many desirable properties of plastic-type polymers based on these monomers, has led to widespread research aimed at incorporating these materials into surface coatings.

Thermoplastic and thermosetting polymer systems have been developed that are based solely, or predominantly, on the vinyl or acrylic monomers (discussed in Chapters 6 and 10). Alternatively, the existing surface-coating vehicles have been modified by the vinyls and/or acrylics with the aim of combining the properties of the two polymer species. In this latter class, the most commonly chosen coating polymers have been based on oils or alkyds. These combined vehicles offer the possibility of uniting the desirable application and wetting properties of the oil or alkyd with the strength, chemical, and weather resistance of the vinyl or acrylic moiety.[1,2,3] In some resins, the vinyl or acrylic polymer may lower the overall cost.

Vinyl or acrylic modified oils and alkyds have been prepared by three distinct methods. These are as follows.

1. Blending of the two polymer entities.

2. Copolymerization, or attempted copolymerization, involving the unsaturation of the fatty acid.

3. Chemical combination of the polymers by other functional groups, usually resulting in an ester or ether linkage between the components.

BLENDING OF THE TWO POLYMER ENTITIES

The simple blending of a vinyl or acrylic polymer with a vegetable oil or an alkyd resin, while attractive from a production viewpoint, presents certain technical problems. Early work in this field repeatedly led to incompatible products, especially when polymers designed for use as plastics were used. In retrospect this incompatibility is not surprising, since the poly-(styrene) or poly(methyl methacrylate) commonly used had molecular weights that were high by surface-coating standards (up to 500,000); it is

117

now recognized that lower molecular weight would enhance the likelihood of a compatible system. It appears that addition polymers with molecular weights 7000 to 10,000 are more desirable for compatibility,[2] but even with these systems, specific oil or alkyd formulations are required.

The lack of compatibility in most of these blended systems is a specific illustration of the general conclusions of Flory[4] that "incompatibility of chemically dissimilar polymers is observed to be the rule and compatibility is the exception." Further, Flory has shown that the mixing process of a ternary system of two polymers and a mutual solvent is governed by the thermodynamic equation:

$$\Delta G = \Delta H - T \Delta S$$

where T = absolute temperature
ΔH = change in internal heat of the system that results from inter-molecular interactions during mixing
ΔS = entropy change during mixing
ΔG = free energy change during mixing

With macromolecules, ΔS will be slight, usually small and positive. Most polymers mix with accompanying heat absorption (ΔH positive), and this term invariably outweighs the term $T \Delta S$ owing to the small magnitude of ΔS. As a consequence, ΔG is positive, and incompatibility between macromolecules is often observed. One exception, which is predicted by the above equation, is the mixing of polymers containing polar groups that react to give negative ΔH's and, hence, a negative free energy change during mixing.

By applying the principle of polar-polar interaction to vinyl and/or acrylic blends with oils and alkyds, it is possible to arrive at compatible mixtures. For example, the use of vinyl or acrylic copolymers containing monomers with basic nitrogen atoms (vinyl pyridine, acrylamide, dimethylaminoethyl methacrylate) gives homogeneous mixtures with alkyds because of the interaction of the basic and acidic functions of the two polymer entities.[5] Alternatively, acidic groups may be incorporated into the addition polymer by the use of unsaturated acids, and the basic nitrogen is built into the alkyd component by the use of an appropriate polyol such as trimethylolaminomethane.[6]

$$CH_2OH$$
$$|$$
$$NH_2-C-CH_2OH$$
$$|$$
$$CH_2OH$$

trimethylolaminomethane

Polymers and copolymers of butadiene which have been depolymerized to a low molecular weight are compatible with specific alkyd resins based on unsaturated fatty acids.[40] Oxidized long oil caster oil phthalic alkyds and vinyl acetate-ethylene copolymers are claimed to yield compatible mixture which give excellent wet adhesion when used as a component in a latex paint.[41]

Technically, the preparation of modified oils and alkyds by a blending process has a number of limitations. Where compatibility is achieved by the use of low molecular weight and polar vinyl or acrylic entities, the advantages of these polymer systems are lost, since their mechanical properties, and weather and chemical resistance, are a function of molecular weight. Therefore, the blended systems often have properties inferior to either of the original components unless other reactive or functional groups are included in the polymers to ensure some cross-linking reactions during film formation. Methods by which this may be done are considered in the chapter on thermosetting acrylics.

MODIFICATIONS INVOLVING THE UNSATURATION OF THE FATTY ACID

The original difficulties experienced with the blending of vinyls and/or acrylics with oil or alkyd polymers suggested that chemical combination of the two entities would be desirable, and should minimize problems of compatibility and subsequent chemical cross-linking. With unsaturated fatty acids, the double bonds or the active methylene groups may copolymerize with the vinyl or acrylic monomer. Many modified oils or alkyds are prepared by carrying out the addition polymerization in the presence of the oil or alkyd.

Vinylated products prepared by the above procedures are far more compatible than simple blended mixtures, and from this it has been inferred that copolymerization occurs. However, recent studies suggest that this view is not entirely correct, although it is difficult to generalize; the monomer structure and the conditions of polymerization can influence the amount of copolymerization.

Conjugated fatty acids appear to form copolymers and Diels-Alder type adducts with styrene, even in the absence of polymerization initiators.[2,7] The adduct probably has the structure:

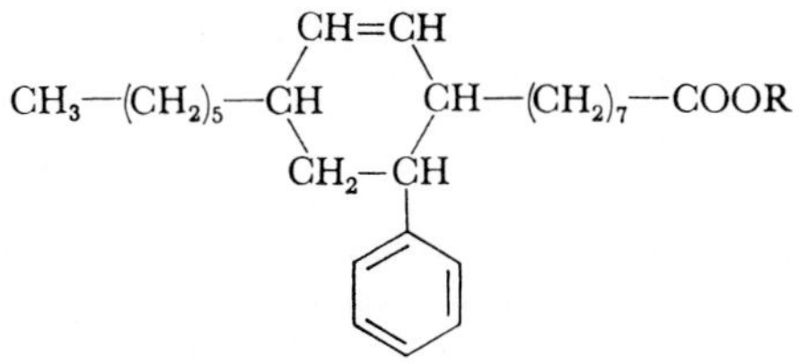

$$CH_3-(CH_2)_5-CH \underset{\displaystyle CH_2-CH}{\overset{\displaystyle CH=CH}{\big\langle \quad \big\rangle}} CH-(CH_2)_7-COOR$$

The copolymer from an α-eleostearate and styrene has, on an average, a ratio of styrene:fatty ester of 4.75:1 (mole), and contains approximately 50 styrene units per molecule. The loss of one double bond per fatty acid molecule, and the absence of a conjugated diene structure[2] suggests a 1:4 addition of the styrene as shown below.

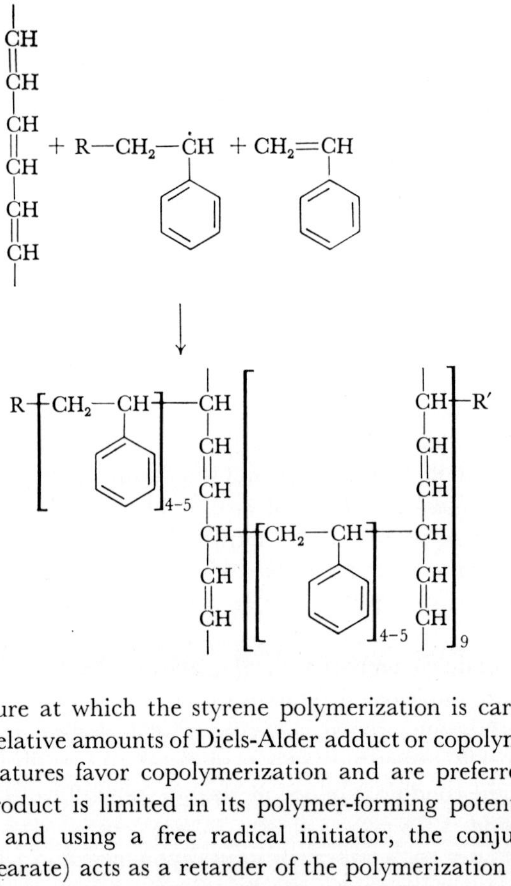

The temperature at which the styrene polymerization is carried out can influence the relative amounts of Diels-Alder adduct or copolymer formed;[8] higher temperatures favor copolymerization and are preferred, since the Diels-Alder product is limited in its polymer-forming potential.[2] At low temperatures, and using a free radical initiator, the conjugated triene (methyl eleostearate) acts as a retarder of the polymerization of styrene or acrylonitrile. It has been suggested[9] that a resonance-stabilized allylic system (as shown opposite) is responsible for this effect.

Modification of conjugated oils or alkyds by methyl methacrylate follows a substantially similar course to that of styrenation.[10]

The reaction of nonconjugated oils and alkyds with vinyl or acrylic monomers depends on the reaction conditions and on the particular monomer chosen. Most of the published literature on attempted copolymerization of nonconjugated fatty acids refers to styrene.[9] With this monomer, temperatures of at least 100°C are necessary to induce some reactivity into the fatty acid molecule. Even so, the nonconjugated esters behave only as retarders for the polymerization;[11] the predominant reaction is one of

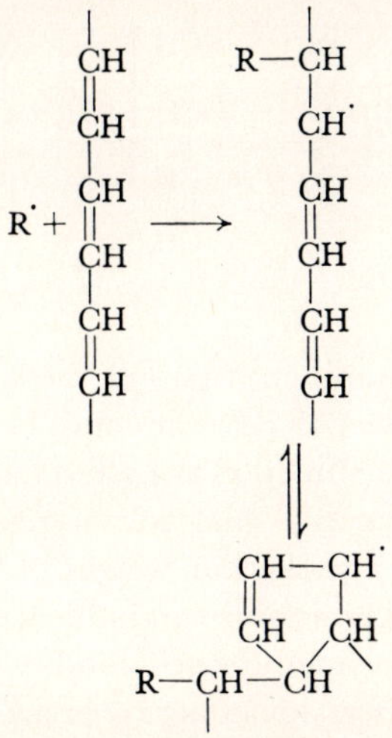

poly(styrene) formation, with the molecular weight being controlled by chain transfer to the active methylene group of the oil.

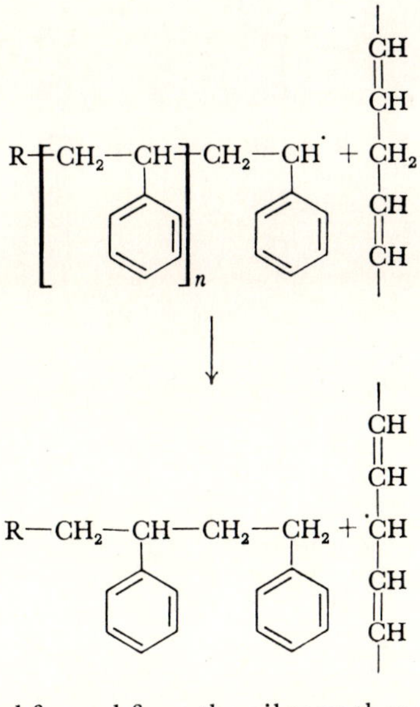

The resonating radical formed from the oil may then initiate the formation of a graft copolymer or may undergo dimerization.

Therefore, a styrenated nonconjugated oil or alkyd, produced by polymerization of the monomer in the presence of the other polymer species, consists essentially of poly(styrene), unchanged oil or alkyd, some graft

$$
\begin{array}{c}
|\\ CH\\ \|\\ CH\\ |\\ \dot{C}H \\ |\\ CH\\ \|\\ CH\\ |
\end{array}
\; + \; n\,CH_2{=}\underset{\underset{Y}{|}}{\overset{\overset{X}{|}}{C}}
\;\longrightarrow\;
\begin{array}{c}
|\\ CH\\ \|\\ CH\\ |\\ CH\\ |\\ CH\\ \|\\ CH\\ |
\end{array}
\!\!\left[\!-CH_2-\underset{\underset{Y}{|}}{\overset{\overset{X}{|}}{C}}-\!\right]_{n-1}\!\!CH_2-\underset{\underset{Y}{|}}{\overset{\overset{X}{|}}{\dot{C}}}-
$$

copolymer, and dimerized oil entities. For these systems to be compatible, the molecular weight of the poly(styrene) must be low, and graft copolymer, which would be expected to function as a solubilizing agent, must be present. The observation by Redknap[2]—that poly(styrene) formed in the presence of methyl linolenate has a molecular weight of 7080—supports the above contentions. The choice of initiator can influence, among other things, the proportion of graft copolymer formed, and it has been suggested that tertiary butyl peroxide offers advantages over azodiisobutyronitrile because of the formation of a more reactive radical which more readily attacks the oil and favors graft copolymer formation.

$$
\dot{R} \;+\;
\begin{array}{c}
|\\ CH\\ \|\\ CH\\ |\\ CH_2\\ |\\ CH\\ \|\\ CH\\ |
\end{array}
\;\longrightarrow\; RH \;+\;
\begin{array}{c}
|\\ CH\\ \|\\ CH\\ |\\ \dot{C}H\\ |\\ CH\\ \|\\ CH\\ |
\end{array}
$$

(from initiator)

$$
n\,CH_2{=}\underset{\underset{Y}{|}}{\overset{\overset{X}{|}}{C}}
$$

$$
\begin{array}{c}
|\\ CH\\ \|\\ CH\\ |\\ CH\\ |\\ CH\\ \|\\ CH\\ |
\end{array}
\!\!-\!\!\left[\!-CH_2-\underset{\underset{Y}{|}}{\overset{\overset{X}{|}}{C}}-\!\right]_{n-1}\!\!CH_2-\underset{\underset{Y}{|}}{\overset{\overset{X}{|}}{\dot{C}}}-
$$

This is, therefore, an additional mechanism by which graft copolymers can be formed.[3,12]

The behavior of monomers, other than styrene, depends upon their structure and upon their tendency to enter into copolymerization reactions with oil-type molecules. Methyl methacrylate is similar to styrene and yields essentially poly(methyl methacrylate) with possibly some graft copolymer.[10] On the other hand, acrylonitrile, because of the polar nature of its substituent group, forms a copolymer with linseed oil type esters, and compositions with 60 to 80 wt. % of the oil are possible.[9]

Various techniques have been developed to control the relative amounts of homopolymer and graft copolymer formed in the modification of oils and alkyds. The use of iodine to inhibit the homopolymerization is claimed to favor copolymerization reactions;[13] the conversion of *cis* to *trans* isomers facilitates vinyl or acrylic modification.[14] The use of mixtures of monomers, chosen so that one of the possible propagating radicals is reactive toward linoleate or linolenate, is another means of favoring graft copolymerization; one of the monomers is often acrylonitrile.[15] The conversion of the non-conjugated to a conjugated olefin by the usual catalysts (page 60), prior to modification with the monomer, is often practiced with diene oils; with trienes, conjugation can lead to retarding effects because of possible cyclization (page 120). As an alternative to isomerizing the oil, mixtures of conjugated and nonconjugated oils—for example, wood oil/linseed oil— have been used in the original alkyd manufacture.[16]

More recently formulations have been developed in which the alkyd component has incorporated in it groups which should enhance copolymerization with the vinyl or acrylic monomer. For example, an alkyd has been described[42] which contains Rianene fatty acid (with about 50% conjugated double bonds), maleic anhydride residues and trimethylolpropane diallyl ether residues all of which should enhance the subsequent copolymerization with a monomer such as methyl methacrylate (a formulation of this type is given in Appendix).

In the preparation of a modified alkyd, the vinyl/acrylic monomer may be reacted with the oil, and the product converted via a monoglyceride into the alkyd. Alternatively, the monoglyceride may be modified and then reacted with the dibasic acid to give the modified alkyd, or the preferred alkyd may be directly modified with the monomer. These alternative processes often stem from the need to avoid patent restrictions.

The modified vehicles have average molecular weights higher than the original oil or alkyd, so that they tend to be intermediate between the usual thermoplastic and thermosetting polymer systems. One of the major uses for modified alkyds has been in fast air-drying industrial finishes, since the initial tack-free stage of these resins is virtually only a solvent release. Subsequently, oxidation of the film takes place. However, the amount of oxygen taken up by the modified alkyd is less than would be predicted from the oil length.[17] This is readily understandable, since the modification process uses either the conjugated unsaturation or the active methylene groups of the oil.

As a consequence, finishes based on modified oils and alkyds may give trouble when a second coat is applied, since the period during which the coating will be swollen by solvent is relatively long. It is common to find that recoating is possible after either a short or long drying period, corresponding to complete solubility, or insolubility, of the polymer in the solvent of the second coat. The reduced oxygen uptake may be advantageous in minimizing the number of groups sensitive to chemical attack, but the disadvantages resulting from the reduced cross-linking are solvent sensitivity and poor weather resistance. It should be noted that the presence of low molecular weight addition polymer in the final film defeats one of the original aims, that is, improvement in durability.

METHODS OF CHEMICALLY COMBINING THE POLYMER ENTITIES
(ESTERIFICATION, ETHERIFICATION AND ACETAL FORMATION)

The methods discussed above for the modification of oils and alkyds impose certain limitations on the use of the vehicle. This arises from the presence of considerable amounts of vinyl and/or acrylic homopolymer. The nature of the reaction involved in the modifying process means that the fully saturated fatty acids and oils can not be successfully vinylated or acrylated. The use of these saturated oils is particularly desirable where highly durable and discoloration-resistant finishes, usually of the stoving type, are required. Therefore, methods that do not rely on the unsaturated centers of the fatty acid molecules have been sought for the modification of oils and alkyds. Usually, the two polymer entities are linked by ether or ester bonding. A recent technique involves the combination of oil and acrylic components by acetal type linkages.[43]

Some "compatible" vinylated and/or acrylated non-drying oil alkyds have been prepared by the conventional method of polymerizing the monomer in the presence of the preferred alkyd. However, it has been shown that the presence of carbon-carbon unsaturation, usually as a maleate or fumarate resulting from impurities in the phthalic anhydride, or as unsaturated acids deliberately added to the alkyd, was responsible for some copolymerization and compatibility.[18-26]

Other methods of combining vinyl or acrylic polymers with alkyds rely on using a vinyl (acrylic) copolymer that contains groups capable of being esterified by either the hydroxyl, or carboxyl, of the alkyd. These methods are formally similar to the use of α-β unsaturated acids in the alkyd (as described above), but the methods differ in that the esterification reaction is conducted after the addition polymerization. This difference in the order in which the various chemical steps are carried out gives a number of important technical benefits[26–32] which include the following:

1. Fully saturated oils may be used.

2. Where unsaturated oils are required, the double bonds or active methylene groups are not lost in the modification process and are available for subsequent reactions.

3. The addition polymerization is carried out in the absence of the glyceride oil, thus eliminating difficulties resulting from the inhibition of the free radical polymerization by the natural antioxidants of the oil.

4. The functionality of the copolymer and, hence, the number of chemical links to the alkyd component, is controllable.

5. Differences in the reactivity ratios of the comonomers may be allowed for.

6. The addition prepolymer can be fully characterized before reaction with the alkyd.

7. The molecular weight of the addition polymer is often higher than that formed by the alleged "copolymerization" with the oil unsaturation (for example, 30,000 as against 8000). Despite this increase in molecular weight, the final resins have comparable viscosities, probably because of the greater chemical homogeneity of the polymer prepared by esterification.

The type of chemical group present in a saturated fatty acid based alkyd (hydroxyl and carboxyl) means that anhydride, epoxy, acid, and hydroxyl groups are used in the addition polymer. The esterification reactions involved can be summarized as follows.

Reaction 1

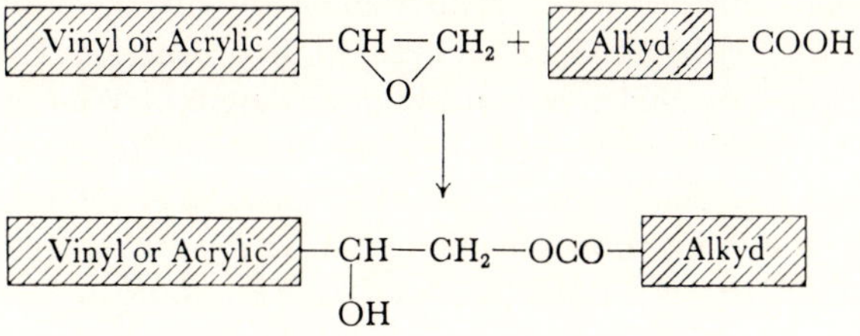

$$\boxed{\text{Vinyl or Acrylic}}-CH-CH_2 + \boxed{\text{Alkyd}}-COOH$$

with the epoxide oxygen bridging CH and CH2, giving

$$\boxed{\text{Vinyl or Acrylic}}-CH-CH_2-OCO-\boxed{\text{Alkyd}}$$

where the first CH bears an OH group.

Reaction 2

$$\boxed{\text{Vinyl or Acrylic}}-OH + \boxed{\text{Alkyd}}-COOH$$

$$\boxed{\text{Vinyl or Acrylic}}-O-CO-\boxed{\text{Alkyd}}$$

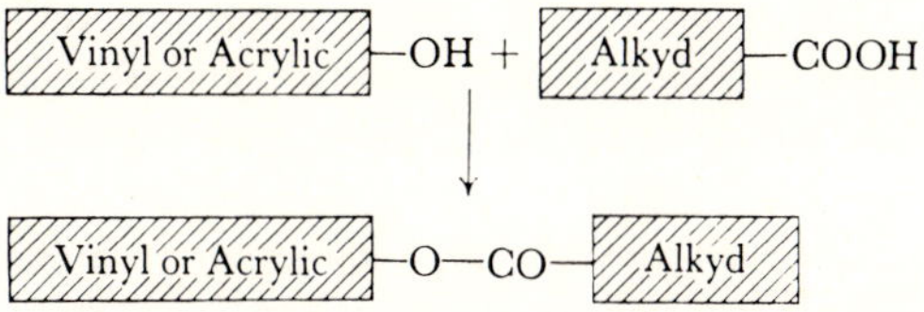

Reaction 3

$$\boxed{\text{Vinyl or Acrylic}} \overset{\displaystyle -C\overset{O}{\underset{O}{\diagdown}} }{\underset{\displaystyle -C\overset{\diagdown}{\diagup}O}{}} O \;+\; \boxed{\text{Alkyd}}-\text{OH}$$

$$\downarrow$$

$$\boxed{\text{Vinyl or Acrylic}} \underset{\displaystyle \overset{|}{\text{COOH}}}{-\text{COO}-} \boxed{\text{Alkyd}}$$

Reaction 4

$$\boxed{\text{Vinyl or Acrylic}}-\text{COOH} \;+\; \boxed{\text{Alkyd}}-\text{OH}$$

$$\downarrow$$

$$\boxed{\text{Vinyl or Acrylic}}-\text{COO}-\boxed{\text{Alkyd}}$$

The number of functional groups present in the resin components will naturally govern the extent of chemical combination. Since the comonomers with hydroxy, epoxy, anhydride, or acid groups are usually more expensive than the predominant backbone monomer (styrene, methyl methacrylate), it is desirable to keep the functional groups to a minimum. High functionality gives problems because of the likelihood of gelation, whereas low functionality increases problems of poor compatibility. In practice, functional groups are present in the vicinity of 4 mole %. Typical monomers containing the required functional groups are listed on page 268. Reactions 1 and 3 generally take place at appreciable rates at approximately 150°C. Under these conditions, further self-condensation of the alkyd, and/or thermal depolymerization of the vinyl, are minimized. Both reactions (1 and 3) can be readily followed by acid or anhydride group determinations. However, in some instances, it may be preferable to use acid or hydroxyl groups in the vinyl copolymer, particularly if the reactivity ratios favor a more distributed polymer, or if the functional monomer is comparatively cheap. With hydroxyl or carboxyl functional groups, it is more difficult to control the amount of vinyl to alkyd combination, since self-condensation of the alkyd groups is a competing reaction. It has been suggested that this problem can be overcome by reacting the alkyd components in a stepwise manner with the preformed acid copolymer. For example, a coconut oil monoglyceride can be esterified with a methyl methacrylate-methacrylic acid copolymer, a reaction that can be followed by acid value determinations; additional glycerol and phthalic anhydride are then added to build up the alkyd structure.

The above methods of preparing vinyl and acrylic modified alkyds, besides enabling non-drying oils to be used, also offer advantages when applied to the drying oil systems.[28,30] Drying oil modified alkyds prepared by this method oxidize and dry more readily, and are less susceptible to recoating problems, than the "copolymerized" vehicles discussed on page 115. This results from the chemical combination of the two polymer systems by ester links; these do not use up the potentially autoxidizable centers of the fatty acid molecule. Methacrylated alkyds prepared via esterification reactions will tolerate weaker solvents than "copolymerized" resins of the

$$
\begin{array}{l}
\text{Acrylic} \left\{
\begin{array}{l}
-COOH \\[2mm]
-COOH
\end{array}
\right.
\quad
\begin{array}{l}
CH_2OH \\
| \\
+ \; CHOH \\
| \\
CH_2-OCOR
\end{array}
\\[6mm]
\downarrow \\[4mm]
\text{Acrylic} \left\{
\begin{array}{l}
-COOCH_2 \\
\quad | \\
\quad CH-OH \\
\quad | \\
\quad CH_2-OCOR \\
-COOCH_2 \\
\quad | \\
\quad CH-OH \\
\quad | \\
\quad CH_2OCOR
\end{array}
\right.
\end{array}
$$

$$
\downarrow \quad + \quad
\begin{array}{c}
\text{(phthalic anhydride)} \\
C_6H_4(CO)_2O
\end{array}
\quad + \quad
\begin{array}{l}
CH_2OH \\
| \\
CHOH \\
| \\
CH_2-OCOR
\end{array}
$$

$$
\text{Acrylic} \left\{
\begin{array}{l}
-COOCH_2 \\
\quad | \\
\quad CH-OCO-\text{(C}_6\text{H}_4\text{)} \\
\quad | \\
\quad CH_2OCOR \\
-COOCH_2 \\
\quad | \\
\quad CH-OCO-\text{(C}_6\text{H}_4\text{)} \\
\quad | \\
\quad CH_2OCOR
\end{array}
\right.
\quad
\begin{array}{l}
COO-CH_2 \\
\quad | \\
\quad CH-OCO-\text{(C}_6\text{H}_4\text{)} \\
\quad | \\
\quad CH_2OCOR \\
COO-CH_2 \\
\quad | \\
\quad CH-OCO-\text{(C}_6\text{H}_4\text{)} \\
\quad | \\
\quad CH_2OCOR
\end{array}
$$

same overall composition, and are less likely to give compatibility problems or low gloss on pigmentation. These properties are probably a reflection of the greater homogeneity of the esterified polymer, and its small content of free poly(methyl methacrylate).

Alternative methods of combining the properties of vinyl (acrylic) and oil-type polymers involve the formation of unsaturated esters or ethers of the fatty acid or fatty alcohol. Vinyl esters of fatty acids can be prepared by reacting the fatty acid with vinyl acetate in the presence of suitable catalysts, such as mercuric acetate and sulfuric acid.[33,34] These fatty acid vinyl esters can be homopolymerized or copolymerized with, for example, vinyl acetate, vinyl chloride, or acrylic esters using conventional free radical initiation.

Such copolymers are, in effect, internally plasticized vinyls (acrylics) and have the properties expected from these systems. Air-drying compositions are possible, if sufficient unsaturated fatty acid ester is present, and these systems undergo similar autoxidation reactions to a drying oil.

$$R-COOH + CH_2=CH$$
$$|$$
$$OCOCH_3$$

$$\Big\downarrow H_2SO_4 \text{ or } Hg(OCOCH_3)_2$$

$$CH_2=CH \quad + CH_3COOH \nearrow$$
$$|$$
$$OCOR$$

$$\Big\downarrow \begin{array}{c} CH_2=CH \\ | \\ OCOCH_3 \end{array}$$

$$-CH_2-CH-CH_2-CH-CH_2-CH-$$
$$| \qquad\qquad | \qquad\quad |$$
$$OCOR \qquad OCOCH_3 \quad OCOR$$

$$(R = \text{fatty acid})$$

The preparation of vinyl ethers involves either reaction of the fatty alcohol with acetylene or an interchange reaction with a simple vinyl ether.[35-38]

$$ROH + CH\equiv CH \longrightarrow CH=CH_2$$
$$|$$
$$OR$$

$$ROH + CH=CH_2 \longrightarrow CH=CH_2 + C_2H_5OH \nearrow$$
$$| \qquad\qquad\qquad |$$
$$OC_2H_5 \qquad\qquad OR$$

$$(R = \text{fatty acid chain})$$

Vinyl ethers are susceptible to cationic polymerization that is usually initiated with Friedel-Craft catalyst systems using, for example, boron trifluoride, ferric chloride, or stannic chloride.[35-39]

$$n CH_2=CH \xrightarrow[\substack{SnCl_4 \\ \text{and} \\ \text{cocatalysts}}]{BF_3} CH_3-CH{\Big[}CH_2-CH{\Big]}CH=CH$$
$$| \qquad\qquad\quad | \qquad\qquad | \quad |$$
$$O \qquad\qquad\quad O \qquad\quad O \qquad O$$
$$| \qquad\qquad\quad | \qquad\qquad |_{n-2} \quad |$$
$$R \qquad\qquad\quad R \qquad\quad R \qquad R$$

Unsaturated fatty alcohol vinyl ethers prepared by these methods still have essentially all the original fatty acid unsaturation[35] and, consequently, the polymers are readily susceptible to autoxidation. The films obtained in the presence of the normal driers have good alkali resistance because the backbone polymer does not contain ester links.[39] A range of copolymers can also be prepared with the simple vinyl ethers—notably ethyl, butyl, and isobutyl. These copolymers offer a wide range of desirable properties. The vinyl backbone is modified by the flexible fatty acid side chain and, since the original molecular weight and functionality are much higher than in an oil, the copolymers dry rapidly to give flexible, solvent-resistant films which are acid and alkali resistant.

Similar reactions can be applied to the direct esterification of functional vinyl or acrylic copolymers with fatty acids or fatty alcohols. For example, a styrene-allyl alcohol copolymer can be esterified with soya fatty acids and this polymer used as a component of a metal primer.[44]

A new approach to the chemical combination of oil and vinyl or acrylic entities has been presented by Sampath and Rheineck[43] who used a vegetable oil derivative which contained aldehyde groups. By the use of the methyl acetal of the aldehyde it was possible to control the reaction with various hydroxyl containing polymers including alkyds, styrene allyl-alcohol copolymers, and hydroxy terminated styrene-butadiene polymers. The general reaction sequence used to prepare these resins is shown below for a monoaldehyde derivative of a vegetable oil. In principal the di- and trialdehydes could also be used but the greater functionality leads to problems with control of the reaction.

$$
\begin{array}{ccc}
\underset{\displaystyle |}{CH_2-O-\overset{\displaystyle O}{\overset{\displaystyle \|}{C}}-(CH_2)_7-CHO} & & \underset{\displaystyle |}{CH_2-O-\overset{\displaystyle O}{\overset{\displaystyle \|}{C}}-(CH_2)_7-CH{<}^{OCH_3}_{OCH_3}} \\[2ex]
\underset{\displaystyle |}{CH-O-\overset{\displaystyle O}{\overset{\displaystyle \|}{C}}-R} & CH_3OH & \underset{\displaystyle |}{CH-O-\overset{\displaystyle O}{\overset{\displaystyle \|}{C}}-R} \\[2ex]
& \longrightarrow & \\[1ex]
CH_2-O-\underset{\displaystyle O}{\underset{\displaystyle \|}{C}}-R & -H_2O & CH-O-\underset{\displaystyle O}{\underset{\displaystyle \|}{C}}-R
\end{array}
$$

where R = fatty acid residue.

Monaldehyde oil   Monoaldehyde oil dimethyl acetal.

$$
HO-\boxed{\text{Polymer}}-OH
$$

$$
\underset{\displaystyle |}{\underset{\displaystyle CH-O-C-R}{\underset{\displaystyle |}{CH-O-\overset{\displaystyle O}{\overset{\displaystyle \|}{C}}-(CH_2)_7-\overset{\displaystyle OCH_3}{\overset{\displaystyle |}{CH}}-\boxed{\text{Polymer}}-OH+CH_3OH}}}
$$

$$
CH_2-O-\underset{\displaystyle O}{\underset{\displaystyle \|}{C}}-R
$$

Further reaction of the hydroxyl groups of the polymer with other aldehyde oil molecules results in a further increase in the molecular weight.

REFERENCES

1. Brown, W. H., and T. J. Miranda, *Offic. Dig. Federation Soc. Paint Technol.*, **36** (**475**), 92 (1964).
2. Redknap, E. F., *J. Oil Colour Chemists' Assoc.*, **43**, 260 (1960).
3. Solomon, D. H., *J. Oil Colour Chemists' Assoc.*, **45**, 88 (1962).
4. Flory, P. J., *Principles of Polymer Chemistry*, Cornell University Press, New York, 1953, p. 555.
5. Johnson, A. L., M. Morf, and J. K. Whiteley, Canadian Industries Ltd., Brit. Pat. 877,236, Ger. Pat. 1,118,911, U.S. Pat. 2,964,483 (1960).
6. Brockman, F. J., J. D. Murdock, and N. Nelan, Canadian Industries Ltd., Ger. Pat. 1,122,647 (1962).
7. Crofts, J., *J. Appl. Chem. (London)*, **5**, 88 (1955).
8. Dyck, M., *Farbe Lack*, **67**, 148 (1961).
9. Mayo, F. R., and C. W. Gould, *J. Am. Oil Chemists' Soc.*, **41**, 25 (1964).
10. Rohm and Haas Co., *Tech. Bull.*, SP-20 1/62, The Methacrylation of Alkyd Resins.
11. Harrison, S. A., and W. E. Tolberg, *J. Am. Oil Chemists' Soc.*, **30**, 114 (1953).
12. Merrett, F. M., *Trans. Faraday Soc.*, **50**, 759 (1954).
13. Shell International Research MIJ, Brit. Pat. 866,485 (1961).
14. Establishments Robbe Freres, French Pat. 1,163,234 (1958).
15. Daniel, J. H., and R. T. Corkum, American Cyanamid Co., Can. Pat. 580,622 (1959).
16. Inshaw, J. L., and J. S. Wakely, Imperial Chemical Industries Ltd., Brit. Pat. 866,598 (1961).
17. Bevan, E. A., M. J. Heavers, and W. R. Moon, *J. Oil Colour Chemists' Assoc.*, **40**, 745 (1957).
18. Kovacs, L., and D. Charlesworth, *J. Oil Colour Chemists' Assoc.*, **46**, 47 (1963).
19. Meeske, J., and D. Leganis, Reichhold Chemicals, Brit. Pat. 670,119 (1952).
20. Walus, A. N., E. I. du Pont de Nemours & Co., Can. Pat. 597,841 (1960).
21. Christenson, R. M., and H. A. Vogel, Pittsburgh Plate Glass Co., Can. Pat. 595,981 (1960).
22. Konen, J. C., and R. A. Boller, Archer Daniels Midland Co., U.S. Pat. 2,877,194 (1959).
23. Christenson, R. M., Pittsburgh Plate Glass Co., U.S. Pat. 2,939,854 (1960).
24. Christenson, R. M., and H. A. Vogel, Pittsburgh Plate Glass Co., U.S. Pat. 2,919,254 (1959).
25. Yeoman, F. A., Westinghouse Electrics Corp., U.S. Pat. 2,590,668 (1952).
26. Shechter, L., and J. Wynstra, *Am. Chem. Soc. Paint, Plastics Printing Ink. Chem.*, Paper presented at the New York Meeting, **14**, 165 (1954).
27. Solomon, D. H., *J. Oil Colour Chemists' Assoc.*, **45**, 88 (1962).
28. Fletcher, J. R., D. P. Kelly, and D. H. Solomon, *J. Oil Colour Chemists' Assoc.*, **46**, 127 (1963).
29. Hopwood, J. J., C. Pallaghy, and D. H. Solomon, *J. Oil Colour Chemists' Assoc.*, **47**, 289 (1964).
30. Gunning, R., and D. H. Solomon, *J. Oil Colour Chemists' Assoc.*, **47**, 319 (1964).
31. Solomon, D. H., and P. J. Wigney, *J. Oil Colour Chemists' Assoc.*, **48**, 440 (1965).

32. Solomon, D. H., and J. J. Hopwood, *J. Oil Colour Chemists' Assoc.*, in press; Zimmerman, R. L., E. R. Moore, and D. M. Pickelman, *Offic. Dig. Federation Soc. Paint Technol.*, **37**, 393 (1965).

33. Asahara, T., and M. Tomita, *J. Oil Chemists Soc. Japan* **1**, 76 (1952), via *J. Am. Oil Chemists' Soc.*, **29**, 439 (1952).

34. Shono, T., and C. S. Marvel, *J. Polymer Sci.*, **1**, 2067 (1963).

35. Teeter, H. M., W. J. Schneider, and L. E. Gast, U.S. Secretary of Agriculture, U.S. Pat. 2,889,309 (1959).

36. Dufek, E. J., R. A. Awl, L. E. Gast, J. C. Cowan, and H. M. Teeter, *J. Am. Oil Chemists' Soc.*, **37**, 37 (1960).

37. Gast, L. E., W. J. Schneider, J. L. O'Donnell, and H. M. Teeter, *J. Am. Oil Chemists' Soc.*, **37**, 78 (1960).

38. Mustakas, G. C., M. C. Raether, and E. L. Griffin, *J. Am. Oil Chemists' Soc.*, **37**, 68 (1960).

39. Brekke, O. L., and L. D. Kirk, *J. Am. Oil Chemists' Soc.*, **37**, 568 (1960).

40. Bailey, B. E., V. C. O. Dando, The International Synthetic Rubber Company Ltd., Brit. Pat., 1, 164, 815, (1969).

41. Kennedy, Richard J., du Pont de Neumours, E. I., and CO. U.S. Pat., 3, 547, 849 (1970).

42. Reichbold—Albert Chemie. Ger. Pat., 1, 295, 816 *via* Ger. Pat. Abs. 1969, 9, No. 26, Gp 1, 2.

43. Sampath, P. R., and A. E. Rheineck, *J. Paint Technol.*, **41**, 17, (1969).

44. Sahni, A. P., Monsanto Co., U.S. Pat., 3,551, 368 (1970).

CHAPTER 5

UNSATURATED POLYESTERS AND HIGH SOLIDS COATING SYSTEMS

One of the ultimate aims in formulating surface-coating polymer systems is to eliminate, or reduce to a minimum, the use of organic solvents as aids to film application or formation. Two distinct approaches to this problem have been used. First, water has been used as the solvent, or carrier, for the polymer (Chapter 11) and second, the components of the polymer system have been chosen so that they all—or nearly all—have film-forming potential. The chemical reactions involved in the second approach are considered in this chapter.

Polymer systems that contain only film-forming components are sometimes referred to as "100% solids" or "solventless coatings." However, in practice, there is usually some loss of the volatile reactive solvent, oxidation products, or stabilizers; therefore, "high solids coatings" is a more apt description of the formulations. Possible advantages of these compositions are reduced atmospheric pollution, increased film builds with a consequent reduction in volume change and stress development during film formation, simplification of the painting operations with respect to the number of coats necessary, and economic gains related to the elimination of volatile expensive solvents.

Chemically, two broad principles have been used in the formulation of high solids coatings. Solutions of relatively high molecular weight polymers (approximately 3000) in polymerizable monomeric solvents of low molecular weight (approximately 150) are one type of composition used. These are typified by the unsaturated polyesters, that is, a blend of a relatively high and low molecular weight film former. Alternatively, polymers of an intermediate molecular weight (approximately 1000), with viscosity characteristics such that no solvent is required for satisfactory application, can be used. These generally dry by autoxidation. High solids coatings based on epoxy resins, or isocyanates, are discussed in Chapters 7 and 8, respectively.

UNSATURATED POLYESTERS

Strictly speaking, any polyester containing carbon-carbon double bonds is an unsaturated polyester and, without further limitations, drying oil

133

alkyds would come within this class of polymers. However, the term unsaturated polyester has come to be associated with polymer systems where the unsaturation arises from the dibasic acid or the polyol, and is generally contained in the backbone polymer chain, not in a side chain. A more clear-cut means of distinguishing this type of polymer system from an alkyd is given by the effect of oxygen on the film-forming reactions. Whereas the drying or curing of an alkyd resin is promoted by oxygen, that of an unsaturated polyester is inhibited,[1] unless it is specifically modified with allyl ether groups (discussed on page 146).

In general, unsaturated polyester compositions consist of a predominantly linear unsaturated backbone polymer, thinned in a polymerizable monomer. An "activator" or "promoter" is also present. Immediately prior to use, an organic peroxide or hydroperoxide is added (that is, the system is marketed as two components) and this, in conjunction with the activator, forms a redox system. The redox system generates the free radicals necessary for the curing or cross-linking reaction. Once the components are mixed, the polymer solution has a limited stability or pot-life.

Briefly, the steps occurring in the cure of an unsaturated polyester composition are the following:

1. The generation of free radicals.

2. The initiation of polymer chains by attack of the free radicals on the monomer or backbone polymer unsaturation.

3. Copolymerization and cross-linking of the polymer chains with the unsaturated backbone polymer and polymerizable monomer.

The final structure can be represented as follows:

$$
\begin{array}{c}
R \\
| \\
[M]_X \qquad\qquad\qquad\qquad\qquad [M]_Z \\
| \qquad\qquad\qquad\qquad\qquad\qquad\quad | \\
U\!-\!G\!-\!S\!-\!G\!-\!S\!-\!G\!-\!U\!-\!G\!-\!U \\
| \\
[M]_Y \\
| \\
-\!G\!-\!U\!-\!S\!-\!G\!-\!U\!-\!G\!- \\
| \\
[M]_X
\end{array}
$$

where R = initiating free radical

M = polymerizable monomer

G = glycol residue

U = unsaturated dibasic acid residue

S = saturated dibasic acid residue

X, Y, Z = number of monomer units in the cross-linking segment

The components of an unsaturated polyester each serve a distinct yet interrelated chemical function that can be modified or controlled by formulation or processing conditions.

THE UNSATURATED POLYESTER BACKBONE POLYMER

The backbone polymer is conventionally a polyester in which by far the most common source of carbon-carbon double bonds is maleic anhydride. This choice is dictated by commercial considerations, and the simplified processing when using anhydrides, since it is necessary to remove only one molecule of water to form the diester. Consequently, systems based on maleic anhydride will be considered in order to illustrate the chemical reactions involved in polyester formation.

The polyester formed by condensing a glycol (such as ethylene glycol) with maleic anhydride, is too highly functional in the subsequent cross-linking reactions; this often leads to films with poor mechanical properties. Therefore, this functionality is controlled by the partial replacement of the maleic anhydride with other dibasic acids or anhydrides that do not contain a double bond that will react with the radical derived from the initiator or polymerizable monomer. For example, maleic/phthalic anhydride ratios of from 1/3 to 3/1 are commonly used. The choice of the non-reactive dibasic acid also influences the properties of the final coating. Isophthalic acid gives backbone polymers with a higher melting point than o-phthalic acid or anhydride (Table 5.1), and this offers some advantages

TABLE 5.1

EFFECT OF REPLACING o-PHTHALIC BY ISOPH-THALIC ACID ON THE MELTING POINT OF POLY-ESTERS[2]

Molar Ratio Isophthalic/o-Phthalic	Melting Point (Ball and Ring) °C
0	66–69
1.0	80–82
3.0	84–88
∞	78–81

where the film needs to be polished.[2] More styrene is required to reduce the isophthalate polymers to a given viscosity. This may offer commercial advantages where styrene is cheaper than the backbone polymer. The higher melting point of the isophthalate polymers has also enabled them to be marketed as solid resins that can be dissolved in styrene when required. This has reduced storage problems.

Aliphatic dibasic acids, such as adipic acid or dimer fatty acids, are generally associated with improved flexibility of the polymer system.

The esterification of maleic anhydride results in isomerization of the double bond; the *cis*-maleate goes to the *trans*-fumarate configuration. This isomerization, the degree of which can be readily estimated by infrared spectroscopy,[3] is favored by higher temperatures and low acid number in the resin.[4] The amount of isomerization varies, but virtually complete formation of the fumarate is possible.[5] The change from maleate to fumarate, while of basic importance to the copolymerization tendency of the backbone polymer (see page 140), also results in a higher melting point polymer that usually requires more styrene to reach a given viscosity.[6] The isomerization of maleate to fumarate in polyesters increases with increasing maleate/saturated dibasic acid ratio and is reported to be higher in *ortho*phthalic than in *iso*phthalic polyesters.[18]

The choice of the polyol used in preparing the polyester also influences the maleate to fumarate change; 2:3 butylene glycol, 1:2 propylene glycol, and ethylene glycol catalyze the conversion to a greater extent than non-vicinal glycols, such as 1:3 or 1:4 butylene glycol.[4] Therefore, the choice of polyol and/or reaction conditions are methods of controlling *cis-trans* isomerization. The polyol can also influence the melting point of the polyester,[2] as shown in Table 5.2.

TABLE 5.2

EFFECT OF 50% MOLAR REPLACEMENT OF PROPYLENE GLYCOL BY CYCLIC DIOLS [2]

Glycols	Acid (in Equimolar Proportions)	Melting Point (Ball and Ring) °C
Propylene glycol	Isophthalic:maleic	78–87
Cyclohexane-1:2-diol	Isophthalic:maleic	115–117
Isosorbide	Isophthalic:maleic	115–118
Hydrogenated bisphenol-A	*o*-Phthalic:maleic	99–101

The amount of water evolved during the condensation of the dibasic acid/polyol mixture is frequently greater than would be predicted from the degree of esterification as measured by acid value. Side reactions of the hydroxyl groups are responsible for this additional water of reaction. Internal dehydration of vicinal glycols can lead to the corresponding aldehyde, or ketone, as shown by the formation of propionaldehyde from 1:2-propylene glycol.[7]

$$CH_3\!-\!\underset{\underset{\textstyle OH}{|}}{CH}\!-\!\underset{\underset{\textstyle OH}{|}}{CH_2} \xrightarrow{-H_2O} \left[CH_3\!-\!CH\!=\!\underset{\underset{\textstyle OH}{|}}{CH} \right] \rightleftharpoons CH_3\!-\!CH_2\!-\!CHO$$

This reaction can be minimized by an initial low temperature (60 to 70°C) condensation which favors half-ester—and possibly some diester—formation, before heating to the final polymerization conditions.[7]

Polyether formation is also a common side reaction. It is catalyzed by strong acids (notably maleic and isophthalic[8]), but can be minimized in some formulations—for example, those using *o*-phthalic/maleic anhydride mixtures—by adding the maleic anhydride late in the condensation stage when the hydroxyl concentration is low. Other aspects of the late addition of maleic anhydride are considered on page 139. The formation of ethers results in resins with higher acid numbers at a given viscosity and also introduces hydrophilic groups into the backbone polymer.

Even though a diol/dibasic acid mixture is superficially bifunctional and should, therefore, lead only to linear polymers, the use of maleic anhydride as part of the dibasic acid can lead to chain branching particularly in the early stages of the esterification. Alkyl-substituted ethylene glycols, for example, 2:3-butylene glycol and 1:2 propylene glycol, add across the maleate (or fumarate) double bond under mild reaction conditions (24 hours at 100°C, 3 hours at 130°C, and a final temperature of 180°C).[9] Glycols containing two primary hydroxyl groups, for example, 1:4 butylene glycol, 1:3 propylene glycol, and ethylene glycol[9] do not add under these conditions. With more forcing initial reaction conditions (temperatures of 150°C), even ethylene glycol will add to the double bond; this reaction is favored by high temperatures, and by high free hydroxyl content of the reaction mixture, as shown in Table 5.3, and it increases with time.[10]

TABLE 5.3

EFFECT OF HYDROXYL CONCENTRATION ON THE ADDITION OF ETHYLENE GLYCOL TO MALEATE OR FUMARATE UNSATURATION [10]

Ethylene Glycol:Maleic Anhydride	Addition to C=C after 5 Hours (%)
1.1:1	10.5
1.2:1	12.0
4.0:1	24.0

A possible structure for the addition product of a glycol and maleic (or fumaric) acid is shown below.[9,10] Part of the evidence to support this structure is the presence of ether bands in the infrared spectrum of the dibasic acids recovered from a polyester by saponification.

$$CH_2\text{—}OH \quad CH\text{—}COO^- \qquad HO\text{—}CH_2\text{—}CH_2\text{—}O\text{—}CH\text{—}COO^-$$
$$CH_2\text{—}OH \ + \ CH\text{—}COO^- \ \longrightarrow \qquad\qquad\qquad\quad CH_2\text{—}COO^-$$

Adducts of this type have a potential functionality of three, but the absence of gelation during the preparation of many backbone polyester compositions, shown to contain glycol-unsaturated acid adducts, suggests[10] that some internal cyclization to a monofunctional lactone takes place.

$$HO-CH_2-CH_2-O-CH-COOR \quad \longrightarrow \quad + \ R'OH$$

$$\underset{\displaystyle CH_2-COOR''}{|}$$

The tendency to lactonize is related to the glycol structure;[11] for example, the cyclohexane-1:2-diol adduct shows a greater tendency to lactonize than that derived from ethylene glycol.

The addition of glycols to the unsaturated polyester backbone means that the double bond content of the polymer will vary from the calculated figure, and this will upset attempts to calculate copolymerization parameters. Also, failure to control the processing conditions closely could lead to variations in the extent of this reaction, and in the subsequent film-forming properties. Processing procedures that withhold the maleic anhydride until late in the reaction help to minimize adduct formation.

The molecular weight of the backbone polyester molecule is generally controlled by the degree of esterification as measured by acid number and viscosity. The hydroxyl/carboxyl group ratio governs the possible molecular weight and, in practice, it is usual to use a slight excess of hydroxyl groups. This reduces the effective functionality of the polyol to slightly less than two. Acid values of 20 to 40 and number-average molecular weights of 1000 to 3000 are typical values[2] for a polyester.

The carboxyl and hydroxyl end groups of the backbone polymer have been claimed to adversely affect the properties of a polyester composition. Methods for removing one, or both, of these groups are available. One process involves reaction with an isocyanate to give urea or urethane linkages from the carboxyl and hydroxyl groups, respectively.[12] This reaction, in conjunction with acid-value titration, has been developed as a quantitative method for determining hydroxyl groups in polyesters.[13] Similarly, reaction of the polyester with an excess of a diisocyanate[14] yields polymers with free isocyanate groups, which are claimed to improve the adhesion to some substrates. This may be related, at least in part, to the possibility of chemically linking the polyester composition to substrate hydroxy groups (for example, wood).

$$\boxed{\text{Polyester}}-OH \; + \; OCN-R-NCO$$

$$\downarrow$$

$$\boxed{\text{Polyester}}-OCONH-R-NCO$$

$$\downarrow$$

$$R-NH-CO-O-\boxed{\text{Polyester}}$$

$$|$$
$$NH$$
$$|$$
$$CO$$
$$|$$
$$O$$
$$|$$

Wooden substrate

Alternatively, the hydroxyl number of the polyester may be reduced by reaction with diketene[15] to give acetoacetic ester type end groups and improved properties.

$$\boxed{\text{Polyester}}-OH \; + \; CH_2{=}C{=}O$$

$$\downarrow$$

$$\boxed{\text{Polyester}}-OCOCH_2-\underset{\underset{O}{\|}}{C}-CH_3$$

It has been indicated above that adding the maleic anhydride after the other dibasic acid has been partially condensed with the polyol can reduce the loss of carbon unsaturation in the backbone polymer and, in the case of o-phthalic acid or anhydride, can minimize polyether formation. Other advantages also result from this method of preparation, which gives a different molecular distribution with the cross-linkable carbon double bonds situated predominantly at the chain ends.[16] This difference in molecular distribution gives polymers with higher softening temperatures, more viscous solutions in styrene, and better heat distortion properties.[16] When the resin is prepared by condensing the maleic anhydride with the glycol and then adding the other dibasic acid, the double bonds are located at the center of the polymer chain, and this gives vastly inferior properties.[17]

The formation of the unsaturated polyester backbone polymer by the chain polymerization of anhydride/olefin oxide has been described. This process is claimed to reduce costs by the use of the olefin oxide in place of glycols, and to reduce processing times. An example of this chain polymerization is the copolymerization of tetrabromophthalic anhydride, maleic anhydride, and propylene oxide.[19]

GENERAL REFERENCES

Boenig, H. V., *Unsaturated Polyesters*, Elsevier Publishing Co., Amsterdam, 1964.

REFERENCES

1. Seaborne, L. R., *Rept. Progr. Appl. Chem.*, **42**, 473 (1957).
2. Jenkins, V. F., A. Mott, R. J. Wicker, *J. Oil Colour Chemists' Assoc.*, **44**, 42 (1961).
3. Chiang, Mu-Tong, and E. G. Bobalek, *Offic. Dig. Federation Soc. Paint Technol.*, **31**, 1287 (1959).
4. Okita, T., and S. Oishi, *J. Chem. Soc. Japan, Ind. Chem. Sect.*, **58**, 315 (1955); via *Chem. Abstr.*, **50**, 4030 (1956).
5. Hayes, B. T., and R. F. Hunter, *Chem. Ind. (London)*, **1957**, 559.
6. Batzer, H., and B. Mohr, *Makromol. Chem.*, 369th. Comm. (1952).
7. Robins, R. G., *Australian J. Appl. Sci.*, **5**, 187 (1954).
8. Brown, R., H. Ashjian, and W. Levine, *Offic. Dig. Federation Soc. Paint Technol.*, **33**, 539 (1961).
9. Knodler, S., W. Funke, and K. Hamann, *Makromol Chem.*, **53**, 212 (1962).
10. Ordelt, Z., *Makromol. Chem.*, **63**, 153 (1963).
11. Ordelt, Z., and F. Ciganek, *Chem. Prumysl.*, **14**, 141 (1964).
12. Dannenbaum, H., Dr. Beck & Co., Ger. Pat. 1,058,731 (1959).
13. Dreher, B., *Farbe Lack*, **67**, 703 (1961).
14. Nischk, G., E. Müller, and L. Goerden, Farbenfabriken Bayer Akt.-Ges, Ger. Pat. 940,018 (1956).
15. Lonza Elektrizitatswerke und Chemische Fabriken Akt.-Ges., Brit. Pat. 815,843 (1959), via *Chem. Abstr.* **53**, 23086 (1959).
16. Carlston, E. F., G. B. Johnson, F. G. Lum, D. G. Huggins, and K. T. Park, *Ind. Eng. Chem.*, **51**, 253 (1959).
17. Szayna, A., *Ind. Eng. Chem., Prod. Res. and Develop.*, **2**, 105 (1963).
18. Gupta, S. K., and R. T. Thampy, *Makromol. Chem.*, **139**, 103 (1970).
19. Anon. *Modern Plastics*, **42** (No. 12), 22 (1969).

THE POLYMERIZABLE SOLVENT

The choice of the polymerizable solvent or monomer for use in an unsaturated polyester formulation depends on a number of technical requirements and on economic conditions. The monomer must be a solvent for the backbone polymer, and must give solutions with acceptable solids/ viscosity relationships. In addition, it should have a low vapor pressure so as to minimize losses due to evaporation when the film is applied. Finally, it should copolymerize readily with the unsaturated centers of the backbone polymer to give the desired final film properties. The most widely used monomers that satisfy these requirements are styrene, vinyl toluene, methyl methacrylate, and some allyl ethers. To illustrate the chemical reactions taking place during the cross-linking stage, the case of styrene will be considered in more detail, but the general principles apply to other monomers. Special considerations related to the allyl ethers are further discussed on page 149.

The tendency of the growing polystyryl chain to react with the backbone polymer will be governed, in part, by the type of unsaturation present. The reactivity ratios of maleic ester and styrene ($r_M = 0.005$, $r_S = 6.52$) are very different from those for fumaric ester and styrene ($r_F = 0.07, r_S = 0.3$); it can be seen that the fumarate favors copolymer formation. The degree of maleate to fumarate isomerization that takes place during the backbone polymer formation, therefore, will be of prime importance, and will markedly affect film properties. For example, the fumarate gives a much harder coating with less flexural strength than does the maleate.[1] Furthermore, as the polymerization or curing reaction proceeds, a viscous partly cross-linked structure forms that limits movements of the backbone polymer. As a result, some of the maleate or fumarate is not available for copolymerization.[2] Therefore, in using the copolymer equation to calculate the film composition, it is necessary to use modified values for the reactivity ratios, and these are available for some systems.[2] The number of styrene molecules in each cross-link has been estimated by infrared spectroscopy; values of one to three styrene units are usual.[3]

The polymerization of styrene and other vinyl or acrylic monomers is exothermic, so that considerable heat is generated in the cross-linking reaction; this is generally described in terms of the peak exotherm temperatures. The liberated heat can be an advantage or a disadvantage; it may facilitate the generation of free radicals from the initiator system or, in cases where large volumes are involved (for example, in certain moulded articles), heat dissipation may become difficult. This may be controlled by using a monomer that liberates less heat on polymerization (for example, α-methyl styrene[4]) or by the use of mercaptans as chain-transfer agents.[4] However, other properties may be affected by these changes. Alternatively, adjustment of the rate of radical generation by control of the redox components will govern the curing rate and the exotherm.

REFERENCES

1. Parker, E. E., and E. W. Moffett, *Ind. Eng. Chem.*, **46**, 1615 (1954).
2. Tokarev, A. V., and S. S. Spasskii, *Zh. Fiz. Khim.*, **33**, 554 (1959); via *Chem. Abstr.*, **53**, 20900 (1959).
3. Hayes, B. T., W. J. Read, and L. H. Vaughan, *Chem. Ind. (London)*, **1957**, 1162.
4. Bossu, B., *Ind. Plastiques Mod.* (Paris), **10** (1), 46 (1958); via *Chem. Abstr.*, **52**, 5874 (1958).

THE INITIATING SYSTEMS

The temperature to which an unsaturated polyester can be heated to initiate the curing reaction is governed by the loss of polymerizable monomer and, to a smaller extent, by the sensitivity of some substrates to heat. Therefore, ambient or slightly elevated temperatures are most commonly used. While it is possible to initiate polymerization by the slow decomposition of peroxides at temperatures in the vicinity of 60°C, by far the most common method is to use a redox system that will operate at ambient temperatures.

The components of the redox system are kept separate until just prior to use and, hence, the term two-pack or two-component system is applied to unsaturated polyesters and similar compositions. Electron beam and radiation curing of unsaturated polyesters are discussed in Chapter 13.

Redox initiator systems can be divided into three types according to the mechanism by which they operate.[1] The first group involves reaction between an oxidizing and reducing agent and the formation of one free radical, for example,

$$ROOH + Me^n \longrightarrow RO^. + Me^{n+1} + OH^- \qquad (5.1)$$
$$ROOH + Me^{n+1} \longrightarrow ROO^. + Me^n + H^+ \qquad (5.2)$$
$$Me^{n+1} + RH \longrightarrow Me^n + R^. + H^+ \qquad (5.3)$$

where Me = metal. These systems involve the ions of the transition metals, and reaction is accompanied by a change in their valency state. The second type of redox system relies on the formation of two radicals from bimolecular reaction between the oxidizing and reducing agent.

$$ROOH + AH \longrightarrow RO^. + H_2O + A^.$$

In the third group are reactions which do not lead directly to free radicals, but which form intermediate complexes that dissociate or decompose to free radicals. The systems most commonly used generally belong to the first and third classes,[1] and the components used in polyesters are a cobaltous salt and a hydroperoxide, or dimethyl aniline and a peroxide. It should be noted that some of these reactions are analogous to those occurring during autoxidation, where hydroperoxides are decomposed by metal salts (page 52), and also that redox reactions are useful in initiating low temperature emulsion polymerizations (page 297). As is the case with autoxidation, the effectiveness of metals in decomposing hydroperoxides is approximately in the order of their activity in oxidation-reduction systems (Co > Mn > Ni > Fe), but this effectiveness depends to some extent on the hydroperoxide.[1,2]

The cobaltous salts commonly in use are of the longer chain organic acids (octoates, naphthenates), since these are soluble in the hydrocarbon medium. A widely used hydroperoxide is "methyl ethyl ketone peroxide." This product is a mixture of the isomers shown below, and the actual composition will depend on the source of supply.[2]

Reactions of the types shown in Eqs. 5.1 to 5.3. occur between the cobaltous ion and the hydroperoxide, and it can be seen that the hydroperoxide acts as an oxidant in reaction 5.1, while in 5.2 it behaves as a reductant. It can be seen from Eqs. 5.1 to 5.3 that the concentration of cobalt salt is important and will control the degree of each reaction. Only trace amounts are required and 1 mole of cobaltous ion has been claimed to decompose 1000 moles of hydroperoxide.[1] The cobalt is alternatively oxidized to cobaltic

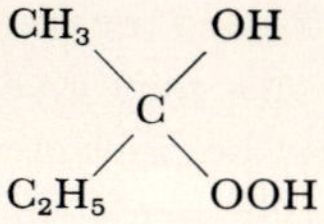

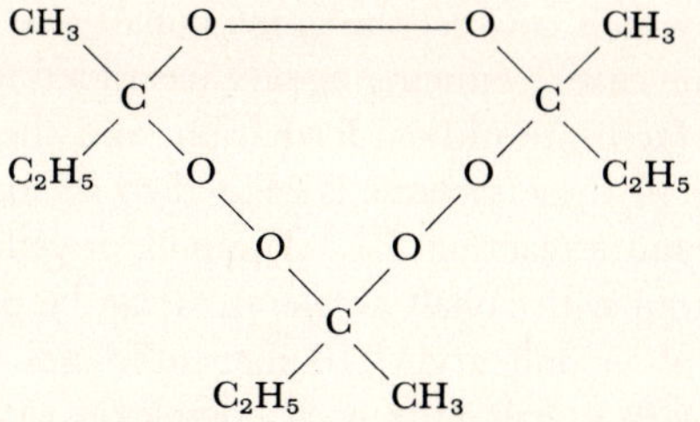

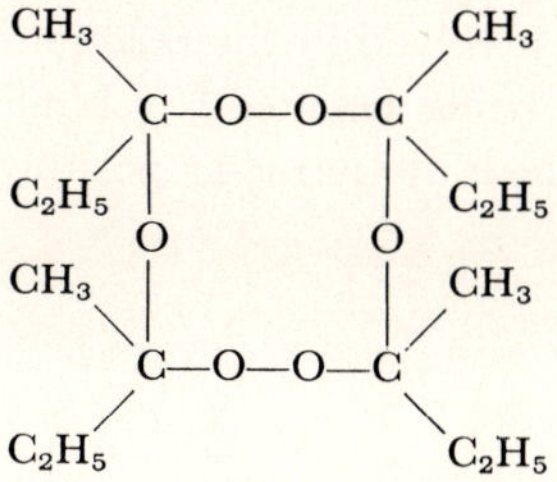

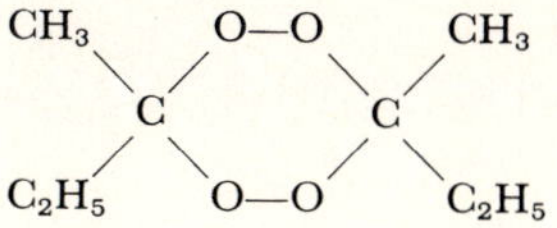

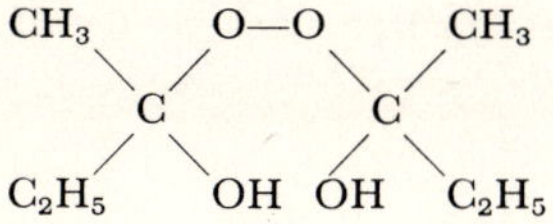

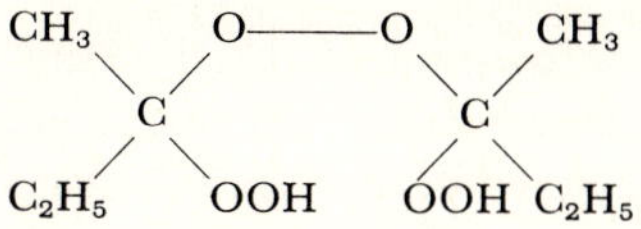

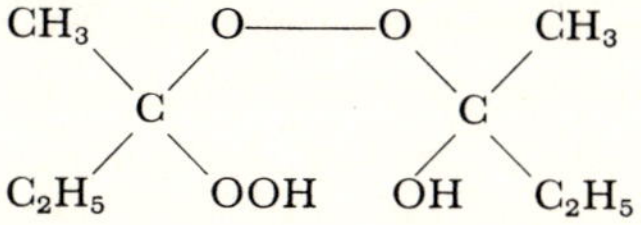

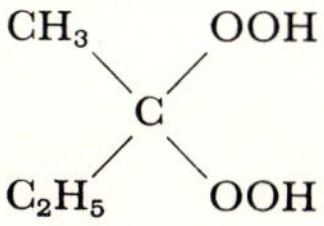

$$\begin{array}{ccc} CH_3 & & OH \\ & \diagdown \!\!\! C \!\!\! \diagup & \\ C_2H_5 & & OOH \end{array}$$

$$\begin{array}{c} CH_3 \quad O \!\!-\!\!\!-\!\!\! O \quad CH_3 \\ \diagdown C \diagup \qquad \diagdown C \diagup \\ C_2H_5 \quad O \qquad O \quad C_2H_5 \\ O \qquad O \\ \diagdown C \diagup \\ C_2H_5 \quad CH_3 \end{array}$$

$$\begin{array}{cccc} CH_3 & & & CH_3 \\ \diagdown C\!-\!O\!-\!O\!-\!C \diagup \\ C_2H_5 \;\big| & & \big|\; C_2H_5 \\ O & & O \\ CH_3 \;\big| & & \big|\; CH_3 \\ \diagup C\!-\!O\!-\!O\!-\!C \diagdown \\ C_2H_5 & & & C_2H_5 \end{array}$$

$$\begin{array}{ccc} CH_3 & O\!-\!O & CH_3 \\ \diagdown C \diagup \quad\quad \diagdown C \diagup \\ C_2H_5 & O\!-\!O & C_2H_5 \end{array}$$

$$\begin{array}{ccc} CH_3 & O\!-\!O & CH_3 \\ \diagdown C \diagup \quad\quad \diagdown C \diagup \\ C_2H_5 & OH \;\; OH & C_2H_5 \end{array}$$

$$\begin{array}{ccc} CH_3 & O\!-\!\!\!-\!\!\! O & CH_3 \\ \diagdown C \diagup \quad\quad \diagdown C \diagup \\ C_2H_5 & OOH \quad OOH & C_2H_5 \end{array}$$

$$\begin{array}{ccc} CH_3 & O\!-\!\!\!-\!\!\! O & CH_3 \\ \diagdown C \diagup \quad\quad \diagdown C \diagup \\ C_2H_5 & OOH \quad OH & C_2H_5 \end{array}$$

$$\begin{array}{ccc} CH_3 & & OOH \\ & \diagdown \!\!\! C \!\!\! \diagup & \\ C_2H_5 & & OOH \end{array}$$

and reduced to cobaltous, but toward the completion of the polymerization, the concentration of cobaltic salts may give rise to an undesirable green coloration in the film. This may be prevented by the addition of reducing agents such as hydroxy carboxylic acids (tartaric, citric, malic),[3] or thioamide promoters,[4] which convert the green cobaltic to the pink cobaltous salts. In fact, in some cases, reducing agents are added to the cobalt/hydroperoxide system to facilitate radical formation, and the accelerating effect of dimethyl aniline in these systems is related to its ability to reduce the cobaltic ion and promote reaction 5.1.[5] The pink or yellow color sometimes formed by resins cured with cobalt accelerators can be prevented by the use as accelerators of alkyl and aryl 1,10-phenantholines or $2,2^1$-bipyridines either as mixtures with cobalt salts or as complexes with the cobalt salts[16] (see also page 56).

The exact mechanism involved in the reaction of dimethyl aniline with benzoyl peroxide is still being debated,[6–11] but it appears that dimethyl anilino and benzoyl radicals are formed,[5] possibly as follows:

$$CH_3\text{--}N(CH_3)\text{--}C_6H_5 + C_6H_5\text{--}C(=O)\text{--}O\text{--}O\text{--}C(=O)\text{--}C_6H_5 \longrightarrow [\text{Complex}]$$

$$\longrightarrow CH_3\text{--}N(CH_3)\text{--}C_6H_4{}^{\cdot} + C_6H_5\text{--}COOH + C_6H_5\text{--}C(=O)\text{--}O^{\cdot}$$

(or an isomer)

Substituents in the para position of the dimethyl aniline influence the reaction—electron-withdrawing groups (halogen, aldehyde, carboxyl, nitro, nitrile) decrease the rate, while electron-donating substituents (methyl, phenyl, hydroxyl, methoxyl, amine) accelerate the rate of peroxide decomposition.[5] These observations have been used to increase the pot-life of polyester compositions by the use of less reactive amines.[12]

Dimethyl aniline reacts with organic acids similar to those present in the backbone polymer to form free radicals,[13,14] Margaritova and Evstratov[14] have proposed the following mechanism for this reaction.

$$\text{--N:} + RCOOH \rightleftharpoons \text{--}\overset{+}{N}\text{--}CORO\bar{H} \rightleftharpoons \text{--}\overset{+}{N} + \bar{O}H + RCO^-$$

Then

$$-\overset{|}{\underset{|}{N}}{}^{+} + \bar{O}H + RCOOH \rightleftharpoons -\overset{|}{\underset{|}{N}}{}^{+}\bar{O}COR + H_2O$$

$$\downarrow$$

$$-\overset{|}{\underset{|}{N}}: + RCOO\cdot$$

The rate of this reaction increases with the dissociation constant of the acid,[14] and could be a possible cause of storage instability in unsaturated polyesters.

Photochemical initiation of the curing reaction is only mentioned in a limited number of patents, but would appear to have considerable potential in the polyester field. In principle, this method involves irradiation of the applied coating with light of a given wavelength. This is absorbed by a photochemically active compound (for example, certain benzophenones or benzoin ethers) with a consequent generation of free radicals either from their own decomposition, or from the induced breakdown of other compounds. For example, it has been claimed that a diacetyl peroxide-benzophenone containing polyester is stable for at least 6 days, but gels in 25 minutes under the action of ultraviolet light.[15] (See also Chapter 13.)

REFERENCES

1. Dolgoplosk, B. A., and E. I. Tiniakova, *J. Polymer Sci.*, **30**, 315 (1958).
2. *The Role of Peroxides in Curing Polyester Resins and Their Influence on the Physical Properties of Reinforced Plastics*, by Novadel Ltd., London, 1960, p. 30.
3. Beck Koller & Co. (England) Ltd., Brit. Pat. 966,661 (1964).
4. Dominion Rubber Co., Can. Pat. 574,602 (1963).
5. Stanko, N. G., *Lakokrasochnye Materialy i ikh Primenenie*, **3**, 14 (1961).
6. Bond, W. B., *J. Polymer Sci.*, **22**, 181 (1956).
7. O'Driscoll, K. F., and E. N. Richezza, *J. Polymer Sci.*, **46**, 211 (1960).
8. Fayadh, J. M., D. W. Jessop, and G. A. Swan, *Proc. Chem. Soc.*, **1964**, 236.
9. Horner, L., *J. Polymer Sci.*, **18**, 438 (1955).
10. Imoto, M., and K. Takemoto, *J. Polymer Sci.*, **19**, 579 (1956).
11. Walling, C., and N. Indictor, *J. Am. Chem. Soc.*, **80**, 5814 (1958).
12. Sargent, E. H. G., Drayton Research, Ltd., Brit. Pat. 941,660 (1963).
13. Uehara, R., *Bull. Chem. Soc., Japan*, **31**, 685 (1958).
14. Margaritova, M. F., and S. D. Evstratov, *Vysokomol. Soed.*, **3**, 390 (1961).
    Margaritova, M. F., and K. A. Rusakova, *Polymer Science U.S.S.R.*, **11**, 3116 (1969).
15. Herbig-Haarhaus, French Pat. 1,300,582
16. Dun, A., and D. B. Fox, Australian Pat. 291, 032 (1969).

STABILIZATION OF POLYESTER SOLUTIONS

Once the components of the redox system are present in the unsaturated polyester solution, the composition has a limited pot-life or stability. From a practical point of view, increase of the pot-life without loss of the curing rate would greatly assist in the application and possible uses of polyesters.

This may be brought about by either a modification of the method of using the polymer system, or by the use of chemical additives or stabilizers.

The methods involving application techniques generally aim at separating the components of the redox system until the polymer system has been applied to the surface. For example, a twin-headed spray gun may be used with the peroxide (or hydroperoxide) as one feed and the polyester/promoter solution as the other. Alternatively, a base or ground coat of either a polyester, or a thermoplastic polymer, containing the initiator may be followed by a top coat of polyester and promoter. The redox components come into contact by diffusion and, to facilitate this, the coating is built up of a number of thin alternate coats of each component. This type of process offers a further advantage where the substrate is sensitive to organic peroxides, since the first coat can be applied peroxide free.[1]

Conventional free radical inhibitors (phenols, quinones) have been used to extend the pot-life, but these generally adversely effect the curing reaction particularly at room temperature. However, a number of compounds have been claimed to act as stabilizers at room temperature and to have little effect on the cure at elevated temperatures.[2] Volatile inhibitors have been sought, and for these to be effective, they must be lost rapidly from the applied film, so that polymerization will commence before significant loss of styrene occurs. Ideally, a volatile inhibitor extends the pot-life with no effect on the curing reaction. Ketoximes[3] are commonly used as volatile inhibitors, and the concentration of acetoxime required to give a pot-life of 8 hours, and yet not seriously effect the curing time, is shown in Table 5.4.[4]

TABLE 5.4

RELATIONSHIP BETWEEN OXIME CONCENTRATION AND GEL TIME FOR VARIOUS CATALYST SYSTEMS AT 23°C[4]

Peroxide Type	Peroxide Concentration (%)	Cobalt Concentration (%)	Acetoxime Concentration for Gel Time of 8 Hours (%)
Methyl ethyl ketone peroxide	2.5	0.06	1.0
Cyclohexanone peroxide	4	0.06	0.7
Cumene hydro-peroxide	5.6	0.06	0.3

All concentrations calculated on 100 parts of resin. Peroxide concentrations were chosen to give the same concentration of peroxide oxygen as determined iodiometrically.

It has been claimed that the oximes of aldehydes and ketones function by forming coordination complexes, thereby inactivating the cobalt ion, and preventing oxidation-reduction type reactions.[4] The complex formation is accompanied by a color change from violet to brown.[5]

$$Co^{3+} + 6 \begin{array}{c} R \\ | \\ C{=}NOH \\ | \\ R' \end{array} \rightleftharpoons \left(Co \left[\begin{array}{c} R \\ | \\ C{=}NOH \\ | \\ R' \end{array} \right]_6 \right)^{3+}$$

Ketoximes are, therefore, not effective stabilizers for peroxide/dimethylaniline catalyst polyesters.

Certain volatile solvents (particularly methanol and ethyl acetate) have a stabilizing action in the bulk on polyester compositions containing hydroperoxide, but have virtually no retarding action on the curing of a film,[6] since most of the solvent is lost by volatilization. The stabilizing action of these solvents is possibly related to their ability to form complexes with hydroperoxides,[7] or to act as reducing agents. Similarly, this postulate could explain the improved properties resulting from the removal of the terminal hydroxyl groups of the backbone polymer (see page 50), since these also would be capable of complexing with and stabilizing the hydroperoxide.

REFERENCES

1. H.E.L.I.C., French Pat. 1,314,547 (1963).
2. Parker, E. E., Pittsburgh Plate Glass Co., U.S. Pat. 2,570,269 (1951).
3. Mott, A., C. Richardson, and R. J. Wicker, Howards of Ilford Ltd., Brit. Pat. 848,826 (1960).
4. Jenkins, V. F., A. Mott, and R. J. Wicker, *J. Oil Colour Chemists' Assoc.*, **44**, 42 (1961).
5. Giesen, M., *F.A.T.I.P.E.C.*, *Congr. 7*, **1964**, 340; via *Chem. Abstr.*, **61**, 12199 (1964).
6. Zvonar, V., and S. Pokorny, *Chem. Prumysl*, **9**, 664 (1959).
7. Khan, N. A., *Pakistan J. Sci.*, **12**, 95 (1960).

AIR INHIBITION

The cure of an unsaturated polyester which contains styrene is inhibited by the oxygen present in the air. As a result, a coating is formed with a soft or tacky surface and a hard cross-linked bottom layer. The oxygen influences the curing reaction by adding to the growing polystyryl radicals to form a relatively stable peroxide, which prevents further polymerization. Oxygen will add to a polystyryl radical between 1,000,000 and 20,000,000 times as fast as a styrene monomer and, hence, marked inhibition occurs.[1]

Air inhibition may be minimized by either the mechanical exclusion of oxygen, or by the incorporation into the polyester of oxygen-sensitive

$$R \left[CH_2-CH \right]_n CH_2-CH + O_2 \longrightarrow R \left[CH_2-CH \right]_n CH_2-CH-O-O^{\cdot}$$

chemical groups. These compete for the oxygen and thereby reduce, or eliminate, interference to the styrene polymerization.

Materials used to form mechanical barriers must be soluble in the polyester resin solution, but must be readily precipitated or exuded out of the film once the cross-linking reactions begin. Various compounds in amounts from 0.001 to 1% have been used for this purpose and some of these are listed in Table 5.5.

TABLE 5.5

MATERIAL USED TO OVERCOME AIR INHIBITION OF STYRENE CONTAINING UNSATURATED POLYESTERS

Material Added	Reference
Wax	Pittsburgh Plate Glass Co., Can. Patent 596,375
Paraffin wax	Dominion Rubber Co., Can. Patent 533,413
Long chain fatty acid esters, e.g. oleyl linoleate	Sichelwerke, Ger. Patent 1,121,330
Stearyl stearate	Beck, Koller & Co., Brit. Patent 850,762
Diels-Alder anthracene adduct	Farbenfabriken Bayer, Akt.-Ges. Ger. Patent 1,106,901
Wax and an organic onium bentonite	Reichhold Chemicals, U.S. Patent 3,014,001

The use of wax, or waxlike materials, has the additional advantage of reducing the loss of styrene by evaporation.[2] In the absence of wax, up to 60% of the total styrene[2] may be lost, but with the wax only 10% is lost. Control of the styrene loss, while of considerable economic importance, vitally affects the final coating composition and film properties.

Two of the major problems associated with wax-type polyesters are the need to remove the wax after curing by buffing or polishing the film and, in some cases, the development of poor adhesion to the substrate. This occurs because the wax tends to migrate to the coating-substrate, as well as to the coating-air, interface.

An alternative mechanical method of separating the polyester surface from the air is to apply, to the wet undried polyester film, a topcoat of a system that is not oxygen sensitive. Generally, polyurethanes with free isocyanate groups have been used for this purpose,[3] since these can react with the hydroxyls of the polyester, and thus, chemically unite the two coats. An extension of this principle has been to apply a thin coat of a reactive isocyanate polymer to give substrate adhesion, then the polyester coating and, finally, a mist coat of polyurethane or other resin (epoxy fatty acid esters) to exclude oxygen.[4] A single polymer composition incorporating these principles has been suggested[5] in which the unsaturated polyester backbone polymer is replaced by an unsaturated polyurethane with terminal isocyanate groups.

Chemical methods of overcoming air-inhibition utilize groups that react readily with oxygen and are preferably contained in low molecular weight, or highly functional, molecules. Various allyl ethers are commonly used as a partial replacement for styrene; the methylene group of the allyl ether is activated by both the carbon-carbon double bond and the ether oxygen, and can effectively compete with the polystyryl radicals for oxygen molecules. The allyl ethers also exhibit a strong tendency to copolymerize with styrene or maleate/fumarate unsaturation. The majority of allyl ethers of interest are compatible with the polyester solution, and the incorporation of these compounds by blending presents few difficulties. Typical compounds in use are the allyl ethers of polyols[6-9] or methylol melamines.[10-11]

Alternatively, the allyloxy grouping may be chemically incorporated into the polyester backbone polymer by partial replacement of the glycol component with mono- or diallyl ethers of a triol. Typical compounds used are trimethylolethane monoallyl ether (**1**) and trimethylolpropane diallyl ether (**2**).

$$CH_3-\underset{\underset{CH_2-O-CH_2-CH=CH_2}{|}}{\overset{\overset{CH_2OH}{|}}{C}}-CH_2OH \qquad C_2H_5-\underset{\underset{CH_2-O-CH_2-CH=CH_2}{|}}{\overset{\overset{CH_2OH}{|}}{C}}-CH_2OCH_2-CH=CH_2$$

(1)   (2)

Allyl ethers of partially polymerized polyols, such as tetra glycerol, have also been used as a blended or reacted component.[12] The reaction of allyl glycidyl ether with free carboxyl groups of the backbone polymer has been used to chemically incorporate the allyloxy group.[13]

The allyl ether groups react with oxygen in a similar manner to other active methylene-containing compounds—by the loss of the allylic hydrogen and the formation of the resonance-stabilized allyl ether radical. A hydro-

$$\boxed{Polyester}-COOH + CH_2-CH-CH_2-O-CH_2-CH=CH_2$$
$$\searrow O \swarrow$$
$$\downarrow$$
$$\boxed{Polyester}-COO-CH_2-CH-CH_2-O-CH_2-CH=CH_2$$
$$\underset{OH}{|}$$

peroxide is formed by the addition of oxygen and the abstraction of a hydrogen atom.

$$CH_2=CH-CH_2-O-R \xrightarrow{O_2} CH_2=CH-\overset{\cdot}{CH}-O-R + \cdot OOH$$

$$\downarrow O_2$$

$$CH_2=CH-\underset{OOH}{\overset{|}{CH}}-O-R$$
$$+ \qquad \xleftarrow{\qquad + CH_2=CH-CH_2-O-R \qquad} \qquad CH_2=CH-\underset{OO\cdot}{\overset{|}{CH}}-O-R$$
$$CH_2=CH-\overset{\cdot}{CH}-O-R$$

Decomposition of the hydroperoxide, particularly when catalyzed by cobalt salts, can produce radicals of the type shown below:

$$CH_2=CH-\overset{\cdot}{CH}-O-R \qquad CH_2=CH-\underset{O\cdot}{\overset{|}{CH}}-O-R \qquad CH_2=CH-\underset{OO\cdot}{\overset{|}{CH}}-O-R$$

These may copolymerize with either styrene or the double bonds of the backbone polymer. Therefore, the oxygen has been consumed to produce active radicals and not stable polystyryl peroxide-type compounds.

Other compounds readily susceptible to oxygen have been used, and these include wood oil,[14] squalene, or polybutadiene.[15] The incorporation of chlorinated acids[16] or rosin[17] also overcomes air-inhibition.

REFERENCES

1. Mayo, F. R., *Ind. Eng. Chem.*, **52**, 614 (1960).
2. Jenkins, V. F., A. Mott, and R. J. Wicker, *J. Oil Colour Chemists' Assoc.*, **44**, 42 (1961).
3. Berger, Jenson & Nicholson, Ltd., Brit. Pat. 883,211 (1961).
4. Abbott, A. C. (Jr.), and Mary G. Brodie, Sherwin-Williams Co., U.S. Pat. 2,993,807 (1961).
5. Solomon, D. H. To be published.
6. Gumlich, W., P. Kranzlein, and G. Bohm, Huls, Chemische Werke Akt.-Ges Ger. Pat. 1,019,421 (1957).
7. Maker, W. J., Glidden Co., U.S. Pat. 2,852,478 (1958).
8. Imperial Chemical Industries Ltd., Can. Pat. 574,069 (1963).
9. Nischk, G., and H. Meckback, Farbenfabriken Bayer Akt.-Ges. Ger. Pat. 1,020,428 (1957).
10. Ishida, E., Union Varnish Co., Japan Pat. 10,093 (1956); via *Chem. Abstr.*, **52**, 21256 (1958).

11. CIBA Ltd., Ger. Pat. 1,122,255 (1961).
12. Behar, R., A. Cahn, M. Dubren, and R. Aussedat., Societe Francaise Duco, Brit. Pat. 947,701 (1964).
13. Delius H., and W. Becker, Reichhold Chemicals Inc., U.S. Pat. 3,006,876 (1961).
14. Scott, K. A., Distillers Co. Ltd., Brit. Pat. 847,532 (1960).
15. Holfort, H., Meknert and Veeck, Kom-Ges., Ger. Pat. 1,133,124 (1962).
16. Lysy, J., French Pat. 1,281,895; *Continental Paint Resin*, **1962**, No. 65, 5.
17. Raichle, K., and C. Nichaus, Farbenfabriken Bayer Akt.-Ges., Ger. Pat. 1,129,688 (1962).

HIGH SOLIDS SYSTEMS

The availability of acrolein and its derivatives in commercial quantities at attractive prices, coupled with the successful use of allyl ethers as a partial replacement for styrene in unsaturated polyesters, has stimulated the development of polymer systems based predominantly on acrolein.[1] These polymers generally dry by oxidation and are formally similar to the vegetable oils which, it should be remembered, are possibly the oldest, and the most widely used, high solids coating.

As a logical extension to the use of small amounts of allyl ethers in polyester compositions, systems in which allyl ethers have replaced entirely the styrene, or other polymerizable solvent, have been developed.[2] These rely on metal-catalyzed oxidation and subsequent copolymerization to bring about the drying of the film.

Acrolein derivatives of molecular weight approximately 1000 have been produced during the search for highly reactive molecules with suitable application characteristics. The high functionality gives much faster drying rates than the vegetable oils, and the oxygen active center is generally present in a short side chain. Hence, any volatile oxidation product will be of a much lower molecular weight (C_1–C_3) than those resulting from the vegetable oils (C_8–C_{12}). Consequently, the loss in film weight will be less than with oils.

Technically, the most promising air-drying polymers prepared from acrolein appear to be the vinyl dioxolanes,[3] although their commercial release is doubtful until a successful cost-property-volume balance is established.[4] The preparation of these compounds relies on the ready formation of cyclic acetals with 1:2 or 1:3 glycols, and may be illustrated for the specific case of 1:2:6-hexane-triol (**3**), itself synthesized from acrolein. The hydroxy vinyl dioxolane (**4**) can then be incorporated into a low molecular compound, or polymer, by alcoholysis with the appropriate alkyl ester.

The —CH group of the vinyl dioxolane is activated by two ether oxygens and the carbon double bond and is, therefore, much more susceptible to

$$2 \; CH_2\text{---}CH_2\text{---}CH_2\text{---}CH_2\text{---}CH\text{---}CH_2 \qquad 2CH_2\text{---}CH_2\text{---}CH_2\text{---}CH_2\text{---}CH\text{-----}CH_2$$

(with OH, OH OH on the left structure **(3)**, and the right structure **(4)** bearing an epoxide with CH—CH=CH$_2$)

$$\text{(3)} + \underset{\overset{|}{2\;CH}}{CHO}\;\;\underset{CH_2}{\overset{\|}{}} \quad\longrightarrow\quad \text{(4)} \quad (CH_2)_n \underset{COOR}{\overset{COOR}{\diagup\diagdown}}$$

$$CH_2\text{---}CH\text{---}(CH_2)_3\text{---}CH_2OCO\text{---}(CH_2)_n\text{---}COOCH_2\text{---}(CH_2)_3\text{---}CH\text{---}CH_2$$

(each end bearing the epoxide with CH—CH=CH$_2$)

$$+\; 2ROH$$

oxidative attack than the —CH$_2$ of an allyl ether. Infrared spectroscopy
and chemical studies have shown that during the drying process, the cyclic
acetal and vinyl unsaturation are lost, and that acrylated and hydroxylated
structures appear.[3] This is consistent with attack at the —CH groups as
indicated below:

$$\boxed{Polymer}\text{---}CH\text{---}CH_2 \;\longrightarrow\; \boxed{Polymer}\text{---}CH\text{---}CH_2$$

(left structure: O—CH—CH=CH$_2$ acetal ring; right structure: O—C—OOH with CH=CH$_2$)

$$\downarrow$$

$$\boxed{Polymer}\text{---}CH\text{---}CH_2$$

(with O, O ring and C—O· + ·OH, CH=CH$_2$)

$$\boxed{Polymer}\text{---}CH\text{---}CH_2$$

(with O, O· and CO—CH=CH$_2$)

(that is, an acrylate)

The intermediates shown above undergo polymerization to form a
cross-linked film.

Similar dioxolanes are prepared[5] by reductive coupling of two molecules of acrolein, followed by acetal formation. This gives a highly functional reactive monomer, 2:4:5-trivinyl-1:3-dioxolane.

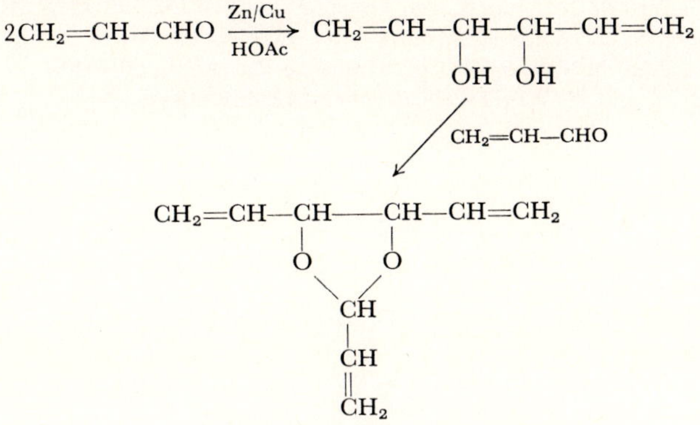

(2:4:5-trivinyl-1:3-dioxolane)

A corresponding compound lacking the dioxolane ring, acrolein diallyl acetal, showed much less reactivity with oxygen.

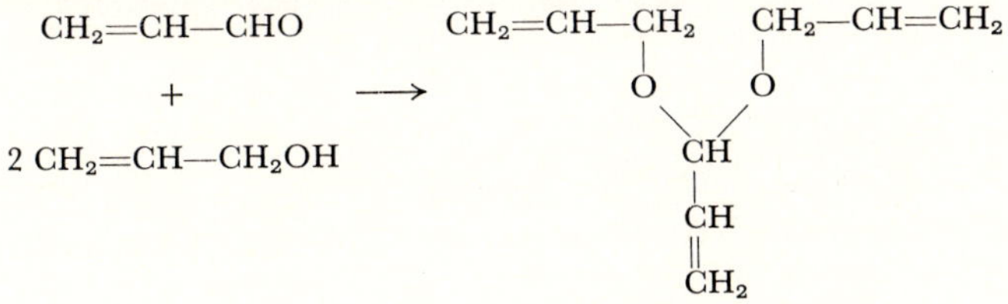

(acrolein diallyl acetal)

Other means of incorporating acrolein residues into polymers for coatings have been based on the ready formation of acetals, or diacetals, with pentaerythritol or similar polyfunctional alcohols. These derivatives may be used as such (for example, pentaerythritol diallyl acetal), polymerized through the carbon unsaturation,[6] or condensed into polyesters through residual hydroxyl groups.

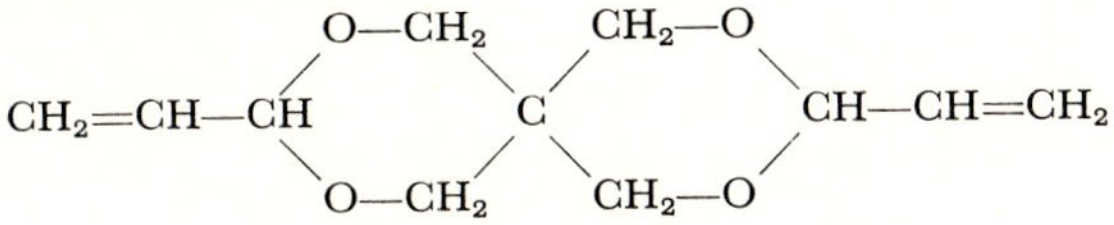

(pentaerythritol diallyl acetal)

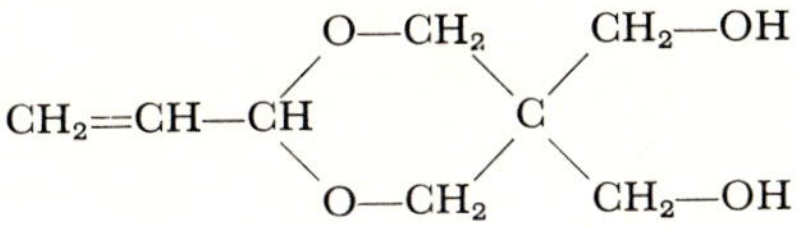

(pentaerythritol monoallyl acetal)

REFERENCES

1. Smith, C. W., ed., *Acrolein*, John Wiley & Sons, New York, 1962.
2. Chatfield, H. W., *Paint Technol.*, **26 (4)**, 17 (1962).
3. E. I. du Pont de Nemours & Co., French Pat. 1,210,192 and 1,257,246.
4. Hochberg, S., *J. Oil Colour Chemists' Assoc.*, **48**, 1043 (1965).
5. E. I. du Pont de Nemours & Co., S. African Pat. Appl., 60/4030.
6. Schulz, H., and H. Wagner, *Angew. Chem.*, **62**, 105 (1950).

CHAPTER 6

THERMOPLASTIC CELLULOSE, ACRYLIC, AND VINYL COATINGS

Coatings based on thermoplastic polymers are used mainly as furniture, general industrial, and automotive lacquers. The overriding virtues of lacquer-type formulations are their simplicity and virtually foolproof application and drying properties (see page 2). The higher formulation costs are offset by relatively trouble-free production and the ease of rectification, or polishing, to remove film imperfections. These coatings are particularly suited to industrial spray painting where, because of the facilities available or the large areas involved, the problems of dirt collection during

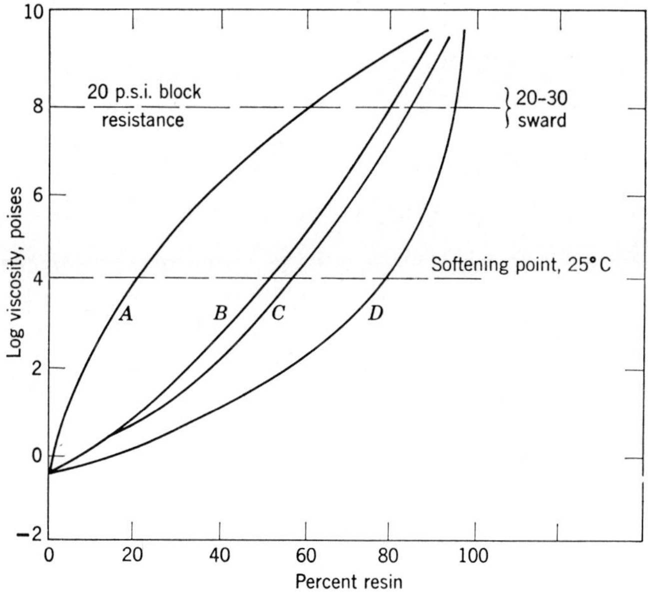

Figure 6.1. Relationship between viscosity and solids content of resin solutions. *A*, $\frac{1}{2}''$ nitrocellulose (T_g 53°C), *B*, phenolic ester gum; *C*, maleic ester gum; and *D*, rosin (T_g 27°C). [1,2]

drying become significant. Also, large volumes of paint present in circulating systems are stable; problems associated with viscosity increase—sometimes experienced with enamels—are few.

The rapid drying of thermoplastic lacquer polymers is related to their glass-transition temperatures, which are usually above room temperature;[1] the tack-free dry stage is reached rapidly, because the viscosity of the polymer solution is high at a comparatively low solids content, as shown in Figure 6.1.[2] The ability of these films to dry rapidly should not be confused with solvent release properties.

REFERENCES

1. Burrell, H., *Offic. Dig. Federation Soc. Paint Technol.*, **34**, 131 (1962).
2. Los Angeles Paint and Varnish Production Club, *Offic. Dig. Federation Soc. Paint Technol.*, **20**, 3 (1948).

CELLULOSE DERIVATIVES

Cellulose is a polymer composed of individual anhydroglucose units linked at the 1 and 4 positions through glucosidic bonds with a β-configuration. The structure of this polymer is shown below where n varies from 50 to 5000.[1]

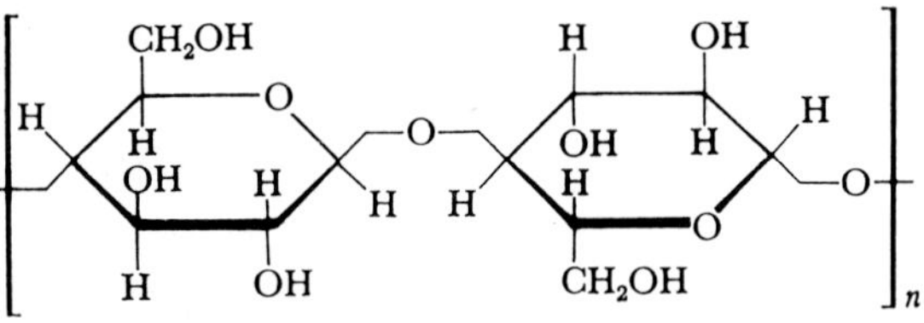

$$\left[\begin{array}{c} CH_2OH \quad\quad H \quad\quad OH \\ H \quad O \quad\quad H \quad O \\ H \quad OH \quad H \quad OH \quad H \\ OH \quad H \quad\quad H \quad H \quad H \\ H \quad OH \quad\quad CH_2OH \quad O \end{array} \right]_n$$

Cellulose is relatively insoluble as a result of the strong hydrogen bonds between molecules. This limits the accessibility of some groups and, in many chemical reactions, the polymer molecules do not behave as expected by analogy with simple molecules.

The chemical derivatives of cellulose used in the coatings industry are the result of reaction of some of the hydroxyl groups to produce esters or ethers. This reduces hydrogen bonding and the hydrophilic nature of the cellulose, and results in the formation of polymers soluble in organic solvents. During these chemical treatments, the backbone cellulose chain is partially degraded; the resultant lower molecular weight polymers have enhanced solubility but have reduced flexibility and mechanical properties. Consequently, it is necessary to reach a compromise between the molecular weight or degree of polymerization (number of anhydroglucose units per chain) and the solution properties of the polymer.

Cellulose Nitrate. Nitration of cellulose is usually carried out with a mixture of nitric and sulfuric acid under carefully controlled conditions. The resultant product is a cellulose nitrate, which is often referred to incorrectly as nitrocellulose. The difference between a nitrate and nitro compound is shown in the formula below.

$$CH_3—O—NO_2 \qquad\qquad CH_3—NO_2$$

methyl nitrate   nitromethane

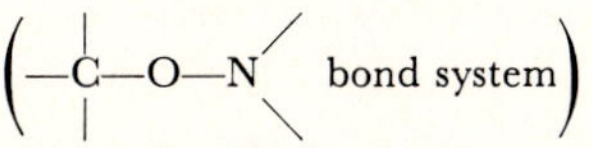

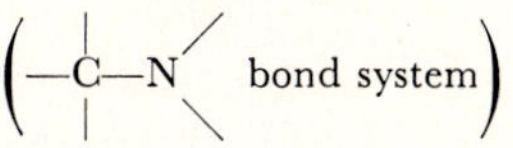

$$\left(—\overset{|}{\underset{|}{C}}—O—N\overset{/}{\underset{\backslash}{}}\ \text{bond system}\right) \qquad \left(—\overset{|}{\underset{|}{C}}—N\overset{/}{\underset{\backslash}{}}\ \text{bond system}\right)$$

The reaction conditions control the amount of chain degradation and the extent of nitration of the hydroxyl groups of the cellulose. Therefore, a range of polymers is available; the properties and uses of typical cellulose nitrates are shown in Table 6.1.[2,3]

TABLE 6.1

PROPERTIES AND USES OF VARIOUS GRADES OF CELLULOSE NITRATE[2]

Nitrogen Content %	Field of Application	Common Solvents	Degree of Polymerization
10.7–11.2	Plastics, lacquers	Ethyl alcohol	175–200
11.2–12.3	Cellulose nitrate lacquers	Ethyl alcohol, methyl alcohol, ethyl, butyl and amyl acetates, acetone, methyl ethyl ketone	500–600
12.0–13.5	Explosives	Acetone	3000–5000

The structure of cellulose nitrate, suitable for use as a surface coating, is shown in Figure 6.2. A nitrogen content of 10.7 to 12.2% corresponds to the

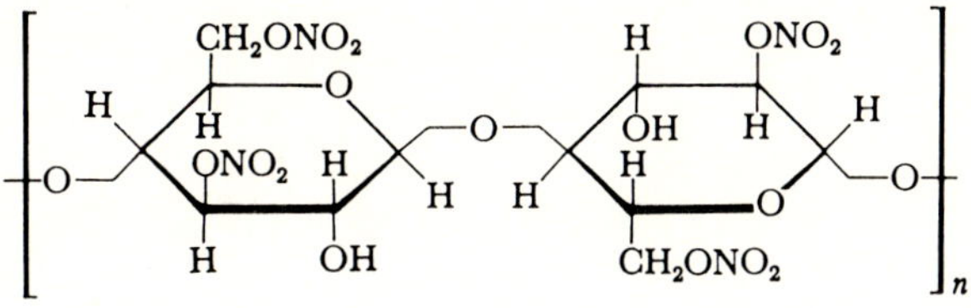

Figure 6.2. Possible structure of cellulose nitrate suitable for surface coatings.

substitution of 2 to 2.25 hydroxyl groups per anhydroglucose residue. The precise location of the hydroxyl groups nitrated is not completely established, since not all groups are equally available; the nitrogen figure, or degree of substitution, should be regarded as a statistical average.[4] The value of n in Figure 6.2 is related to the molecular weight of the cellulose nitrate and to its viscosity in a given solvent. It is usual to specify a cellulose nitrate in terms of the viscosity as determined under standard conditions. The lower molecular weight grades give much higher solids at a given viscosity and are used for wood finishes; they are not recommended where high film strength and durability are required. The higher molecular weight grades are preferred for automotive lacquers.[4]

Cellulose nitrate has a T_g of 53°C,[5] and it is necessary to add an external plasticizer to give films of acceptable flexibility, adhesion, and gloss. Suitable plasticizers are the simple, or polymeric, esters of aliphatic or aromatic acids. The sebacates and adipates give better low temperature flexibility and greater elasticity than the aromatic plasticizers, but the aromatic plasticizers stabilize the cellulose nitrate against ultraviolet light.[6] Mixtures of plasticizers are frequently used to give a lacquer with balanced properties. Resins are added to cellulose nitrate to increase the gloss and adhesion of the film. The solids content at application viscosity is then higher, since the resins have lower molecular weights than the cellulose nitrate. The naturally occurring resins (shellac, ester gum, etc.) are now being superseded by alkyd resins, which are based on either drying or nondrying oils. The drying-oil alkyds impart greater film strength and resistance to embrittlement on heat aging than do the nondrying-oil alkyds; antioxidants must be added to the drying-oil alkyds to prevent cross-linking. If cross-linking does occur, there will be a period during which the lacquer film will be susceptible to lifting during recoating; this period corresponds to incomplete oxidation of the oil and only partial solubility of the resin in the solvents of the final coat.[6]

Cellulose nitrate is compatible with other polymers including poly(vinyl acetate), poly(vinyl acetal), and the poly(methacrylates). It is used in combination with these resins where pale color, ultraviolet resistance, and/or nonyellowing properties are required.

Organic Esters of Cellulose. Cellulose acetate, the most widely used organic ester of cellulose,[7] is manufactured by the acetylation of cellulose by acetic anhydride and suitable catalysts. A solution of high molecular weight cellulose triacetate is formed, and the backbone polymer is then degraded (under controlled conditions) to the required molecular weight. The resulting cellulose triacetate, or primary acetate as it is sometimes called, is partially hydrolyzed to the secondary acetate, which has an acetate content equivalent to the esterification of 2.4 hydroxyl groups per anhydroglucose unit.[8] The molecular weight of the polymer is specified by viscosity measurement under controlled conditions. Cellulose propionate and butyrate are manufactured in a similar manner, but these esters give rather soft films of low strength. Consequently, mixed esters have been prepared to overcome the poor dimensional stability of the acetate and the softness of the higher esters. The most common mixed ester is cellulose-acetate-butyrate with 2.2 acetyl, 0.6 butyryl groups, and 0.2 free hydroxyl groups per anhydroglucose residue, and this is used as the predominant film former or as a blend with other polymers.

Organic Ethers of Cellulose. The hydroxyl groups of cellulose are capable of etherification, which is usually carried out by treating an alkaline suspension of cellulose with the corresponding alkyl halide.

$$R\text{---}ONa + R'Cl \longrightarrow ROR' + NaCl$$
$$(R = \text{cellulose})$$
$$(R' = \text{alkyl, etc.})$$

Ethyl cellulose is the most widely used ether, and this polymer is perhaps the most versatile of the organic-soluble cellulose derivatives.[9] The types most commonly used in surface coatings are prepared by the controlled hydroylsis of the triethyl cellulose; they have between 2.15 and 2.60 ethoxyl groups per anhydroglucose residue. Ethyl cellulose is compatible with a wide range of other polymers, plasticizers, and solvents.[9]

REFERENCES

1. Ott, E., and H. G. Tennent in E. Ott, H. Spurlin, and M. Grafflin, eds., *High Polymers 5, Cellulose,* second ed., Interscience Publishers, Inc., New York, 1963, p. 6.
2. Barsha, J. in E. Ott, H. Spurlin, and M. Grafflin ed. *High Polymers 5, Cellulose,* second ed., Interscience Publishers, Inc., New York, 1963, p. 713.
3. Billmeyer, F. W., *Textbook of Polymer Chemistry,* Interscience Publishers, Inc., New York, 1957, p. 359.
4. Hall, F. C., in J. B. G. Lewin, collator, *Paint Technology Manual, Part 1, Non-Convertible Coatings,* Chapman and Hall, London, 1961, p. 13.
5. Burrell, H., *Offic. Dig. Federation Soc. Paint Technol.,* **34**, 131 (1962).
6. Geilenkirchen, W., *Bull. Schweiz Verein Lack Farben-Chem.,* **1951** (No. 16), 19.
7. Malm, C. J. and G. D. Hiatt, in E. Ott, H. Spurlin, and M. Grafflin, ed., *High Polymers 5, Cellulose,* second ed., Interscience Publishers, Inc., New York, 1963, p. 763.
8. Fisher, J. W., in J. B. G. Lewin, collator, *Paint Technology Manual, Part 1, Non-Convertible Coatings,* Chapman and Hall, London, 1961, p. 168.
9. Bridle, P. F., in J. B. G. Lewin, collator, *Paint Technology Manual, Part 1, Non-Convertible Coatings,* Chapman and Hall, London, 1961, p. 187.

ACRYLIC POLYMERS AND COPOLYMERS

Acrylic lacquers have now almost exclusively replaced cellulose nitrate finishes in the painting of new automobiles in the United States;[1] a similar trend is evident in most other countries. The acrylic lacquers are based on a hard poly(methyl methacrylate) polymer ($T_g = 105°C$) that is suitably plasticized.

The poly(methyl methacrylate) is usually prepared by a free radical solution polymerization, using benzoyl peroxide as the initiator and toluene/acetone as the solvent. Temperatures of 90 to 110°C are used, the reaction being conducted under a slight pressure.[2] Initiator concentrations of 0.2 to 1.0%—by weight of the monomer—control the molecular weight of the polymer within the limits of 55,000 to 105,000 which correspond to relative viscosities of 1.148 to 1.183 when measured on a solution of 0.5g polymer in 100 cc. ethylene dichloride (ASTM D445-46T). The most common polymer has a molecular weight of approximately 90,000. Polymers of this type give lacquers with outstanding gloss retention on prolonged exposure. Further improvements in the durability of the polymer may be obtained by techniques that avoid the introduction of chemical groups capable of absorbing sunlight and initiating photo-degradation.[3] These groups arise from benzoyl peroxide residues, from termination by disproportionation, and from oxidation; they are minimized by using azodiisobutyronitrile as the initiator and Woods metal as an oxygen scavenger.[3]

Polymers with average molecular weights greater than 105,000 tend to cobweb or form long filaments when applied by spray at commercially acceptable solids contents, whereas low molecular weight polymers result in poor film properties and durability. Furthermore, the improvement in gloss retention when the molecular weight is increased above 105,000 is proportionally small, and is more than offset by the reduced solids at application viscosity.

The molecular weight distribution of the poly(methyl methacrylate) is related to the solution and film properties of the coating composition. The low molecular weight fraction decreases the film strength and mechanical properties, whereas the higher fractions cause application difficulties. Consequently, at a given average molecular weight, a narrow molecular distribution is preferred. Conversely, at a given tendency to cobweb, a narrower molecular weight distribution permits the use of a higher average molecular weight and, therefore, gives better gloss retention.[4] On the other hand, the addition of small amounts (0.02 to 0.05%) of extremely high molecular weight polymer (relative viscosity 6.0 to 9.0) to conventional acrylic compositions facilitates melt-in of overspray, presumably by maintaining an open structure in the applied film.[5]

The processing conditions used in the preparation of the polymer influence the molecular weight distribution of the poly(methyl methacrylate). Techniques in which all the monomer, solvent, and initiator are heated together, under pressure, give a narrower distribution than the commonly used alternative process—gradual addition of monomer and initiator to refluxing solvent.[6] Aqueous[6,7] or nonaqueous emulsion polymerization of the monomer also gives a narrower distribution of molecular species. The non-aqueous technique is particularly suited where the polymer is eventually required as a solution in organic solvents; it is only necessary to remove some of the diluent by distillation and then dissolve the concentrated dispersion in suitable solvents.

Poly(methyl methacrylate) requires plasticization to improve cold crack resistance, adhesion to undercoats, solvent release properties, and flexibility. The external plasticizers used include butyl benzyl phthalate[1] and linear polymeric phthalates derived from coconut oil fatty acids.[8] The solvent blend used for the lacquer is a balanced composition chosen to give acceptable viscosity and evaporation characteristics. To avoid excessive solvent retention by the film, it is necessary to use solvents free from a high boiling tail fraction, and to carefully balance the evaporation rate and molecular size of the component solvents. As the film begins to set or become touch dry, the ability of a solvent molecule to diffuse through the film is very dependent on its molecular size, which must be less than the structural holes in the film for good release characteristics. The external plasticizer assists solvent release by maintaining a fluid film for as long as possible. The drying process is accelerated by baking the film (for example, 1 hour at 90°); this treatment is a sound technical approach, since it also allows shrinkage stresses, caused by the drying process, to be relieved.[9] The recently

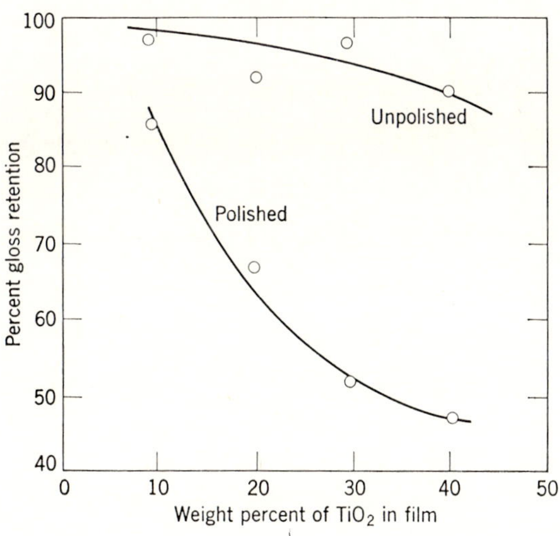

Figure 6.3. Gloss retention of poly(methyl methacrylate) butyl benzyl phthalate plasticized lacquers after 17 months exposure.[1]

introduced bake-sand-bake process is also a technical advance over previous procedures where the baked finish was mechanically polished to remove "orange peel" and to provide the typical mirrorlike showroom finish. Surface imperfections in the film are now removed by a fine sanding operation and the film is then rebaked, often at a higher temperature than the initial bake (up to 140°C), so that it flows to a smooth flaw-free surface. This treatment gives films with lower initial gloss than by mechanical polishing, but the long term durability is better as shown in Figure 6.3.[1]

A relationship has been observed between specular gloss and durability as a function of bake temperature and the T_g of the polymer. The initial specular gloss values at a constant pigment/binder ratio were inversely proportioned to the T_g of the polymer.[26]

Copolymers of methyl methacrylate are used in thermoplastic coatings where internal plasticization or improved adhesion, resistance to degradation, pigment wetting, or corrosion resistant properties are required. The long chain methacrylates, and the acrylates, have lower T_g's than poly(methyl methacrylate), (Table 6.2[9]). The copolymer composition necessary to give

TABLE 6.2

GLASS-TRANSITION TEMPERATURES OF ACRYLIC POLYMERS[9]

Polymer	T_g, °C	Polymer	T_g, °C
Poly(methyl methacrylate)	105	Poly(methyl acrylate)	−9
Poly(ethyl methacrylate)	65	Poly(ethyl acrylate)	−22
Poly(n-butyl methacrylate)	22	Poly(n-butyl acrylate)	−56
Poly(n-hexyl methacrylate)	−5	Poly(2-ethyl hexyl acrylate)	−70

an acceptable T_g for a coating composition can be calculated from the equation (see page 29).

$$\frac{1}{T_g} = \frac{W_1}{T_{g_1}} + \frac{W_2}{T_{g_2}}$$

(copolymer)

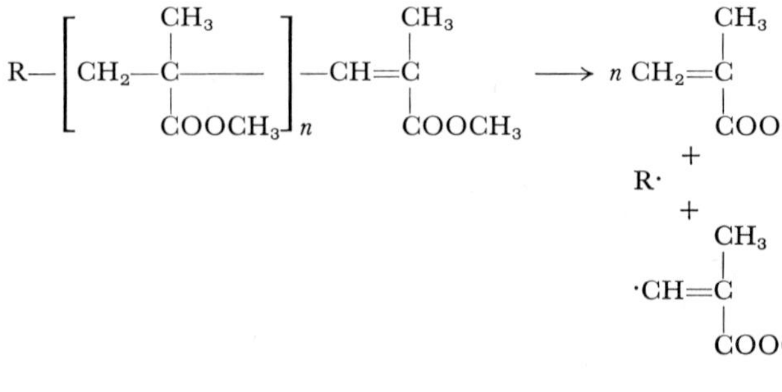

Internally plasticized copolymers of methyl methacrylate are claimed to be less critical in their solvent requirements and to have a greater film hardness at a given flexibility than the externally plasticized homopolymer. Some external plasticizer (butyl benzyl phthalate) is used to assist in solvent release and improve the adhesion to the substrate. Poly(methyl methacrylate) is degraded or "unzipped" by heat or ultraviolet light through a chain mechanism, which is the reverse of the propagation step in polymer formation.

Some other polymers do not "unzip," but are susceptible to an alternative breakdown mechanism such as oxidation at the tertiary carbon atoms.*

$$-CH_2-{}^*\overset{\displaystyle H}{\underset{\displaystyle COOR}{C}}-CH_2-$$

Copolymers of methyl methacrylate with the latter type monomers often exhibit greatly reduced tendencies to degrade; the "unzipping" process is

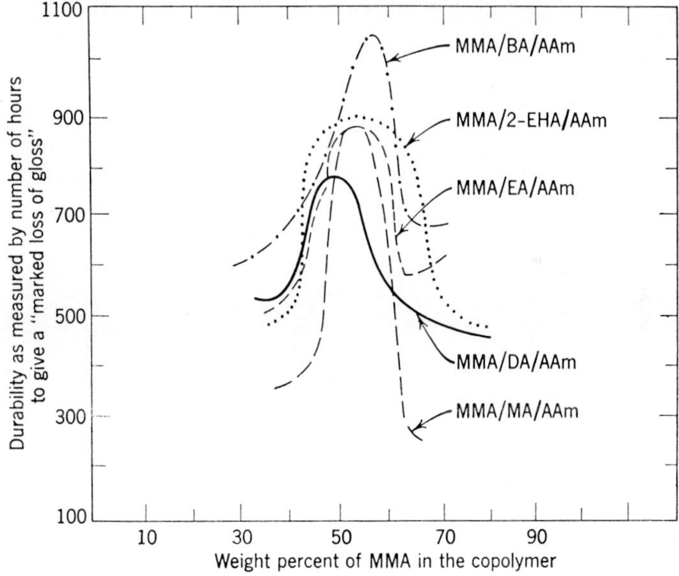

Figure 6.4. Expected relationship between copolymer composition and durability.[11]

MMA = Methyl methacrylate
BA = Butyl acrylate
AAm = Acrylamide
2-EHA = 2-Ethyl hexyl acrylate
DA = Decyl acrylate
MA = Methyl acrylate
EA = Ethyl acrylate

stopped at the comonomer and the oxidative degradation (of the co-monomer) is minimized by the methacrylate. Grassie[10] has shown that 0.24 mole % acrylonitrile, which corresponds to one acrylonitrile unit for each 410 units of methyl methacrylate, substantially reduces the formation of monomer on heat degradation. By analogy with recent theories proposed for thermosetting acrylics,[11] it would be expected that the copolymer composition would be related to exterior durability, and the optimum composition would be characterized by a sharp peak in the durability/composition curves (Figure 6.4). The maxima in these curves are characteristic of the hard monomer (alkyl methacrylates and styrene). At the composition of maximum durability, a copolymer may not give adequate physical properties, and it may be necessary to introduce additional monomers to give a balanced composition satisfying the maximum durability requirements and the film properties. This composition can be calculated from the durability curves of the two individual copolymers. For example, if the durability maximum for copolymer A is at 55% methyl methacrylate and for copolymer B at 70% ethyl methacrylate,[11] then the terpolymer composition of maximum durability lies on the line CD, shown in Figure 6.5.

The comparatively poor pigment and surface-wetting characteristics of poly(methyl methacrylate) can be partially offset by the use of copolymers containing small amounts of polar groups. These groups may be incor-

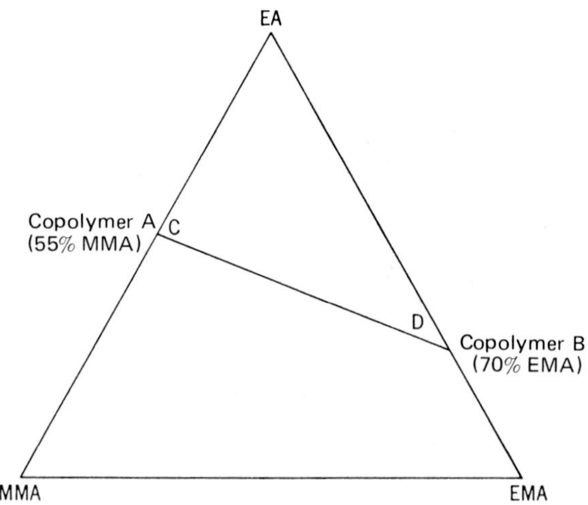

Figure 6.5. Polymer composition of maximum durability from a mixture of three monomers. EA, ethyl acrylate; MMA, methyl methacrylate; and EMA, ethyl methacrylate.[11]

porated by using the appropriate monomer in the initial polymerization stage as shown in Table 6.3.

TABLE 6.3

FUNCTIONAL GROUP MONOMERS

Functional Group	Monomer
Carboxyl	Acrylic or methacrylic acid[12]
Hydroxyl	Hydroxyethyl methacrylate, hydroxypropyl methacrylate or acrylate[13]
Amino	Dimethylaminoethyl methacrylate, tertiary butylaminoethyl methacrylate[14]
Amino-amide	Dimethylaminoethyl methacrylamide[27]
Aziridinyl	2-(1-Aziridinyl) ethyl methacrylate[27]

Alternatively, a functional group may be modified after the copolymerization, as is frequently done with glycidyl methacrylate-containing polymers. The epoxide group can be opened by carboxylic acids, alcohols, amines,[15] ammonia,[15] or phosphoric acids,[2,16,17] as shown in Eqs. 6.1 to 6.4.

$$(6.1) \quad [\text{Polymer}]\!-\!\overset{\triangle}{\underset{O}{}} + \text{RCOOH} \longrightarrow [\text{Polymer}]\!-\!\underset{\text{OH}}{\overset{\text{OCOR}}{|}}$$

$$(6.2) \quad [\text{Polymer}]\!-\!\overset{\triangle}{\underset{O}{}} + \text{ROH} \longrightarrow [\text{Polymer}]\!-\!\underset{\text{OH}}{\overset{\text{OR}}{|}}$$

$$(6.3) \quad [\text{Polymer}]\!-\!\overset{\triangle}{\underset{O}{}} + \text{RNH}_2 \longrightarrow [\text{Polymer}]\!-\!\underset{\text{OH}}{\overset{\text{NHR}}{|}}$$

$$(6.4) \quad [\text{Polymer}]\!-\!\overset{\triangle}{\underset{O}{}} + \text{H}_3\text{PO}_4 \longrightarrow [\text{Polymer}]\!-\!\underset{\text{OH}}{\overset{\text{OPO}_3\text{H}_2}{|}}$$

The reaction of an epoxide polymer with phosphoric acid proceeds via an intermediate, if ketone solvents are present.[17]

$$\text{H}_3\text{PO}_4 + 2\,[\text{Polymer}]\!-\!\overset{\triangle}{\underset{O}{}} + (\text{excess})\ \text{R}\!-\!\underset{\underset{O}{\|}}{\text{C}}\!-\!\text{R}'$$

$$\downarrow$$

$$[\text{Polymer}]\!-\!\underset{O \quad\ O}{\overset{| \quad\ |}{}} + [\text{Polymer}]\!-\!\underset{\text{OH} \quad \text{OPO}_3\text{H}_2}{\overset{| \quad\quad |}{}}$$

$$\underset{\underset{R \quad\ R'}{\diagup \quad \diagdown}}{C}$$

The marked improvement in adhesion conferred by the hydroxy-amino system (Eq. 6.3) is probably related to the formation of a cyclic coordination complex with the substrate.

$$
\begin{array}{c}
\boxed{\text{Polymer}} \\
| \\
CH_2-HC\text{——}CH_2 \\
\quad | \qquad | \\
\quad HO \qquad NH_2 \\
\hline
\text{Substrate}
\end{array}
$$

A polymer composition incorporating an alkyd resin as an internal plasticizer, methyl methacrylate as the durable and hard film former, and hydroxy-amino groups for adhesion has been prepared,[18] as shown below. This resin can be regarded as a particular example of an acrylic modified alkyd (see Chapter 4).

$$
\boxed{\text{Alkyd}}-OH + \underset{\underset{O}{|}}{\overset{HC\!=\!CH}{\underset{OC\quad CO}{}}} \longrightarrow \boxed{\text{Alkyd}}-O-CO-CH\!=\!CH-COOH
$$

heat and free radical initiators

$$
+ CH_2\!=\!\underset{\underset{COOCH_3}{|}}{\overset{\overset{CH_3}{|}}{C}}
\qquad
+ CH_2\!=\!\underset{\underset{COOCH_2-CH-CH_2}{|}}{\overset{\overset{CH_3}{|}}{C}}
$$

$$
-CH_2-\underset{\underset{COOCH_3}{|}}{\overset{\overset{CH_3}{|}}{C}}-CH_2-\underset{\underset{COOCH_2}{|}}{\overset{\overset{CH_3}{|}}{C}}\;\underset{COO-\boxed{\text{Alkyd}}}{\overset{\overset{COOH}{|}}{CH}}-CH-
$$

$$
\underset{H_2C}{\overset{HC}{\diagdown}}O
$$

$$
RNH_2 \downarrow
$$

$$
-CH_2-\underset{\underset{COOCH_3}{|}}{\overset{\overset{CH_3}{|}}{C}}-CH_2-\underset{\underset{COOCH_2}{|}}{\overset{\overset{CH_3}{|}}{C}}\;\underset{COO-\boxed{\text{Alkyd}}}{\overset{\overset{CO\;\;H}{|\;\;|}}{CH}}-CH-
$$

$$
CH-OH
$$
$$
CH-NHR
$$

Acrylic polymers are compatible with cellulose derivatives and the vinyl resins; film-forming compositions of these polymer blends are in use. Cellulose acetate-butyrate is added to acrylic lacquers to overcome solvent crazing while acrylic polymers with low T_g's are sometimes used as the polymeric plasticizers for cellulose nitrate; black acrylic lacquers are of this type[19] with poly(butyl methacrylate) a common plasticizer.

Alkylacrylates, such as methyl ethacrylate

$$CH_2{=}\underset{\underset{\displaystyle COOCH_3}{|}}{\overset{\overset{\displaystyle C_2H_5}{|}}{C}}$$

do not polymerize readily by free radical mechanisms,[20] but it has been shown recently that metallic sodium yields polymers by an anionic mechanism. These polymers could possibly be used as resins for surface coatings.

Acrylic Organosols. The major limitations of solution acrylic polymers as automobile lacquers are the low solids (20 to 30%) at application viscosity and the expensive solvents sometimes necessary to obtain a satisfactory solids/viscosity relationship. Organosols, in which the polymer is present as a dispersion in an organic diluent, overcome these problems; the diluent is usually a hydrocarbon, which is much cheaper than the esters and ketones used for some solution coatings, and the viscosity approximates to that of the continuous (diluent) phase. In practice, it is sometimes necessary to add thickening agents to arrive at a satisfactory viscosity at solids of 50 to 60%. Aqueous emulsions can also be used, but they suffer from limitations such as sensitivity of the substrate to water, high latent heat of vaporization of the diluent phase, lack of control over the evaporation rate of the diluent, and the presence of water-sensitive colloids.

Conventional organosols, particularly of vinyl-type polymers, are made by dispersing polymer particles (prepared by aqueous emulsion or suspension polymerization) in the organic diluent. Acrylic organosols prepared in this manner are unsatisfactory as automotive lacquers because of the difficulties of controlling the particle size and integration of the polymer drops, and because of the low gloss of the resultant films. These problems are related to the presence of the hydrophilic colloid or surfactant used in the original aqueous polymerization, which is now in contact with the diluent phase. These difficulties may be overcome by the preparation of the organosol by an in situ process that involves carrying out a nonaqueous emulsion polymerization using organic-soluble colloids. The kinetics of the dispersion polymerization of methyl methacrylate in n-dodecane have been studied by Barrett and Thomas.[28] They have suggested that the polymerization can be described in terms of a bulk polymerization within the monomer swollen polymer particles. The theoretical expression derived by these workers assumes that all radicals produced in the diluent phase are transferred immediately to polymer particles, monomer swells the polymer particles in partition equilibrium with monomer in the diluent, and the polymerization proceeds within the particle according to the kinetics of bulk polymerization.

Considerable attention has been given to the synthesis of suitable stabilizers for organosols because most of the commonly used stabilizers are designed for use in aqueous emulsion systems. A further difference between aqueous emulsion polymerization and the specific case of methyl methacrylate organosols is that in the organosol formulation the monomeric methyl methacrylate is soluble in the continuous phase. Of the two general methods of stabilizing dispersions, that is by surface ionic charges or by the provision of a steric barrier, the latter was used by Osmond to stabilize acrylic organosols.[21] In the preparation of poly(methyl methacrylate) organosols, a rubber-poly(methyl methacrylate) graft has been used; this may be prepared prior to, or during, the initial stages of the dispersion polymerization by using an initiator that yields free radicals capable of attacking the rubber backbone. Benzoyl peroxide is suitable for this purpose and may abstract a hydrogen atom from, or add across the double bond of, the rubber molecule and allow grafting as shown in Figure 6.6.

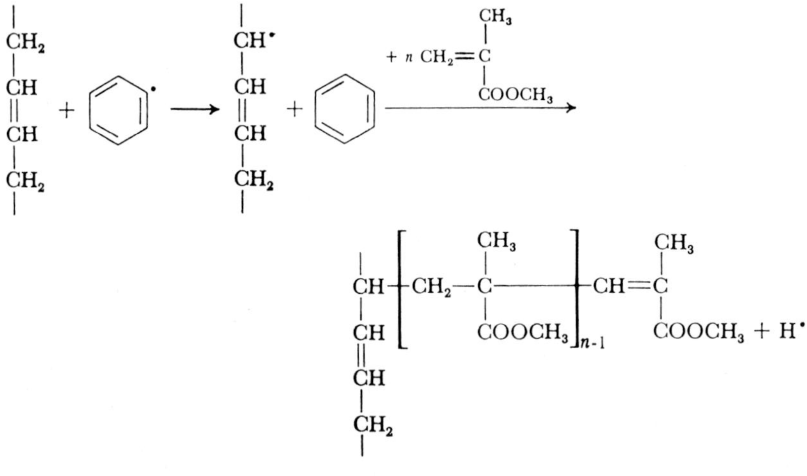

$$
\begin{array}{c}
| \\ CH_2 \\ | \\ CH \\ \| \\ CH \\ | \\ CH_2 \\ |
\end{array}
\; + \; \bigcirc \!\cdot \; \longrightarrow \;
\begin{array}{c}
| \\ CH^{\bullet} \\ | \\ CH \\ \| \\ CH \\ | \\ CH_2 \\ |
\end{array}
\; + \; \bigcirc \quad
\xrightarrow{\; +\, n\, CH_2\!=\!\overset{\displaystyle CH_3}{\underset{\displaystyle COOCH_3}{C}} \;}
$$

$$
\begin{array}{c}
| \\ CH \\ | \\ CH \\ \| \\ CH \\ | \\ CH_2 \\ |
\end{array}
\!\!-\!\!\left[CH_2-\overset{\displaystyle CH_3}{\underset{\displaystyle COOCH_3}{C}} \right]_{n-1}\!\!\!-\!CH=\overset{\displaystyle CH_3}{\underset{\displaystyle COOCH_3}{C}} \; + \; H^{\bullet}
$$

Figure 6.6.   Possible mechanism of graft copolymer formation.

Once sufficient graft copolymer is formed to stabilize the polymer droplets, the polymerization may be continued with a nongrafting initiator[21] (that is, one that forms free radicals which are not sufficiently active to attack the rubber, yet which polymerize the monomer), such as azodiisobutyronitrile.

The understanding of the mechanism of organosol stabilization has progressed considerably over the last few years. From the original concept of a graft-copolymer stabilizer, various copolymers prepared by synthetic procedures which were less haphazard than the grafting process used with rubber have been evaluated. For example the use of copolymers of lauryl

TABLE 6.4
STABILIZERS FOR ORGANOSOLS

Stabilizer	Disperse Polymer Phase	Liquid Medium	Reference
Poly(ethylene-vinyl acetate)	Poly(vinyl acetate)	Cyclohexane	Fr. 1,531, 022 (Union Carbide)
Poly(hydroxystearic acid) graft on poly (glycidyl methacrylate methacrylic acid)	Epoxy resins Maleic anhydride Vinyl acetate/vinyl chloride polymers	Petroleum	Fr. 1,543,838 (I.C.I.)
	Acrylic resins	Petroleum	Br. 1,122,397
Poly(hydroxystearic acid) graft on poly (glycidyl methacrylate -styrene)	Polyester resins Polyurethane resins	Petroleum	Fr. 1,543,838 (I.C.I.)
Poly(vinyl octanoate) graft on poly(methyl methacrylate-allyl methacrylate)	Acrylic resins	Hydrocarbons	Sth. African 68/02930 (du Pont)
Poly(hydroxystearic acid) graft on poly (methacrylic acid- methoxy- [polyethylene oxide] acrylate)	Acrylic resins	Petroleum	Australian Appl. 30079/69 (BALM)
Poly(lauryl methacrylate-methoxy poly [ethylene oxide] acrylate)	Acrylic resins	Petroleum	
Poly($C_8 - C_{10}$ methacrylates) graft on poly(methyl methacrylate- methacrylic acid)	Acrylic resins	Petroleum	Brit. 1,122,397 (I.C.I.)
Poly(methyl methacrylate) graft on poly(vinyl pyrolidone)	Acrylic resins	Alcohols	
Poly(methyl methacrylate) graft on degraded rubber	Acrylic resins	Petroleum	Brit. 941,305 (I.C.I.)
Poly(methacrylic acid) graft on polystyrene	Polystyrene	Alcohols	
Carboxy-terminated poly(ethylhexyl acrylate)-epoxy resin adduct	Unstated	Petroleum	Brit. 1,096,912 (I.C.I.)
Poly(vinyl acetate) graft on poly(ethylhexyl acrylate)	Acrylic resins	Petroleum	Fr. 1,474,058 (du Pont)

methacrylate and glycidyl methacrylate esterified with methacrylic acid is one method by which the number of attachment points for graft copolymerization can be controlled. Thus if this copolymer is polymerized in the presence of methyl methacrylate, a stabilizer consisting essentially of poly (lauryl methacrylate) with a specified number of poly (methyl methacrylate) graft chains is formed. Further developments in stabilizer technology have led to copolymers with polar anchor groups located at either the end of the polymer chain, or at specified intervals along the chain. The general requirement for stabilization is claimed to be a steric barrier 12Å thick; the number of polar anchor groups is related to the strength of the attraction of the group to the particle surface. The strength of the steric barriers has been discussed by Doroszkowski and Lambourne[30] and general reviews of steric stabilization written by Osmond and Wallbridge[31] and by Berryman.[29] Typical stabilizers are listed in Table 6.4.

Poly(methyl methacrylate) organosols require plasticization in the same way as solution coatings. Also, the films are baked at temperatures in excess of the T_g of the plasticized coating compositions to ensure fusion of the individual polymer particles and uniform film formation. At lower temperatures, the films are susceptible to mud-cracking because of poor particle fusion. Consequently, organosols are not used in touch-up lacquers, and the normal solution type acrylics are used for repair, and refinishing, of organosol derived films.

The use of the organosol technique enables coatings to be prepared from polymer systems, such as poly(acrylonitrile), which are insoluble in the common commercial solvents.[22]

Copolymers may be prepared by the in situ organosol technique and can be used where internal plasticization, improved adhesion, or wetting characteristics are required. The same general principles of polymer composition apply to organosols and solution copolymers—with some additional important advantages. By controlling the addition of the respective monomers, it is possible to concentrate functional groups toward the center of the polymer droplet. This technique is useful where the diluent, pigment, or filler may react with the functional group and cause instability to the liquid coating composition. Carboxyl groups are a typical example of such a functional group.[23] An alternative application of this technique is in the preparation of skewed copolymer compositions within the polymer droplet, so that particle fusion is favored. For example, a two-stage polymerization of an acrylonitrile/alkyl acrylate mixture using a charge high in acrylonitrile for stage 1, and a charge low in acrylonitrile for stage 2, gave organosols which, on baking, formed films with properties similar to those obtained from dimethylformamide solution.[24] This can be related to the outer polymer high in acrylate that would fuse readily.

The reactivity of monomers is often different in dispersion and solution polymerizations. The nonaqueous emulsion technique makes it possible to prepare copolymers that are not available by other methods. These copolymers may be used as such or may be converted to solution coatings. Careful selection of the original block or graft copolymer, or the use of monomeric surfactants, should enable copolymer compositions to be controlled to an extent not possible in solution processes. Consequently, new copolymers can be expected from developments in nonaqueous emulsion polymerization. The molecular weight distribution of polymers and copolymers is narrower in dispersion than in solution polymerization, and this gives the property advantages discussed previously on page 154. Thermoplastic acrylic organosols appear likely to eventually supersede the solution acrylic lacquer.

Microgels, which are suspensions of colloidal dimensions (0.05 to 1 micron in diameter) appear to combine the advantages of the solution and organosol coatings.[25] By careful control of the number of cross-links in the microgel particles, the swelling of the microgel in organic solvents has been controlled. The polymer solution has the viscosity characteristics of an emulsion, yet forms a continuous film of a similar type to that obtained from a solution composition. A typical example of a microgel is a polymer of methyl methacrylate and ethylene glycol dimethacrylate (0.1 mole %).[25]

REFERENCES

1. Mercurio, A., *Offic. Dig. Federation Soc. Paint Technol.*, **36 (475)**, 135 (1964).
2. Crissey, L. W., and J. H. Lowell, E. I. du Pont de Nemours & Co., Brit. Pat. 807,895 (1959).
3. Fitzgerald, E. B., and P. F. Stehle, *Ind. Eng. Chem., Prod. Res. Develop.*, **1**, 254 (1962).
4. Elgood, E. J., N. S. Heath, B. J. O'Brien, and D. H. Solomon, *J. Appl. Polymer Sci.*, **8**, 881 (1964).
5. E. I. du Pont de Nemours & Co., Brit. Pat. 848,148 (1960).
6. Solomon, D. H., Unpublished observations.
7. Zimm, B. H., General Electric Co., (U.S.A.), U.S. Pat. 3,001,922 (1955).
8. Christenson, R. M., and K. R. Gosselink, Pittsburgh Plate Glass Co., Can. Pat. 681,279 (1964).
9. Burrell, H., *Offic. Dig. Federation Soc. Paint Technol.*, **34**, 131 (1962).
10. Grassie, N., *The Chemistry of High Polymer Degradation Processes*, Butterworth & Co. Ltd., London, 1956.
11. Graham, N. B., F. R. Crowne, and D. E. MacAlpine, *Chem. Eng. News*, April 12, 1965, 67; *Offic. Dig. Federation Soc. Paint Technol.*, **37**, 1228 (1965).
12. Rohm and Haas Co., Special Products Bulletin, SP-88, 1960.
13. Rohm and Haas Co., Special Products Bulletin, SP-216, 1961.
14. Gusman, S., and S. Melamed, Rohm and Haas Co., U.S. Pat. 2,940,872 (1960).

15. Blake, J., E. I. du Pont de Nemours & Co., U.S. Pat. 2,949,383 (1960); U.S. Pat. 2,949,445 (1960).
    McFadden, R. T., and R. H. Cramm, Dow Chemical Co., U.S. Pat., 3,514,473 (1970).
16. Staicopoulos, D. N., E. I. du Pont de Nemours & Co., Can. Pat. 557,944 (1959); U.S. Pat. 2,868,760 (1959); Ger. Pat. 1,123,066 (1959).
17. Simms, J. A., Paper presented to Am. Chem. Soc., Sept. 1959.
18. Fitch, R., E. I. du Pont de Nemours & Co., Brit. Pat. 857,956 (1961).
19. Sanderson, J. J., E. I. du Pont de Nemours & Co., U.S. Pat. 2,989,492 (1959).
20. Bevington, J. C., and B. W. Malpass, *Trans. Faraday Soc.*, **60**, 1268 (1964).
21. Osmond, D. W. J., and H. H. Thompson, Ger. Pat. 1,093,992, via. *Chem. Abstr.*, **56**, 1611 (1962); Merrett, F. M., *Trans. Faraday Soc.*, **50**, 759 (1954).
22. Yakovlev, A. D., *Lakokrasochnye Materialy i ikh Primenenie*, **1963** (No. 4), 18.
23. Osmond, D. W. J., Imperial Chemical Industries Ltd., Brit. Pat. 958,023 (1964).
24. Victorius, C., *Am. Chem. Soc., Div. Org. Coatings*, St. Louis Meeting, **21**, 237 (1961).
25. Bullitt, O. H., and Hochberg, S., E. I. du Pont de Nemours & Co., Brit. Pat. 967,051 (1964).
26. Brendley, W. H. (Jr.) and A Mercurio, *Am. Paint J.*, **54**, 76 (1970).
27. Nyquist, E. B., and R. H. Yocum, *J. Paint Technol.*, **42**, 308 (1970).
28. Barrett, K. E. J., and H. R. Thomas, *J. Polymer Sci.*, Part A-1, **7**, 2621 (1969).
29. Berryman, D. W., *Proc. and News Aust. Oil and Colour Chemists Assoc.*, **7**, 4 (1970).
30. Doroszkowski, A. and R. Lambourne, *Am. Chem. Soc., Div. Org. Coatings*, Chicago meeting, **30**, 592 (1970).
31. Osmond, D. W. J., and D. J. Walbridge, *International Symposium on Macromolecular Chemistry Toronto—1968*; Interscience, New York, 1970. (*J. Polymer Sci.* **C 30**) Page 381.

VINYL POLYMERS AND COPOLYMERS

The vinyl resins have the general structure:

$$\left[CH_2 - \overset{\displaystyle X}{\underset{\displaystyle Y}{C}} \right]_n$$

In this section, the discussion does not include rubber derivatives, or the acrylic and methacrylic esters (X or Y = —COOR).

Poly(ethylene). The simplest polymer of this general type is poly(ethylene), which is prepared by the polymerization of ethylene with one of the following.

1. Free radical initiators at moderate and very high pressures.
2. Coordination catalysts.
3. Metal oxides.[1]

The structure of the polymer is related to the polymerization conditions; it may range from essentially linear and highly crystalline (high density)

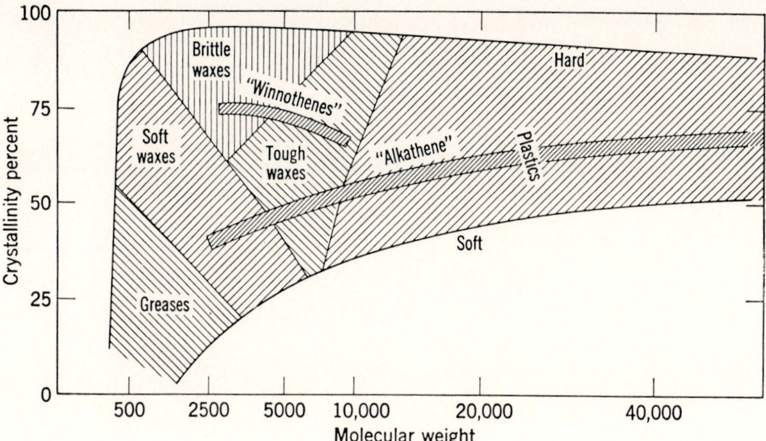

Figure 6.7. Effect of molecular weight and degree of crystallinity on the mechanical properties of poly(ethylene).[2]

to a branched partially crystalline (low density) polymer. The influence of the degree of crystallinity, or the amount of branching, and the molecular weight on the properties of poly(ethylene) is shown in Figure 6.7.[2] Poly(ethylenes) are insoluble in common solvents below 60°C; at higher temperatures, they are soluble in a wide range of solvents. Consequently, little use has been found for poly(ethylene) as a major film former for coatings that are applied at room temperatures. The low density and low molecular weight polymers are used as additives to provide "slip" to coatings, and as temporary protective coatings. The poly(ethylene) is heated above its melting point in the presence of a hydrocarbon and then cooled rapidly to give a stable dispersion, which is used to apply temporary protective coatings to painted articles. These polymers are prepared by using free radical initiators, and the branched structures are of two distinct types.[3] Long chain branching results when a free radical, either from the initiator or a propagating polymer chain, abstracts a hydrogen atom from a dead polymer molecule.

The molecular weight of the branch is often similar to the original

$$R^{\cdot} + R_2-CH_2-CH_2-R_3 \longrightarrow RH + R_2-\overset{\cdot}{C}H-CH_2-R_3 \xrightarrow{n\ CH_2=CH_2}$$

$$R_2-CH-CH_2-R_3$$
$$|$$
$$[CH_2-CH_2\ ]_{\overline{n}}$$

backbone molecule. Short chain branching arises from intramolecular chain transfer to give ethyl and butyl branches as shown.

$$R\!-\!CH_2\!-\!CH_2\!-\!CH_2\!-\!CH_2\!-\!\dot{C}H_2$$

$$\downarrow$$

$$R\!-\!\underset{\cdot}{C}H\!-\!CH_2\!-\!CH_2\!-\!CH_2\!-\!CH_3$$

butyl branch

$$R\!-\!CH_2\!-\!\underset{\cdot}{C}H\!-\!C_4H_9$$

$$\downarrow +CH_2\!=\!CH_2$$

$$R\!-\!CH_2\!-\!CH\!-\!C_4H_9$$
$$|$$
$$\dot{C}H_2\!-\!\dot{C}H_2$$

$$\downarrow$$

$$R\!-\!CH_2\!-\!CH\!-\!\dot{C}H\!-\!CH_2\!-\!CH_2\!-\!CH_3$$
$$|$$
$$CH_2\!-\!CH_3$$

ethyl branch

Apart from the molecular weight and the degree of chain branching, the molecular weight distribution influences the properties of poly(ethylene);[4] the wider the distribution the greater the solubility. Ethylene copolymers with vinyl acetate have increased compatibility with a range of surface coating resins including rosin, rubber derivatives, and chlorinated diphenyls.[5]

Poly(styrene). Styrene can be converted to poly(styrene) by numerous mechanisms and procedures, but for surface-coating polymers, a free radical process gives more controllable conversion rates and molecular weights. Poly(styrene) is a typical thermoplastic lacquer polymer, readily soluble in a wide range of solvents and possessing good chemical resistance. Compared to poly(methyl methacrylate), it yellows badly on prolonged exposure as a result of oxidation that is promoted by ultraviolet light. At 60°C, carboxyl and hydroxyl groups are formed,[6] probably by the following mechanism which is similar to the autoxidation of drying oils.

Ultraviolet light absorbers help to minimize yellowing but, even so, poly(styrene) is not widely used as a coating vehicle. Copolymers in which styrene is the major component are used in thermosetting compositions and, to a much lesser extent, as lacquer coatings. Minor amounts of carboxyl groups overcome the poor adhesion of the homopolymer while maleic anhydride copolymers can be converted to the half-ester, and half-amide, to yield spirit-soluble polymers compatible with tannin.[7] The half-ester of copolymerized maleic anhydride is a cheap means of internal plasticization, since it is equivalent to an acrylate.

$$-CH-CH_2-CH- \quad \xrightarrow[UV]{O_2} \quad \overset{\displaystyle OOH}{-C-CH_2-CH-}$$

$$\overset{\displaystyle O^\bullet}{-C-CH_2-CH-}$$

$$\xrightarrow{RH}$$

$$-\overset{\displaystyle O}{\underset{\displaystyle \parallel}{C}} \quad + \quad {}^\bullet CH_2-CH- \qquad \overset{\displaystyle OH}{-C-}$$

Conventional plasticizers of the phthalate type and blends with rubbers are used as external plasticizers for poly(styrene); random copolymers with acrylates, or graft copolymers with rubber are methods of internally

$$\overset{\displaystyle X}{\underset{\displaystyle Y}{-C-}}HC\underset{OC\diagdown_O\diagup CO}{-CH-} \xrightarrow{ROH} \overset{\displaystyle X}{\underset{\displaystyle Y}{-C-}}\underset{\underset{OR}{CO}}{CH}\underset{COOH}{CH-}$$

(acrylate or substituted
acrylate residue)

plasticizing poly(styrene) compositions. When the polymerization of the styrene is carried out in the presence of rubber, with *tert*-butyl hydroperoxide as the initiator, graft copolymerization takes place, probably as shown overleaf.

Grafted compositions of this type have better mechanical properties than a blend of the two polymers. Styrene-butadiene copolymers are used in emulsion form, and their preparation and properties are considered on page 301.

Fluorine-Containing Polymers. Polymers derived from monomers that contain fluorine are characterized by high thermal stability, chemical inertness, and low solubility in most solvents. Consequently, as solution or dispersion

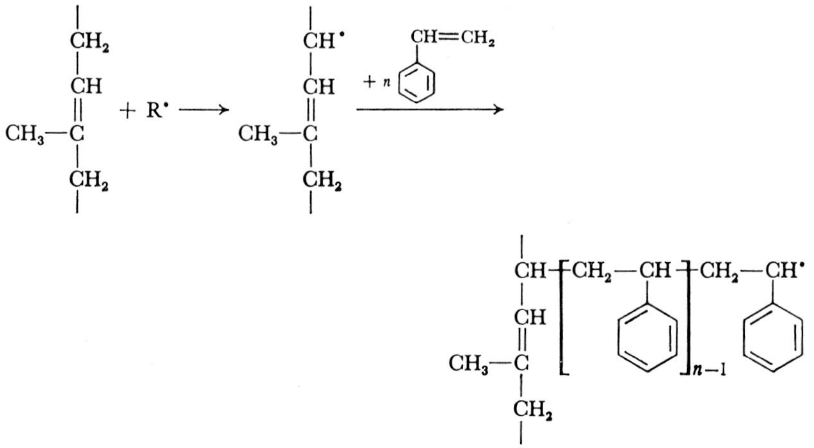

$$
\begin{array}{c}
| \\
CH_2 \\
| \\
CH \\
\| \\
CH_3-C \quad + R^{\bullet} \longrightarrow \\
| \\
CH_2 \\
|
\end{array}
\qquad
\begin{array}{c}
| \\
CH^{\bullet} \\
| \\
CH \\
\| \\
CH_3-C \\
| \\
CH_2 \\
|
\end{array}
$$

coatings, they are of limited use; recent developments in the use of sheet
coatings, and in the factory painting and finishing of building materials,
make these polymers of major interest to the coating industry. Poly(tetra-
fluoroethylene) is prepared by the free radical polymerization of the
monomer at elevated temperatures, in the presence of water.[8] The polymer
is usually applied by techniques similar to those used for molding solid
polymers and ceramics. Aqueous emulsions are available. Poly(vinyl
fluoride) is now commercially available as a plastic film that has high
tensile strength, flexibility, and resistance to outdoor weathering.[9] Useful
film life in excess of 10 years Florida exposure is claimed.[9] These films are
extremely tough and chemically inert, and may possibly be used in the
factory coating of building materials where the polymer sheet is bonded to
the substrate. The polymer is prepared by the free radical polymerization
of the monomer in an aqueous suspension.[10]

Poly(vinyl acetate). Homopolymers of vinyl acetate are prepared by typical
free radical polymerization processes in solution or emulsion. The major
uses of poly(vinyl acetate) are as a paint vehicle in the form of an aqueous
emulsion (see page 294), as an adhesive, and as an intermediate in the
preparation of other polymers. The second-order transition temperature of
30°C, and the tendency toward cold flow even at room temperature,[11]
contribute to the ability of the polymer to act as an adhesive. Vinyl acetate
resins are soluble in ketones, esters, chlorinated hydrocarbons, nitroparaffins,
and the lower boiling aromatic hydrocarbons.[11] The polymers are relatively hydrophilic. They swell and soften on continued immersion in water.
Prolonged exposure to temperatures of 65° to 70°C causes degradation
apparently by loss of acetic acid and the development of a conjugated
unsaturated system that is responsible for discoloration. This mechanism

is similar to that suggested for the decomposition of poly(vinyl chloride) (see page 182).

$$-CH_2-CH-CH_2-CH-CH_2-CH-CH_2-$$
$$\qquad\quad |\qquad\qquad |\qquad\qquad |$$
$$\qquad\quad OAc\qquad\ OAc\qquad\ OAc$$

$$-CH=CH-CH=CH-CH_2-CH- \ + \ HOAc$$
$$\qquad\qquad\qquad\qquad\qquad\qquad |$$
$$\qquad\qquad\qquad\qquad\qquad\quad OAc$$

Copolymers based predominantly on vinyl acetate are used where an internally plasticized polymer system is required. Recent fundamental studies on the vinyl acetate-alkyl maleate and fumarate systems have clearly demonstrated the influence of the comonomer and its distribution along the polymer chain on polymer properties.[12,13] At equal comonomer content, a fumarate ester is a more effective internal plasticizer than the corresponding maleate (Figure 6.8[13]); it has been suggested that rotation about the saturated succinic ester bond is not entirely free, possibly as a result of steric factors and the entrance of the fumarate into the polymer chain prior to isomerization. Under these conditions, the two alkyl side chains of a maleate

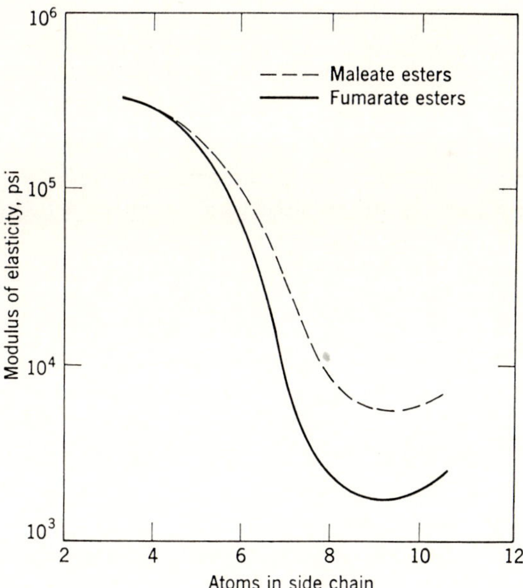

Figure 6.8. Modulus of elasticity of vinyl acetate alkyl maleate and fumarate ester copolymers.[13]

may be located more closely together than the fumarate.[13] The homogeneity of the copolymer has a marked effect on its properties, and the difference between a distributed and nondistributed copolymer may involve a factor of 100. The homogeneous or distributed copolymer is the most effective method of using the comonomer (Figure 6.9).[13] The inclusion of maleic

	Temperature at which Modulus is 135,000 psi	Percent Elongation	Modulus, psi	Tensile Strength, psi
	28.5	2	308,000	5,100
Increasing	24.0	5	64,000	3,090
homogeneity	21.5	14	62,000	2,130
	−4.0	360	57,000	1,590
	−8.0	550	1,630	730

Figure 6.9.   Influence of the distribution of comonomer on the physical properties of a vinyl acetate-di(2-ethylhexyl) fumarate, 80/20 copolymer.[13]

anhydride in vinyl acetate copolymers is a means of improving the adhesion of the copolymer but, in the presence of basic pigments, elimination of acid groups and the formation of unsaturation conjugated with the carboxyl group results in discoloration.[14]

A number of polymer systems are derived from poly(vinyl acetate) by chemical treatments or modifications. Hydrolysis, or alcoholysis, gives poly(vinyl alcohol), and polymers of this type may vary in molecular weight and in the degree of hydrolysis. Poly(vinyl alcohols) are soluble in water or alcohol-water mixtures; they are used for paper and textile coatings where oil and grease resistance are required, and as protective colloids, dispersing and thickening agents in aqueous dispersions. The poly(vinyl acetals) are derived from poly(vinyl alcohol) by reaction, under acidic conditions, with an aldehyde.

$$\begin{array}{c} -CH-CH_2-CH- \\ | \qquad\qquad | \\ OH \qquad\quad OH \\ \\ + R-CHO \end{array} \longrightarrow \begin{array}{c} -CH-CH_2-CH- \\ | \qquad\qquad | \\ O \qquad\qquad O \\ \diagdown \qquad \diagup \\ CH \\ | \\ R \end{array} + H_2O$$

Three common resins are poly(vinyl formal), poly(vinyl acetal), and poly-(vinyl butyral). Poly(vinyl butyral) is used as a safety glass adhesive, in wood sealers and finishes, and in wash primers.[15] In practice, the resins are prepared—without the isolation of the poly(vinyl alcohol), by heating the poly(vinyl acetate) with a water-methanol mixture in the presence of a strong acid, distilling off the methyl acetate, and adding water and the aldehyde.[16] Usually, some acetate and hydroxyl groups remain in the backbone polymer. The analysis of a typical commercial polymer corresponds to[15]

poly(vinyl butyral)   80.7%
poly(vinyl alcohol)   19.0%
poly(vinyl acetate)   0.3%

The residual hydroxyl groups aid solubility in alcohols or glycol ether solvents, and allow thermosetting reactions with urea, melamine, and phenol-formaldehyde polymer systems.

Polymers derived from vinyl ethers are discussed on page 129.

REFERENCES

1. Raff, R. A. V., and J. B. Allison, *High Polymers 11, Polyethylene*, Interscience Publishers Inc., New York, 1956, p. 66.
2. Richards, R. B., *J. Appl. Chem. (London)*, **1**, 370 (1951).
3. Roedel, M. J., *J. Am. Chem. Soc.*, **75**, 6110 (1953); Bryant, W. M. D., and R. C. Voter, *J. Am. Chem. Soc.*, **75**, 6113 (1953); Billmeyer, F. W., *J. Am. Chem. Soc.*, **75**, 6118 (1953).
4. Richards, R. B., *Trans. Faraday Soc.*, **42**, 10 (1946).
5. Union Carbide Co., Bulletin on CoMer VA resins.
6. Grassie, N., *The Chemistry of High Polymer Degradation Processes*, Butterworth & Co. (Publishers) Ltd., London, 1956, p. 236.
7. Gerlich, H., H. Knobloch, and F. Meyer, Badische Anilin and Soda Fabrik Akt.-Ges., Ger. Pat. 1,082,052 (1960).
8. Billmeyer, F. W., *Textbook of Polymer Science*, Interscience Publishers, Inc., New York, 1962.
9. Simril, V. L., and B. A. Curry, *J. Appl. Polymer Sci.*, **4**, 62 (1960).
10. Kalb, G. H., D. D. Coffman, T. A. Ford, and F. L. Johnston, *J. Appl. Polymer Sci.*, **4**, 55 (1960).
11. Union Carbide Co., Bulletin on Vinyl Acetate Resins.
12. Reaville, E. T., and W. F. Fallwell, *Offic. Dig. Federation Soc. Paint Technol.*, **36**, 625 (1964).
13. Cass, R. A., and L. O. Raether, *Offic. Dig. Federation Soc. Paint Technol.*, **36**, 947 (1964).
14. Chiba, T., and T. Yonezawa, *Makromol. Chem.*, **48**, 248 (1961).
15. Union Carbide Co., Bulletin on Vinyl Butyral Resins.
16. Berardinelli, F., Celanese Corp. of America, U.S. Pat. 2,915,504 (1959); Brit. Pat. 823,260 (1959).

VINYL CHLORIDE POLYMERS

Vinyl chloride polymers are commonly referred to as "vinyl resins," although—chemically speaking—the term has a more general meaning. Vinyl chloride is a cheap monomer whose polymers, and copolymers, are widely used as plastics as well as surface coatings. The characteristic properties of these polymers include good color, flexibility, chemical resistance, and freedom from taste; consequently, they are used in coatings designed for beer and food cans, paper, metal foil, and wire coating.

The vinyl resins result from the free radical polymerization of the monomer(s) in either a bulk, solution, suspension, or emulsion process; for surface-coating polymers, the bulk process is rarely used.[1] Solution polymerization is carried out under pressure, to only partial conversion of the monomer mixture, and the polymer is isolated by precipitation, washing, and then drying. Where the reactivity ratios of the monomers do not favor an even distribution of the monomer units in the copolymer, skew feeding can be used to offset the differences in reaction rates of the monomers. Consequently, the solution polymerization gives a more uniform copolymer than the suspension process (Figure 6.10),[1] and the improvement in properties counterbalances the increased processing cost. Suspension polymerization is the most widely used method of making vinyl resins; two major problems associated with this process are the comparatively wide molecular weight and chemical distribution in the polymer, and the presence of a hydrophilic colloid on, or within, the resin droplets. Emulsion polymerization yields smaller polymer particles than the suspension process, and these are used predominantly in the manufacture of organosols and plastisols.

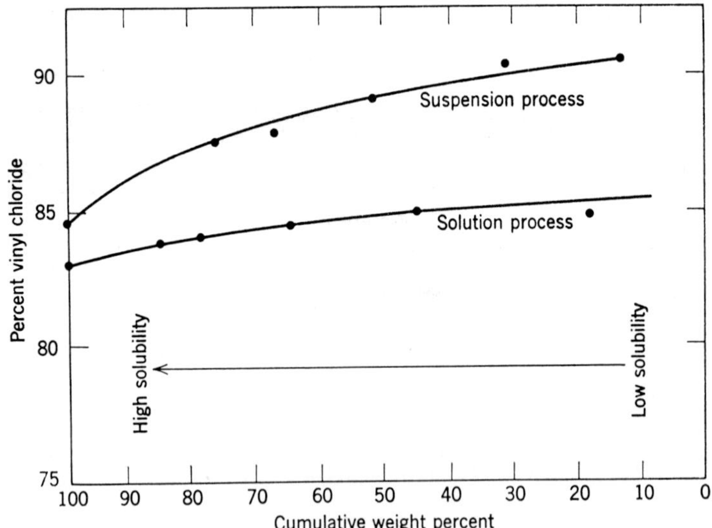

Figure 6.10. Comonomer distribution in vinyl chloride-vinyl acetate polymers.[1]

Vinyl chloride copolymers are soluble in ketones and, in some cases, ketone-hydrocarbon mixtures. For use as solution coatings, a plasticizer, commonly of the phthalate type, is added. Alternatively, the inclusion of an alkyl acrylate in the polymer chain gives an internally plasticized molecule. However, some external plasticizer is desirable, since the vinyls have a stronger tendency to retain solvent than poly(vinyl acetate) or poly(methyl methacrylate).[2] Polar groups are often desirable in a polymer molecule to improve its solubility and adhesive characteristics; hydroxyl groups can be introduced into vinyls by hydrolysis of acetate residues, or some unsaturated acid monomer may be used in the original monomer mixture. As little as 1% methacrylic acid or maleic anhydride promotes adhesion in a vinyl chloride/vinyl acetate (80/19) copolymer.[3] Often, adhesion is improved after baking the film, but stabilizers must be added to prevent polymer degradation under these conditions (see below). High temperatures are also of interest where thermosetting vinyl chloride compositions are in use.

It has been pointed out previously that one of the major disadvantages of thermoplastic solution coatings is the need to restrict the molecular weight of the polymer so as to obtain acceptable solids/viscosity relationships; even so, the solids are low by comparison with enamels. Polymer dispersions in either a diluent (organosol) or plasticizer (plastisol) overcome this problem, and the vinyls are commonly used in one of these forms. Dry polymer that is produced by emulsion polymerization followed by careful drying, is dispersed in the organic liquid; this is generally a mixture of diluent and solvent chosen so as to give a minimum viscosity. At higher diluent contents, the increase in viscosity is related to flocculation of the polymer particles, whereas in the presence of excess solvent, the viscosity increases because of swelling of the polymer particles. Consequently, the viscosity-solvent composition relationship is much more critical than with solution coatings.[1] The type of diluent chosen will also influence the viscosity characteristics of the organosols; an aliphatic hydrocarbon is a stronger precipitant than an aromatic, as shown in Figure 6.11.[1]

The choice of plasticizers is also related to the need to balance the solvent/nonsolvent ratio but, in this case, the two functions may be regarded as being present in the one molecule. For example, didecyl phthalate is roughly equivalent in dispersion potential to a blend of nonane and dimethyl phthalate.[1]

Film formation from organosols is similar to solution coatings, except that it is necessary to coalesce the polymer particles before a uniform film is formed. The presence of plasticizers, particularly those whose attack on the polymer particles increases rapidly with temperature, and the use of elevated temperatures, favors particle fusion as illustrated by Figure 6.12.[1]

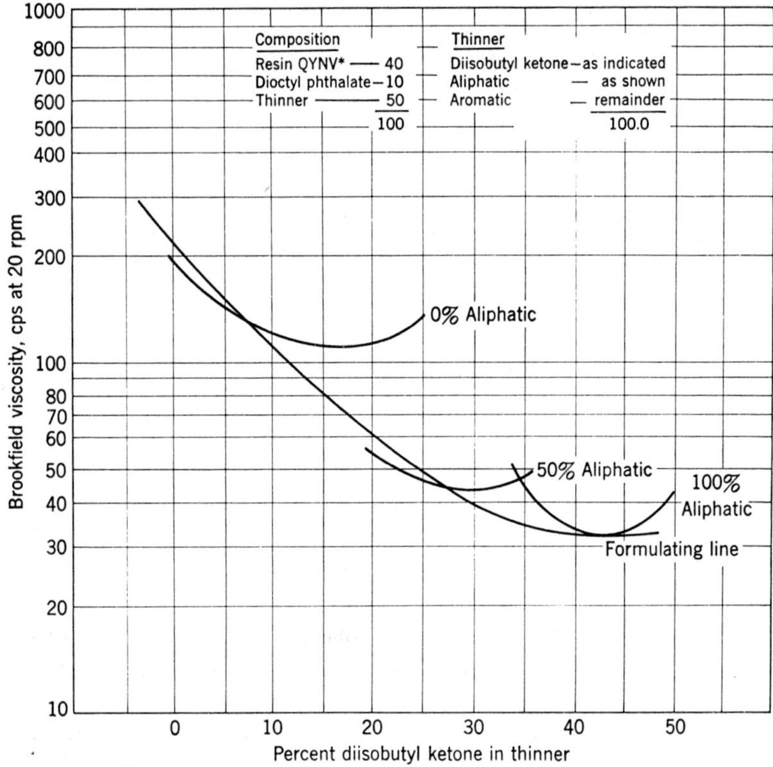

Figure 6.11.   Influence of type of diluent and diluent-solvent ratio on the organosol viscosity.[1]

When vinyl chloride polymers and copolymers are heated, or exposed to ultraviolet light, or to an oxidizing atmosphere, degradation takes place by a "zipperlike" elimination of hydrogen chloride and the formation of a colored polymeric residue. Consequently, it is necessary to add stabilizers

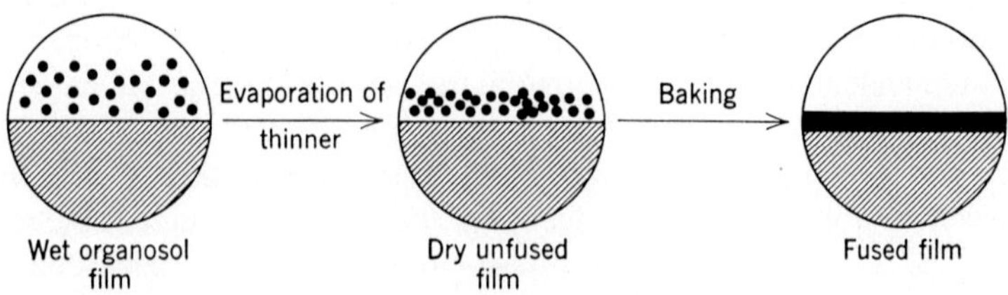

Figure 6.12.   Fusion of organosols.[1]

to enable films to be baked, and then be resistant to normal exposure. The dehydrochlorination is initiated at the active centers α- to unsaturated chain endings or initiator residues, or at chlorine atoms attached to tertiary

$$-CH_2-CH-CH_2-CH- \xrightarrow{\;-HCl\;} -CH=CH-CH=CH-$$
$$\qquad\quad |\qquad\qquad\;\; |$$
$$\qquad\quad Cl\qquad\qquad Cl$$

carbon atoms (these arise from chain branching). Other sites are associated with oxygen-containing groups, incorporated incidentally during the polymerization.[4]

$$-CH_2-CH-CH=CH \qquad\qquad -CH_2-\overset{\overset{\displaystyle CH_2}{|}}{\underset{\underset{\displaystyle Cl}{|}}{C}}-CH_2-$$
$$\qquad\quad |\qquad\qquad |$$
$$\qquad\quad Cl\qquad\qquad Cl$$

active sites

It has been agreed that the dehydrochlorination is autocatalyzed by hydrogen chloride,[5] although recent evidence suggests this is only likely in specific circumstances.[6] Nevertheless, one class of stabilizers is regarded as HCl acceptors, but the evidence shows that the stabilizing action is not necessarily a function of the ability to remove HCl. In fact, some stabilizers have been shown to form chlorides, by reaction with HCl, which actually enhance the decomposition of poly(vinyl chloride).[7] The stabilizing action of organotin compounds, and of group II metal carboxylates, has been related to the interaction between the stabilizer and the more labile chlorine atoms of the polymer.[4,8] When a mixture of two alkali metal laurates are used synergism may result.[13] Barium, cadmium, calcium and zinc laurates, when used alone, allow escape of HCl well before stoichiometric uptake is achieved whereas synergistic mixtures of calcium-zinc and barium-cadmium laurates absorb almost the theoretical quantity of HCl. Cadmium and zinc laurates replace labile chlorine atoms in the backbone polymer by ester groups, reducing formation of long polyene sequences. Equations 6.5 to 6.7 show the reactions involved.

The mechanism proposed in Eq. 6.5 suggests that elimination of carboxylic acids from ester containing polymers (Eq. 6.6) is less likely than the loss of HCl from chlorine containing polymers. Epoxide-containing compounds are used as "HCl-acceptors," but the stabilizing action is not due to reaction with HCl, but rather to the grafting of the epoxide to the double bond, formed by the HCl elimination.[9] Diepoxides can form cross-linked structures with the polymer.

$$\left(\text{R}-\overset{\overset{\textstyle O}{\|}}{\text{C}}-\text{O}\right)_2 \text{M}'' \; + \; \overset{\overset{\textstyle |}{-\text{C}-}}{\underset{\overset{\textstyle |}{-\text{C}-}}{\text{CH}-\text{Cl}}} \tag{6.5}$$

$$(\text{R} = \text{C}_7\text{H}_{15})$$
$$(\text{M}'' = \text{divalent metal})$$

$$\downarrow$$

$$\text{R}-\overset{\overset{\textstyle O}{\|}}{\text{C}}-\text{O}-\overset{\overset{\textstyle -\text{C}-}{|}}{\underset{\overset{\textstyle |}{-\text{C}-}}{\text{CH}}} \; + \; \text{R}-\overset{\overset{\textstyle O}{\|}}{\text{C}}-\text{O}-\text{M}''\text{Cl}$$

$$\text{R}-\overset{\overset{\textstyle O}{\|}}{\text{C}}-\text{O}-\overset{\overset{\textstyle \text{CH}_2}{|}}{\underset{\overset{\textstyle |}{\text{CH}_2}}{\text{CH}}} \; \longrightarrow \; \text{R}-\overset{\overset{\textstyle O}{\|}}{\text{C}}-\text{OH} \; + \; \overset{\overset{\textstyle \text{CH}_2}{|}}{\underset{\overset{\textstyle |}{\text{CH}}}{\text{CH}}} \tag{6.6}$$

$$\overset{\overset{\textstyle -\text{C}-}{|}}{\underset{\overset{\textstyle |}{-\text{C}-}}{\text{CH}-\text{Cl}}} \; + \; (\text{C}_4\text{H}_9)_2\text{Sn} \overset{\text{O}-\overset{\overset{\textstyle O}{\|}}{\text{C}}-\text{R}}{\underset{\text{O}-\underset{\underset{\textstyle O}{\|}}{\text{C}}-\text{R}}{}} \tag{6.7}$$

$$\downarrow$$

$$\text{R}-\overset{\overset{\textstyle O}{\|}}{\text{C}}-\text{O}-\overset{\overset{\textstyle -\text{C}-}{|}}{\underset{\overset{\textstyle |}{-\text{C}-}}{\text{CH}}} \; + \; (\text{C}_4\text{H}_9)_2\text{Sn} \overset{\text{O}-\overset{\overset{\textstyle O}{\|}}{\text{C}}-\text{R}}{\underset{\text{Cl}}{}}$$

$$\left(\text{R} = \text{CH} \overset{\text{C}_4\text{H}_9}{\underset{\text{C}_2\text{H}_5}{}}\right)$$

As an extension to the use of added epoxide stabilizers, copolymers that contain the epoxide groups may be prepared from glycidyl ethers and esters which contain vinyl or allyl unsaturation. These copolymers have superior heat and light stability.[10] Plasticizers that contain epoxide groups act as stabilizer-plasticizers, and monomeric compounds (such as **1**) or polymeric

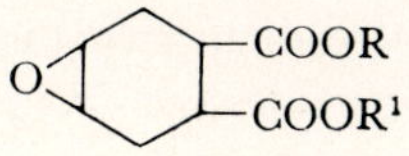

$$\text{(1)}$$

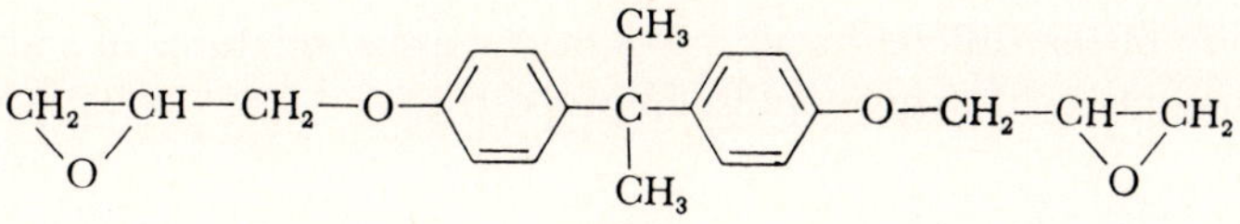

$$\text{(2)}$$

epoxide-esters prepared from **2** are used[11]   (see also page 202 for methods of preparing epoxide-esters).

The theory developed above for the mechanisms of stabilization by the traditional HCl-acceptors shows that the stabilizers function by removing active centers of the polymer chain, whereas the simple acceptance of HCl would leave the chain more prone to further elimination. The second general type of stabilizer-plasticizer also chemically adds to the chain, this time to the conjugated diene structure that results from elimination of HCl. Maleate and fumarate monomeric, and polymeric, esters[12] are the most common stabilizers of this type.

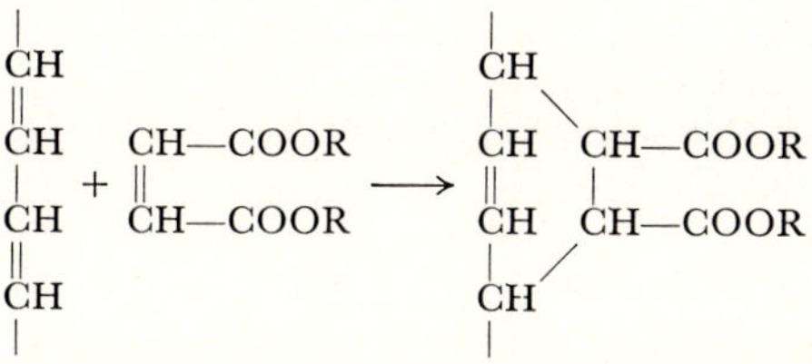

Consequently, the "unzipping" action is stopped by this reaction. It has also been suggested that the liberated HCl adds to the fumarate or maleate double bond.[12]

Organotin complexes, such as dibutyltin bis(monomethyl maleate), thermally dissociate to liberate maleic anhydride, which can then add to the polymeric diene.[8] Other organotin complexes also function as antioxidants,[8] and contribute to the stabilizing of the polymer by reducing oxidative degradation. Conventional antioxidants can also be used. Degradation by ultraviolet light is controlled by the use of ultraviolet light absorbers and/or pigments.

Therefore, stabilizers act by a number of mechanisms that include the following:

1. Replacement of labile chlorine by more stable groups.
2. The addition to the olefin double bonds formed in the polymer.
3. Absorption of HCl.

4. Addition across conjugated diene structures.

5. Destruction of hydroperoxides, or oxygen-sensitive groups, in the polymer.

6. Absorption of ultraviolet radiation.

Because of the different stabilizing mechanisms, mixtures of stabilizers are commonly used to give optimum properties.

REFERENCES

1. McKnight, W. H., *Offic. Dig. Federation Soc. Paint Technol.*, **36 (475)**, 152 (1964).
2. Hansen, C. M., *Offic. Dig. Federation Soc. Paint Technol.*, **37**, 57 (1965). Union Carbide Co., Vinyl Resins for Solution Coatings.
3. Lonza Elektrizitatswerke und Chemische Fabriken Akt-Ges., Swiss Pat. 273,079 (1951); via. *Chem. Abstr.*, **46**, 774 (1952). Union Carbide Co., Vinyl Solution Resin VMCH.
4. Frye, A. H., and R. W. Horst, *J. Polymer Sci.*, **45**, 1 (1960).
5. Fox, V. W., J. G. Hendricks, and H. J. Ratti, *Ind. Eng. Chem.*, **41**, 1774 (1949).
6. Druesdow, D., and C. F. Gibbs, *Modern Plastics*, **30**, 123 (1953).
7. Danyushevskii, A. S., *Plasticheskie Massy*, **1961**, 35.
8. Frye, A. H., R. W. Horst, and M. A. Paliobagis, *J. Polymer Sci.*, *Part A*, **2**, 1801 (1964).
9. Hopf, P. P., and B. D. Sully, *Paint Manuf.*, **30**, 312 (1960).
10. Ellingboe, E. K., E. I. du Pont de Nemours & Co., U.S. Pat. 2,589,237 (1952); Ravve, A., and J. T. Khamis, Continental Can. Co., Inc., U.S. Pat. 3,100,758 (1963).
11. Shokal, E. C., and C. A. May, Shell Development Co., U.S. Pat. 2,728,781 (1955).
12. Roser, C. F., G.м.в. H., Brit. Pat. 811,946 (1959).
13. Briggs, G., and N. F. Wood, *J. Appl. Polymer Sci.*, **15**, 25 (1971).

CHAPTER 7

EPOXY RESINS

Epoxy resins is a term applied to materials in which one or more epoxide groups $\left(\begin{array}{c}-CH-CH_2\\ \diagdown O \diagup\end{array}\right)$ are present in each molecule (Gen. Ref. 1–4). Other functional groups, the most common of which is the hydroxyl group, are often present in the molecule. The chemical reactions involved in the formation of films from the low molecular weight epoxy resins are characteristic of the epoxy group. In this chapter, the methods used to prepare epoxy resins are discussed, and then the reactions of the functional groups—that is, epoxy and hydroxyl groups—are considered.

SYNTHESIS OF EPOXY RESINS

The most widely used epoxy resins are those derived from epichlorhydrin and diphenylolpropane (bisphenol-A) and these are currently marketed by a number of companies. Bisphenol-A is prepared by the condensation of acetone and phenol; the product may be purified by a number of procedures to remove unchanged phenol. Fractional distillation and then extractive crystallization under pressure give bisphenol-A with a melting point of 156.90°C (the pure compound melts at 157.25°C).[1] Another process involves the formation of a complex with *m*- or *p*-cresol, and then purification by vacuum distillation.[2] Reaction of bisphenol-A with epichlorhydrin, in the presence of sodium hydroxide, gives a polymeric epoxide as shown overleaf. The first reaction is the addition of the phenolic hydrogen atom across the epoxide ring of the epichlorhydrin. This is followed by the elimination of hydrogen chloride and the regeneration of the epoxy group. The molecular weight of the polymer is increased by further reaction of these epoxy groups with another bisphenol-A molecule. The simplest epoxy resin of the bisphenol-A type is bisphenol-A diglycidyl ether. This compound results when two moles of epichlorhydrin and one mole of bisphenol-A react, although in practice, it is usual to employ a large excess of epichlorhydrin to minimize the likelihood of higher molecular weight compounds being formed. Since epichlorhydrin is a relatively expensive chemical, it is necessary to recover the unreacted material. As the mole ratio of bisphenol-A to epichlorhydrin increases, the average molecular weight of the epoxy resin rises, and the common commercially available resins have molecular weights corresponding to values of n (in formula **1**) of up to 20.[3] Typical resins and their properties are shown in Table 7.1.[4]

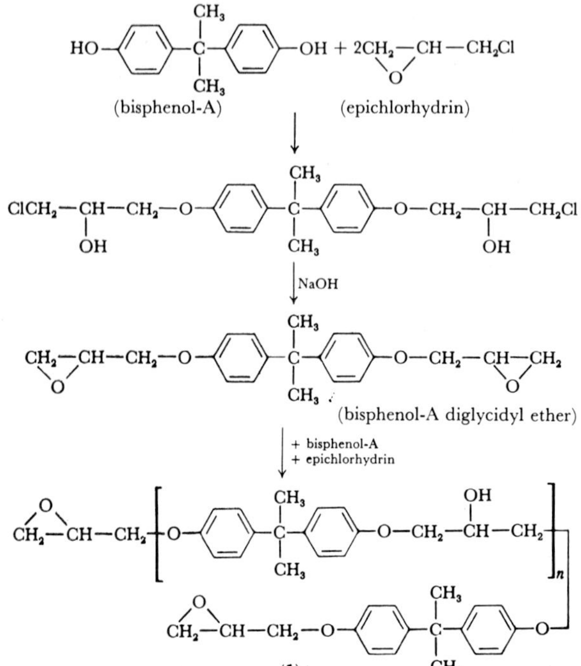

As the value of n increases, more hydroxyl groups are present in the molecule, and the relative percentage of epoxide groups decreases.

Formula **1** depicts the epoxide resins as linear polymers, but considerable chain branching occurs for higher values of n,[3] and structures probably of the type shown on page 190 are developed.

Solid epoxy resins can be made by reacting low molecular weight resins with bisphenol-A. This method, sometimes referred to as the fusion method, has been used as the basis of a novel procedure, described by the Shell Chemical Company,[22] by which it is possible for formulators to prepare their own epoxy resins. The process relies on the use of a special catalyst which selectively promotes the epoxy/phenolic hydroxyl reaction and is totally ineffective in promoting the competing epoxy/epoxy and epoxy/alphatic hydroxyl reactions. The precatalysed proprierty liquid epoxy is marketed as Epon® 829 by the Shell Chemical Company.

The precatalysed epoxy resin/bisphenol-A process for the manufacture of epoxy resins is claimed to yield higher molecular weight esters at short oil length than are possible with the existing commercially available epoxy resins.[22]

The epoxy resins may contain "bound" chlorine which arises from two side reactions. Addition of epichlorhydrin to the chlorhydroxy chain ends

TABLE 7.1

TYPICAL PROPERTIES OF EPOXIDE RESINS [4]

Epikote Resin Grade	Specific Gravity 20°C	Refractive Index at 20°C	Equivalent Weight[a]	Molecular Weight	Number of Repeats (*n* in general formula above)	Viscosity in Poise
Epikote 815	1.139	1.556	83	—	0	7–11
Epikote 828	1.167	1.573	85	380	0	100–150
Epikote 834	1.181	1.583	105	470	0.5	3.8–9.0[c]
Epikote 1001	1.206	1.595[b]	130	900	2.0	0.8–1.7[d]
Epikote 1004	1.156	1.600	175	1400	3.7	4.3–6.3[d]
Epikote 1007	1.147	1.598	190	2900	8.8	17.5–27.0[d]
Epikote 1009	1.190	1.601	200	3750	12.0	36.2–98.5[d]

[a] Grams of resin required to esterify completely 1 gm. mole. of monobasic acid, for example, 280 gm. of C_{18} fatty acid or 60 gm. acetic acid.

[b] At 25°C.

[c] Poise at 25°C of 70% solution in butyl 'Dioxitol'.

[d] 40% solution in butyl 'Dioxitol' at 25°C.

Epikote is a registered trade mark of the Shell Chemical Co.

(present before loss of HCl) results in a structure in which one chlorine atom is not removed during dehydrochlorination.

Chain branching in epoxy resins

(main polymer chain)

$$CH_2\text{---}CH\text{---}\boxed{Polymer}\text{---}CH\text{---}CH_2$$

$$\underset{O}{\diagdown\diagup} \qquad\qquad\qquad \underset{O}{\diagdown\diagup}$$

$$\begin{array}{c} O \\ | \\ CH_2 \\ | \\ CH\text{---}OH \\ | \\ CH_2 \\ | \\ O \quad \text{(branch chain)} \end{array}$$

$$CH_3\text{---}\underset{|}{C}\text{---}CH_3$$

OH

"Bound" chlorine in epoxy resins

$$R\text{---}O\text{---}CH_2\text{---}\underset{\underset{OH}{|}}{CH}\text{---}CH_2Cl + CH_2\text{---}CH\text{---}CH_2Cl$$

$$R\text{---}O\text{---}CH_2\text{---}\underset{\underset{O}{|}}{CH}\text{---}CH_2Cl \longleftarrow \text{Unreactive or "bound" chlorine}$$

$$CH_2\text{---}CH\text{---}CH_2\,(Cl) \longleftarrow \text{Chlorine can be removed as HCl}$$

$$\underset{O(H)}{|}$$

(R = epoxy resin molecule consisting of bisphenol-A units)

Alternatively, the addition of epichlorhydrin to the phenolic group gives a small amount of the compound **2** as a result of the secondary carbon, rather than the primary, of the epichlorhydrin molecule being linked to the phenolic oxygen.

$$R'OH + CH_2\text{---}CH\text{---}CH_2Cl \longrightarrow R'\text{---}O\text{---}CH\overset{\diagup CH_2Cl}{\diagdown CH_2OH}$$

(R' = bisphenol-A chain end)   **(2)**

This chlorine is not removed during dehydrochlorination by sodium hydroxide. There are also small amounts of chain ends with glycol or chlorhydrin structures (Gen. Ref. 3).

$$\begin{array}{ccc} -CH\!\!-\!\!CH_2 & \qquad & -CH\!\!-\!\!CH_2 \\ |\quad\ \ | & & |\quad\ \ | \\ OH\ \ OH & & Cl\ \ \ OH \end{array}$$

Therefore, the generally accepted view—that the higher epoxy resins (n greater than 3) are linear molecules with epoxide groups at either end— needs to be modified. Some branching is present in the molecule, groups other than epoxy are present as some chain ends, and each molecule contains, on an average, less than two epoxy groups (see Table 7.1).

Phenolic derivatives other than bisphenol-A may be used to prepare epoxy resins. The condensation of phenol with formaldehyde gives diphenyl-olmethane or bisphenol-F, whereas the reaction between resorcinol and acetone gives a polymer that can be converted to an epoxide resin, as shown in the following reaction sequence.[5]

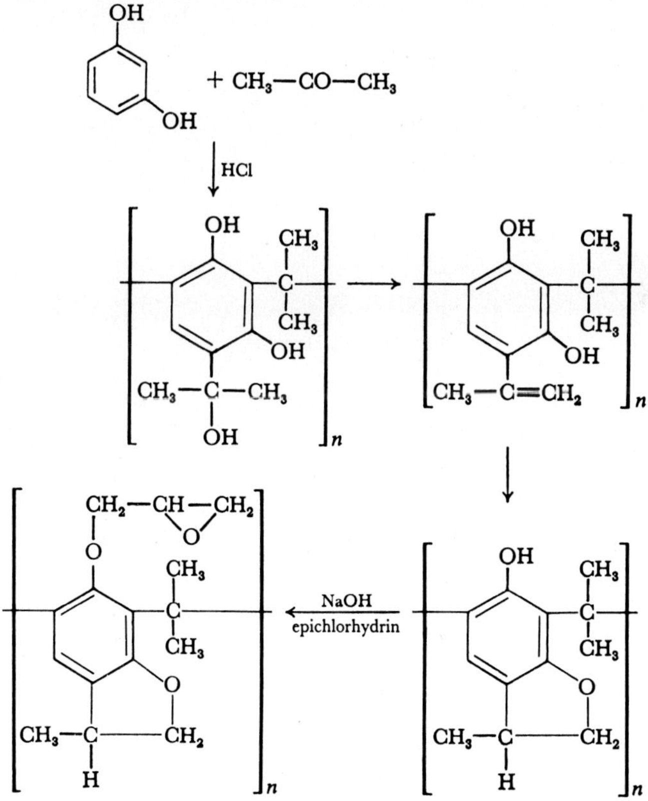

A number of fluorodiglycidyl ethers of the general type shown below have been prepared.

$$CH_2-CH-CH_2-O-R-O-CH_2-CH-CH_2$$
$$\underset{O}{\diagdown\diagup}\qquad\qquad\qquad\underset{O}{\diagdown\diagup}$$

where R = grouping with fluorine substituents.

A typical example of this type is the diglycidyl ether of 2,2-bis (4-hydroxyphenyl hexafluoropropane)

Linear copolymers of the fluorodiglycidyl ethers and diols have been reported.[23]

The phenolic compounds of cashew nut oil, of which the diphenol (**3**) is a predominant component, are also used as intermediates for epoxide resins; Cardolite 7019 (Minnesota Mining & Mfg. Co.) is a commercial resin derived from this source. The long aliphatic chain confers flexibility on these resin systems (Gen. Ref. 1).

(**3**)

Novolac-type phenolic resins are also used in the preparation of epoxide resins (Gen. Ref. 2), and these have structures of the general type:

Aromatic hydrocarbon/formaldehyde condensates, after reaction with phenol, offer a further source of intermediates for conversion to an epoxy-containing polymer.[6] The potential epoxy functionality of a phenol can be increased by first forming an allyl ether, then rearranging this compound to

$$CH_3—(CH_2)_7—CH=CH—(CH_2)_7—COOR$$

$$CH_3—(CH_2)_7—CH\underset{O}{\diagdown\diagup}CH—(CH_2)_7—COOR$$

a ring-substituted allyl phenol and reconverting the liberated phenolic group to an allyl ether. The allyl groups are then epoxidized by peroxides, peracids, etc.[7]

The carbon-carbon unsaturation of vegetable oils, or their derivatives,[8] can also be epoxidized to give compounds of the type shown below:

$$CH_3—(CH_2)_7—CH=CH—(CH_2)_7—COOR$$

$$CH_3—(CH_2)_7—CH\underset{O}{\diagdown\diagup}CH—(CH_2)_7—COOR$$

Similarly, the unsaturation present in polyesters prepared from tetrahydrophthalic acid,[9] or in cyclohexene derivatives, can be converted to an epoxide.

The etherification of epichlorhydrin with simple polyols is an alternative route to epoxy resins, and this can be used to produce low molecular weight polymers free from aromatic unsaturation. Esterification of monobasic fatty acids by epichlorhydrin gives compounds that are monofunctional in terms of epoxide reactivity, and that find use as stabilizer/plasticizers in vinyl resins.

$$R\!-\!COOH + ClCH_2\!-\!CH\!\!-\!\!-\!\!CH_2$$
$$\diagdown O \diagup$$

$$\downarrow$$

$$R\!-\!COO\!-\!CH_2\!-\!CH\!\!-\!\!-\!\!CH_2$$
$$\diagdown O \diagup$$

$$(R = \text{fatty acid})$$

This process also enables tertiary carboxylic acids to be converted to derivatives that are reactive in esterification reactions. For example, the Cardura resins (Shell trademark) are produced from epoxy esters of tertiary C_9–C_{11} acids.

Vinyl or acrylic copolymers of compounds containing both a reactive double bond and an epoxide group are discussed in Chapter 10, and can be represented generally by this formula

$$
\begin{array}{ccc}
X & & X \\
| & & | \\
-CH_2\!-\!C\!-\!CH_2\!-\!C\!-\!\!-\!\!-\!\!-\!\!- \\
| & & | \\
Y & & Z \\
& & | \\
& & CH\!\!-\!\!-\!\!CH_2 \\
& & \diagdown O \diagup
\end{array}
$$

where X, Y, and Z are substituted groups. These polymers behave as highly functional epoxide resins in reactions involving the epoxide group with amines, acids, etc.

FILM FORMATION FROM EPOXY RESINS

The majority of epoxy resins are too low in molecular weight to possess adequate film-forming properties, and must be chemically cross-linked to give acceptable surface coatings. However, recently a high molecular weight bisphenol A-type epoxy—Eponol-55 (marketed by Shell Chemical Co.)—has become available; it is a thermoplastic-type polymer that produces films by solvent evaporation.[3,10] This polymer has a wide distribution of molecular weights (weight average = 300,000; number average 7000) and, in contrast to the more conventional epoxy resins, it has a predominantly linear structure.[3] In addition to its inherent film-forming properties, this epoxy can also be cross-linked via the hydroxyl groups present along the polymer chain. Because of the high molecular weight, the terminal epoxy groups do not form a significant percentage of reactive groups.

Self-Condensation of Epoxy Resins. The epoxide group is susceptible to polymerization, under the action of acidic or basic catalysts, to yield

$$CH_2\overset{O}{\overset{\diagup\diagdown}{-}}CH-R + HX \longrightarrow HO-CH_2-\overset{+}{C}H-R + X^-$$

$$\downarrow \; +\,nCH_2\overset{O}{\overset{\diagup\diagdown}{-}}CH-R$$
$$\downarrow \; +\,H_2O$$

$$HO\!\left[CH_2-CH-O\right]_n\!CH_2-CH-OH$$
$$\underset{R}{\qquad} \underset{R}{\qquad}$$
$$+\ HX$$

Figure 7.1.   Acid-catalyzed polymerization of an epoxy resin.

polyether-type compounds. The general reaction scheme for an acid-catalyzed process is given in Figure 7.1, but the reaction is not necessarily a chain process and may only involve a small number of monomer units. Copolymers of monomeric or polymeric epoxy compounds with vinyl monomers can be prepared—provided the latter are susceptible to cationic polymerization. A simple example is the copolymer formed from styrene and propylene oxide[11] under the catalytic action of boron trifluoride etherate. Amine complexes of boron trifluoride are marketed as latent catalysts for epoxy resins and are claimed to be stable at room temperature, but to react readily at $110°C$[12] when dissociation into boron trifluoride and the amine (both reagents for catalyzing attack on the epoxy group) takes place.

$$R_3N\!:\!BF_3 \underset{\textstyle\rightleftharpoons}{\overset{110°}{}} R_3N + BF_3$$

Tertiary amines can open the epoxide ring and lead to polyether formation, probably by the following mechanism.[13]

$$CH_2\overset{O}{\overset{\diagup\diagdown}{-}}\underset{R}{\overset{|}{C}}H + R_3'N \longrightarrow R_3'N^+-CH_2-\underset{R}{\overset{\overset{\textstyle O^-}{|}}{\overset{|}{C}}}H$$

$$\downarrow \; n CH_2\overset{O}{\overset{\diagup\diagdown}{-}}\underset{R}{\overset{|}{C}}H$$

$$R_3'N^+-CH_2-\underset{R}{\overset{|}{C}}H\!\left[O-CH_2-\underset{R}{\overset{|}{C}}H\right]_n\!O^-$$

$$CH_2\text{—}CH\text{—}[\text{Polymer}]\text{—}CH\text{—}CH_2 \xrightarrow{R_3N} CH_2\text{—}CH\text{—}[\text{Polymer}]\text{—}CH\text{—}CH_2$$

When the epoxy resin contains hydroxyl groups, the tertiary amine can induce a second reaction by catalyzing the formation of an alkoxide ion, as shown overleaf.[14]

The size of the cross-linked polymer that is formed depends on the amine used. DMP-30 (marketed by Rohm and Haas), a very active catalyst, removes approximately seven epoxy groups for each nitrogen in the catalyst.[15] The structure of DMP-30 is shown on page 197.

In other reactions of epoxy resins, tertiary amines are used as catalysts to assist in the opening of the epoxide ring (see page 272). Tertiary amine salts have been used as latent catalysts that give relatively long pot-life at room temperature; at higher temperatures, they react rapidly. DMP-30 tri-2-ethyl hexoate is such an example, and it has been suggested that this salt first dissociates into the amine and acid; the acid is then removed by esterification with some epoxy groups. The tertiary amine initiates the polymerization of other epoxy groups (Gen. Ref. 4).

$$(CH_3)_2N-CH_2 \underset{\underset{\underset{N(CH_3)_2}{|}}{CH_2}}{\overset{\overset{OH}{|}}{\bigcirc}} CH_2-N(CH_3)_2$$

(tri dimethyl aminomethyl phenol)
(DMP-30)

Reaction of Epoxy Resins with Amines. Epoxy resins react readily with primary and secondary amines at room temperature; consequently, it is necessary to separate the epoxy and amine components until immediately prior to, or after, film application. This is done by marketing the compositions as two-pack systems, or by the use of chemical masking agents.

Secondary amines open the epoxide ring and are converted to tertiary amines, which may catalyze further reaction of the epoxy group as discussed above.

$$R'_2NH + R-CH\overset{}{\underset{\diagdown\underset{O}{}\diagup}{}}CH_2 \longrightarrow R'_2N-CH_2-\underset{\underset{OH}{|}}{CH}-R$$

Similarly, a primary amine can react to give a secondary and then a tertiary amine.

$$R'NH_2 + R-CH\overset{}{\underset{\diagdown\underset{O}{}\diagup}{}}CH_2 \longrightarrow R'-NH-CH_2-\underset{\underset{OH}{|}}{CH}-R$$

In commercial formulations, it is usual to use a diprimary amine and, in the epoxide reaction, this is tetrafunctional. Since most epoxy resins are difunctional (or nearly so) in terms of epoxy groups, a cross-linked structure forms with, for example, ethylene diamine.

The simple diamines are relatively volatile compounds that have unpleasant odors and are rather difficult to handle. These properties detract from their use in coating compositions. Higher molecular weight compounds that contain amino groups are, therefore, often used—the two most common types being amine adducts with epoxy resins, and polyamides. Amine adducts are prepared by adding the epoxy resin to an excess of the diamine, so as to favor the formation of an amine-terminated polymer chain.

When this adduct is blended with an epoxy resin, the normal amine-epoxide reaction occurs to give a cured film. The amine adduct is claimed to reduce blushing, odor, and the need for preaging of the blended composition.[4]

$$CH_2\!\!-\!\!CH\!-\!\boxed{Polymer}\!-\!CH\!-\!CH_2 \xrightarrow[\text{(in excess)}]{NH_2-R-NH_2}$$

$$NH_2-R-NH-CH_2-\underset{\underset{OH}{|}}{CH}-\boxed{Polymer}-\underset{\underset{OH}{|}}{CH}-CH_2-NH-R-NH_2$$

Polyamides react with epoxy resins through the terminal amino groups. The commonly used polymers are prepared from dimer fatty acids and a slight excess of an aliphatic diamine; the presence of a highly polar amino group and a long fatty acid portion in the molecule gives the polyamide surface active properties.[16] This results in films with good corrosion resistance and adhesion, when the polyamide is cross-linked by reaction with epoxy resins.[16]

Epoxy-amine compositions that are stable in solution (one-pack systems) are available and, as a general rule, the stability results from the complexing of the amine with another compound. The complex can be decomposed, when cure is required, to yield the free amine. The condensation of an amine with a ketone gives a ketimine and, since this reaction is reversible in the presence of moisture, it is possible to release the amine curing agent on exposure to the atmosphere.

$$\underset{\overset{|}{R'}}{\overset{\overset{R}{|}}{C}}=N-R''-N=\underset{\overset{|}{R'}}{\overset{\overset{R}{|}}{C}} \xrightarrow{\ H_2O\ } 2\ \underset{\overset{|}{R'}}{\overset{\overset{R}{|}}{C}}=O + H_2N-R''-NH_2$$

For example a one pack moisture curing system of Epon 812 and the ketimine N,N[1]-bis (2,4 dimethyl-3-pentylidene) ethylenediamine has been reported to have a stability of >200 days but as a film exposed to the atmosphere the composition dries in 16 hrs.[24]

Alternative methods of forming stable amine-epoxy resin compositions involve the formation of a thermolabile amine-metal coordination complex. Aliphatic amine complexes with zinc or cadmium halides are stable in epoxy resin solutions at temperatures below 40°C; at elevated temperatures (60 to 200°), the complexes are unstable, and they dissociate to yield the free amine that rapidly cross-links the epoxy resins.[17] Complexes of aromatic amines and suitable salts behave in a similar manner to the aliphatic amines.[18] Molecular sieves, which are crystalline metal alumino-silicates with "holes" or pores in their structures, will selectively absorb a range of compounds, including amines. When used to prepare a one-pack stable epoxy system, the molecular-sieve-diamine complex is dispersed in the resin; this mixture is stable for up to one year.[19] When the resin solution is applied as a film, moisture from the substrate (or from the atmosphere) displaces the absorbed amine, which reacts in the normal manner. Besides giving one-pack stability, the molecular sieve enables the amine to be released gradually and, hence, the exotherm of the cross-linking reaction is controlled;[20] therefore, the composition can be used on heat-sensitive materials. An alternative means of releasing the amine from a molecular sieve is to heat it to temperatures of approximately 100°C.

The epoxy-amine system is well suited to the formation of high solids or solventless coatings, because no volatile reaction products are liberated. The epoxy-amine resin combinations, discussed above, have been used as solventless coatings by carefully selecting the resin system to give suitable viscosity characteristics; for example, the low molecular weight liquid epoxy resins have been used with ketimines. Alternatively, where it is more desirable to use a polyamide cross-linking agent, the increased viscosity resulting from the polymeric nature of this amine source can be offset by the use of diluents. These may be nonreactive ester plasticizer-solvents, or reactive diluents, which are usually low molecular weight monoepoxy compounds—such as cresyl glycidyl ether,[21] phenyl glycidyl ether, or allyl glycidyl ether.

GENERAL REFERENCES

1. Skeist, I., *Epoxy Resins*, Reinhold Publishing Corp., New York, 1960.
2. Spitzer, W. C., *Offic. Dig. Federation Soc. Paint Technol.*, **36**, (**475**), 52 (1964).
3. Wismer, M., in E. M. Fettes, ed., *Chemical Reactions of Polymers*, Interscience Publishers, Inc., New York, 1964, p. 905.
4. Lee, H., and K. Neville, *Epoxy Resins*, McGraw-Hill Book Co., New York, 1957, p. 96.

REFERENCES

1. *European Chem. News, 1965,* July, 38.
2. Ges. Fur Teerverwertung, Ger. Pat. 1,153,028 (1963).
3. Marshall, C. D., and G. R. Somerville, *Offic. Dig. Federation Soc. Paint Technol.*, **34**, 286 (1962).
4. Epikote Resins for Paints, Shell Chemical (Aust.), Ltd.
5. Segal, C. L., and J. B. Rust, *Ind. Eng. Chem.*, **52**, 324 (1960).
6. Konininklijke Zwavelzuur Fabriek, Ketjen, N. V., Ger. Pat. 1,139,649 (1962).
7. Dobinson, B., A. J. Duke, K. W. Humphreys, E. Johnston, R. J. Martin, L. S. A. Smith, and B. P. Stark, *Makromol. Chem.*, **59**, 82 (1963).
8. Aelony, D., General Mills, Inc., Can. Pat. 598,317 (1960); Rohm & Haas Co., Brit. Pat. 811,852 (1959).
9. Greenlee, S. O., and J. W. Pearce, S. C. Johnson & Son, Inc., U.S. Pat. 3,113,932 (1963).
10. Shell Chemical (Aust.) Ltd., Technical Bulletin "Eponol 55-B-40."
11. Merten, R., Farbenfabriken Bayer Akt.-Ges., Ger. Pat. 1,083,055 (1960).
12. Firth, F. G., "*Potting and Encapsulation of Electrical Components*," Modern Plastics Encyclopedia, New York, **1957**, 753.
13. Narracott, E. S., *Brit. Plastics*, **26**, 120 (1953).
14. Shechter, L., J. Wynstra, and R. P. Kurkjy, *Ind. Eng. Chem.*, **48**, 94 (1956).
15. Bruin, P., *Chem. Ind. (London)*, **1957**, 616.
16. Wittcoff, H., *J. Oil Colour Chemists' Assoc.*, **47**, 273 (1964).
17. Bishop, R. R., Leicester, Lovell & Co., Ltd., Brit. Pat. 836,695 (1960).
18. Leicester Lovell & Co. Ltd., French Pat. 1,351,709 (1964); via *Chem. Abstr.*, **61**, 8482 (1964).
19. *Chem. Eng.*, **68**, Nov. 27th, 68 (1961).
20. O'Connor, F. M., and D. J. Waythomas, *Offic. Dig. Federation Soc. Paint Technol.*, **34**, 162 (1962).
21. Ott, G. H., *Offic. Dig. Federation Soc. Paint Technol.*, **29**, 1370 (1957).
22. Somerville, G. R., and H. L. Parry, *J. Paint Technol.*, **42**, 42 (1970).
23. O'Rear J. G., J. R. Griffith, and S. A. Reiners, *J. Paint Technol.*, **43**, 113 (1971).
24. Gardner, R Jr. and A. H. Keough, Norton Research Corp., U. S. Pat. 3,547,880 (1970).

Esterification of Epoxy Resins. Carboxylic acids esterify the epoxide group to form the hydroxy ester. This reaction takes place more readily than the hydroxyl-carboxyl esterification and, consequently, in a resin containing both epoxy and hydroxyl groups, the initial reaction involves the epoxide ring.[1]

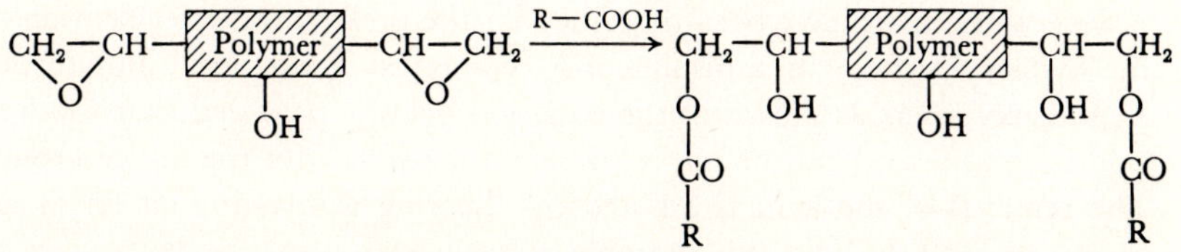

Appreciable reaction rates are obtained at 150°C in the presence of a basic catalyst, usually a tertiary amine. Stannous oxide or hydroxide, or stannous salts of weak acids are reported to be catalysts which reduce the reaction time for the fatty acid–epoxy reaction.[17] Under more forcing conditions, the original, and the newly formed, hydroxyl groups can also be esterified. Fatty acids form a range of epoxy-esters somewhat similar to alkyd resins and, in fact, similar terminology is used in both fields. The epoxy-esters are described in terms of "oil length."

Where drying oil fatty acids are used, the epoxy-esters will dry by autoxidation in the presence of conventional driers. The films are more resistant to alkali than alkyd resins, because the polymer backbone is devoid of ester links. Epoxy resins can be used as a partial replacement for the polyol in an alkyd resin formulation. However, the higher functionality of the epoxy resin may lead to gelation unless a prior "defunctionalizing" process with

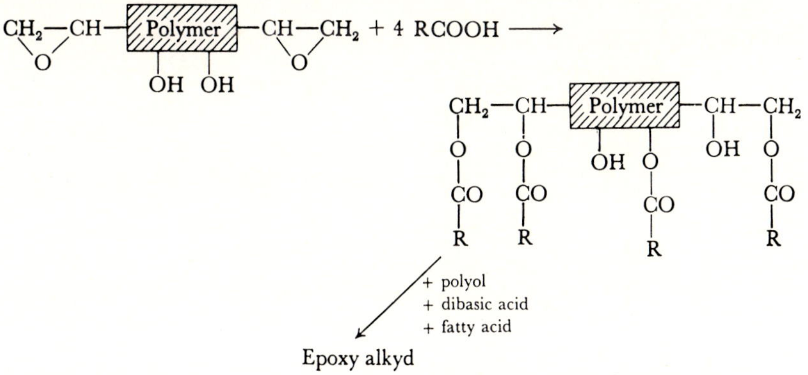

Epoxy alkyd

a monobasic acid is carried out. Sufficient monobasic ester is formed to give a molecule with 2 hydroxyl groups.

The epoxy-acid reaction can be applied to more highly functional and polymeric acids. Polyesters, including alkyd resins, and acid-containing acrylic copolymers are cross-linked by the epoxy resins   (see page 272). Some etherification of the epoxy groups occurs when strong acids are used in place of tertiary amines to catalyze the epoxy-acid reaction.[2]

Several methods have been described for the preferential esterification of the hydroxyl groups in a bisphenol-A type epoxy resin. Acid anhydrides form epoxy esters as shown in the equation below,[3] provided that the free acid is volatile and can be removed before it reacts with the epoxy group. The removal of the acid is assisted by choosing a solvent that forms an azeotrope with the acid (for example, toluene with acetic acid).

$$\underset{O}{CH_2-CH}-[Polymer]-\underset{OH \; OH}{\underset{|}{CH}}-\underset{O}{CH_2} + 2 \; \begin{matrix} CH_3-CO \\ CH_3-CO \end{matrix}\!\!>\!O$$

$$\downarrow$$

$$\underset{O}{CH_2-CH}-[Polymer]-\underset{\underset{CH_3 \;\; CH_3}{\underset{|}{CO} \;\; \underset{|}{CO}}}{\underset{O \quad O}{CH}}-\underset{O}{CH_2} + 2 \; CH_3-COOH\!\uparrow$$

Fatty acid esters of this type are of special interest because they are often a cheap and effective means of plasticizing thermosetting epoxy-polymer systems. The alcoholysis of an epoxy resin with a fatty acid methyl ester occurs readily in the presence of an alkaline catalyst, and the methanol formed in the reaction is easily removed[4] to give an ester that still contains epoxy groups.

$$\underset{O}{CH_2-CH}-[Polymer]-\underset{OH \;\; OH}{\underset{|}{CH}}-\underset{O}{CH_2} + R-C\!\!\begin{matrix} \nearrow O \\ \searrow OCH_3 \end{matrix}$$

$$\downarrow$$

$$\underset{O}{CH_2-CH}-[Polymer]-\underset{\underset{R \qquad R}{\underset{|}{CO} \;\; \underset{|}{CO}}}{\underset{O \quad\;\; O}{CH}}-\underset{O}{CH_2} + CH_3OH\!\uparrow$$

This technique can be applied to the manufacture of epoxy-modified alkyd resins, and offers a means of utilizing the cheaper vegetable oils, rather than the free fatty acids.[5]

Ester interchange of inorganic acids with the hydroxyl group of an epoxy resin is also used to cross-link or to modify epoxy esters. Tetrabutyl titanate, in the presence of a base (piperidine), is used as a curing agent for epoxy resins, and the butyl ester interchanges with the epoxy resin hydroxyl groups[6] to form butanol.

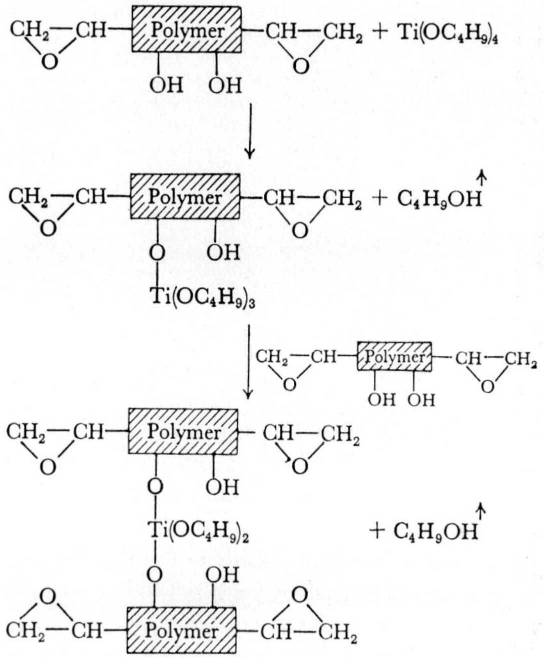

$$CH_2{-}CH{-}[\text{Polymer}]{-}CH{-}CH_2 + Ti(OC_4H_9)_4$$

with OH OH groups below the Polymer, epoxide O rings at each end.

$$\downarrow$$

$$CH_2{-}CH{-}[\text{Polymer}]{-}CH{-}CH_2 + C_4H_9OH\uparrow$$

with $O{-}Ti(OC_4H_9)_3$ and OH below the Polymer.

$$\downarrow \quad CH_2{-}CH{-}[\text{Polymer}]{-}CH{-}CH_2\ (OH\ OH)$$

$$CH_2{-}CH{-}[\text{Polymer}]{-}CH{-}CH_2$$

with $O{-}Ti(OC_4H_9)_2{-}O$ bridging to a second Polymer:

$$CH_2{-}CH{-}[\text{Polymer}]{-}CH{-}CH_2 \qquad + C_4H_9OH\uparrow$$

Additional linking is likely by the formation of molecular complexes or by the catalyzed opening of the epoxide ring.[7]

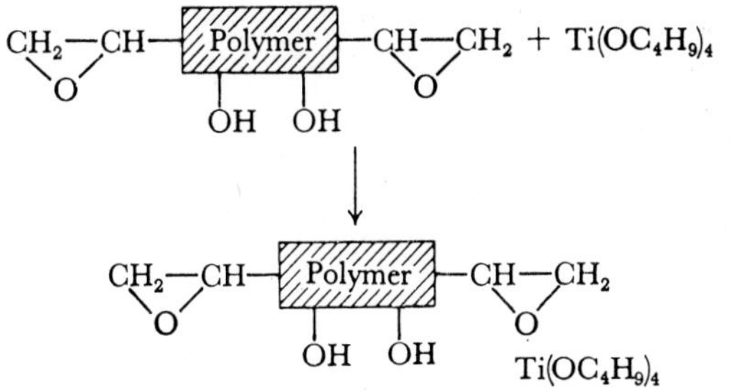

$$CH_2{-}CH{-}[\text{Polymer}]{-}CH{-}CH_2 + Ti(OC_4H_9)_4$$

with OH OH below the Polymer.

$$\downarrow$$

$$CH_2{-}CH{-}[\text{Polymer}]{-}CH{-}CH_2$$

with OH OH below the Polymer and $Ti(OC_4H_9)_4$.

Interchange with silicone resin enables silicone-modified epoxy derivatives to be prepared, as shown in the following equations.[8]

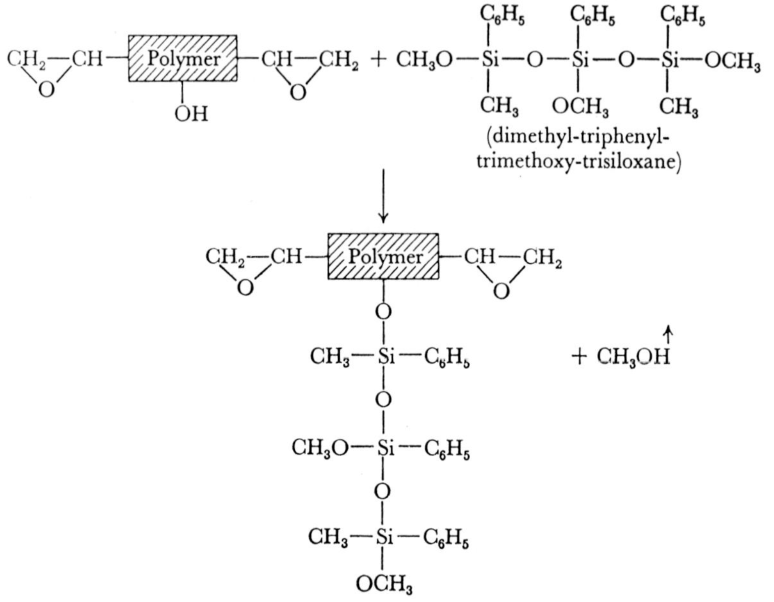

$$CH_2-CH-\boxed{Polymer}-CH-CH_2 + CH_3O-\underset{\underset{CH_3}{|}}{\overset{\overset{C_6H_5}{|}}{Si}}-O-\underset{\underset{OCH_3}{|}}{\overset{\overset{C_6H_5}{|}}{Si}}-O-\underset{\underset{CH_3}{|}}{\overset{\overset{C_6H_5}{|}}{Si}}-OCH_3$$

(dimethyl-triphenyl-trimethoxy-trisiloxane)

$$CH_2-CH-\boxed{Polymer}-CH-CH_2$$

$$CH_3-Si-C_6H_5$$

$$CH_3O-Si-C_6H_5$$

$$CH_3-Si-C_6H_5$$

$$OCH_3$$

$$+\ CH_3OH\uparrow$$

Careful control of the reaction conditions is necessary to prevent gelation. The silicone-modified resin, when cross-linked with a polyamide, gives a film that has much better cold water resistance than the unmodified epoxy resin.[8]

Acid anhydrides react with epoxy resins at elevated temperatures, and this curing system is characterized by long pot-life at room temperature, low exotherm during curing, and low volatility of the components.[9] The curing reactions are catalyzed by amines, alcohols, phenols, and acids.[9,10] The precise reactions occurring between an epoxy resin and an anhydride are still not established, but it appears that, in general terms, the steps shown in Scheme 1 occur.[1,9,11–14]

Reaction 3 is catalyzed by acids and by the anhydride curing agent. Basic catalysts increase the rate of reactions 1 and 2, and etherification is negligible when amine catalysts are used.

In many of the reactions involving an epoxy group, the amine catalyst functions by facilitating the opening of the epoxide ring. However, in the epoxy-anhydride system a more likely mechanism is that proposed by Fischer,[15] and shown in Scheme 2, in which the amine first interacts with the anhydride. Taraka and Kakiuchi[10] also suggest that the amine interacts with the anhydride rather than with the epoxy group.

Scheme 1

1. The anhydride forms a half ester with the hydroxyl group of the resin.

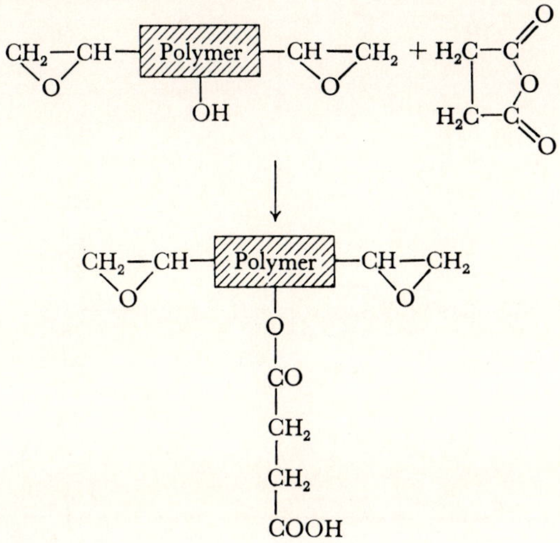

2. Esterification of the acid with the epoxide groups.

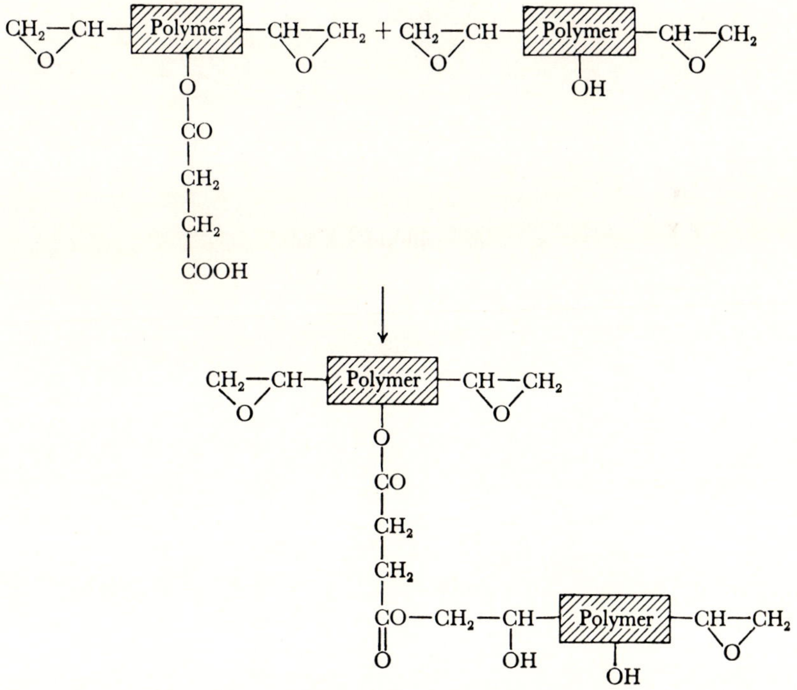

3. Etherification by interaction of the hydroxyl and epoxy groups.

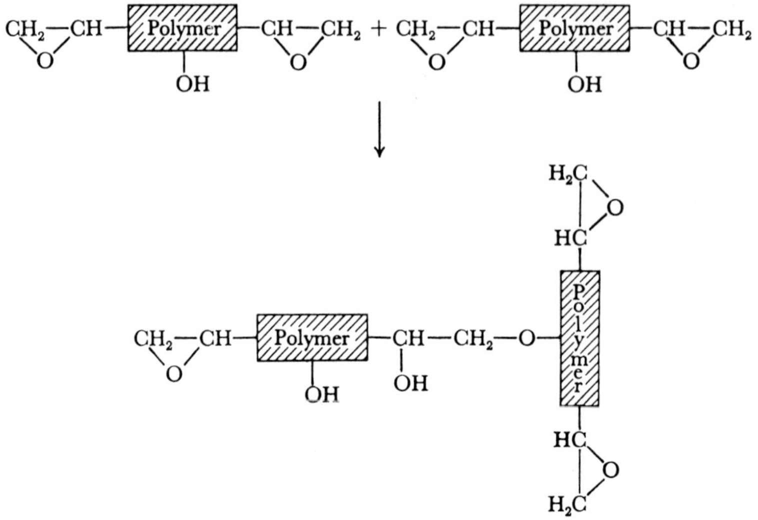

Scheme 2

1. Activation of anhydride by the amine.

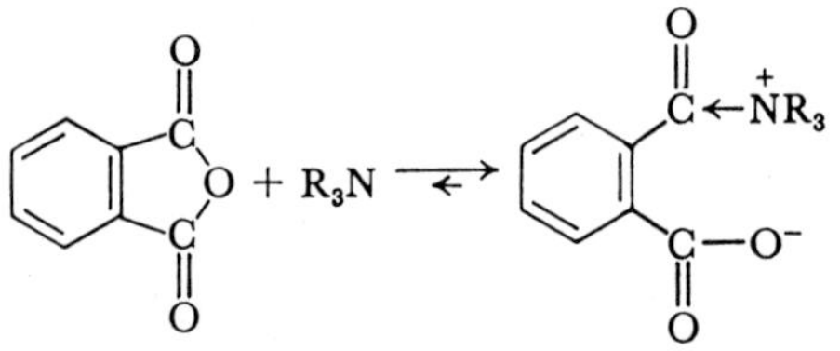

2. Reaction of the carboxyl anion with the epoxide.

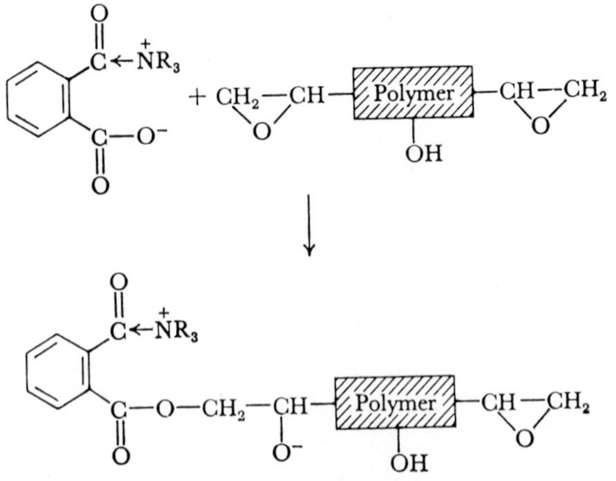

3. Reaction of alkoxide anion with the anhydride.

4. A displacement of the amine by alkoxide ion would also continue polyester formation.

$+ R'O^-$ (R' = a polymeric residue)

$+ NR_3$

Reactions of Epoxy Resins with Thiol, Methylol, Hydroxyl, and Phenolic Groups.
The reaction of a thiol compound with an epoxy resin is one of addition
across the epoxide ring.

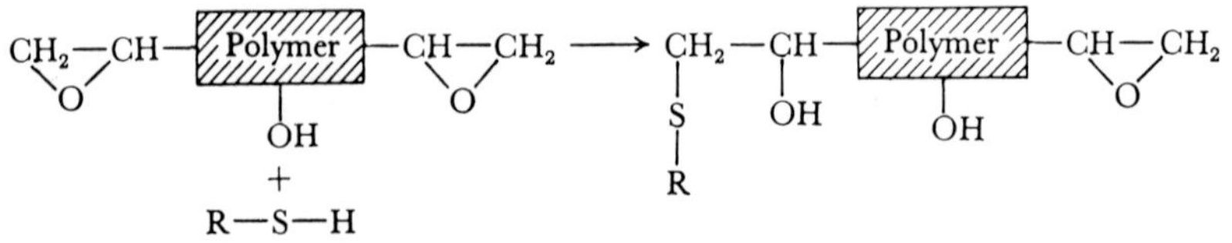

Thiokol resins cross-link epoxy polymers by this mechanism, and the thiol
polymer serves as an effective means of plasticizing the composition.

$$H-S+C_2H_4-O-C_2H_4-O-C_2H_4-S-S+_6 C_2H_4-O-C_2H_4-S-H$$

(Thiokol resin)

The plasticising action of Thiokol resins in epoxies has been attributed
to the penetration of the Thiokol macromolecules into the micropores of the
resin structure. The use of lead catalysts affects the properties by altering
the structural size of the resin units—higher lead concentrations result in
smaller size units with improved tensile and compressive strength.[18] Thiol
alkylene amines, prepared from polyamines and episulphides are useful as
curing agents for epoxy resins; the cured compositions have biological
activity.[19] Thioalkanoic acids have also been used as curing agents for epoxy
resins.[21]

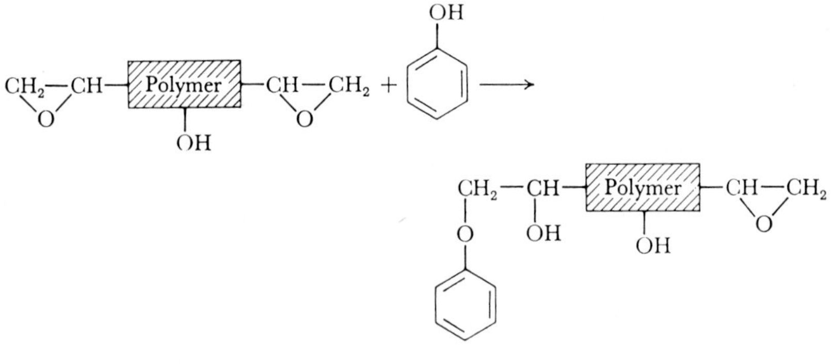

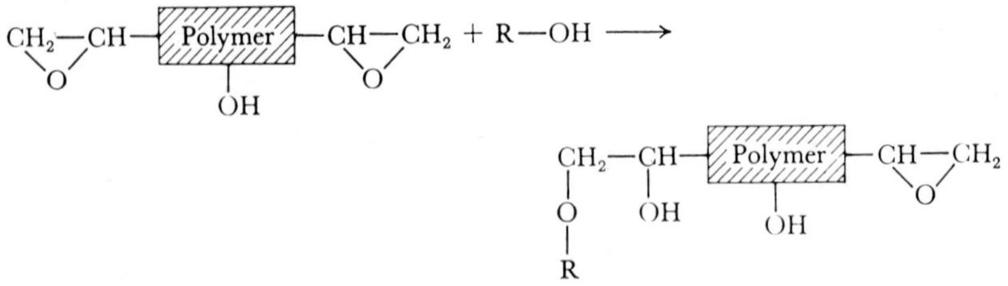

Compositions which cure rapidly at temperatures down to -20°C have been reported and these are based on adducts of the diglycidyl ether of bisphenol-A and low molecular weight dimercaptans. The catalyst used was triethylene diamine.[20] Phenols and alcohols add to the epoxy resin, usually at elevated temperatures, to form ethers.

These reactions are of importance in the cross-linking of epoxy resins by polymeric hydroxy compounds or phenolic resins. In the case of phenolic resins the methylol and methylol ether groups also add to the epoxide group and, in addition, condense with any free hydroxyl groups[16] along the polymer molecule. These reactions are catalysed by acids or by bases. Generally the epoxide/phenol and epoxide/secondary alcohol reactions both respond to acid catalysis which is therefore non-selective. On the other hand alkaline catalysts, such as sodium hydroxide, of N,N dialkyl acid amides selectively catalyze the epoxide/phenol reaction.[22]

(R = H or alkyl)

Partial reaction of an epoxy/phenolic resin blend, before the film is applied, is often carried out to improve the flow properties and to reduce the tendency for pinholing, or cissing (crawling), to occur. The extent of the precondensation is kept to a minimum, because the increased molecular weight of the interacted polymer lowers the solids at application viscosity. Urea- and melamine-formaldehyde resins react with epoxy resins in a similar manner to the phenolic resins, that is, via the methylol or methylol ether groups. These reactions are very slow at room temperature, and the compositions are sufficiently stable to be sold as one-pack formulations. Elevated temperatures (100°C) are used to give acceptable curing rates.

REFERENCES

1. O'Neill, L. A., and C. P. Cole, *J. Appl. Chem. (London)*, **6**, 356 (1956).
2. Hofmann, W., and W. Fisch, VI, *F.A.T.I.P.E.C., Congr.*, **1962**, 243.
3. Shell Chemical (Australia) Pty Ltd., Private communication.
4. Crecelius, S. G., Devoe and Reynolds Co., Inc., U.S. Pat. 24,047 (1955). Shell International Maatschappij N. V., Australian Pat. Appln. 3385/66.
5. Bataafse Petroleum Maatschappij, N. V., Brit. Pat. 858,827 (1961). Turner, R. J., and G. Swift, *F.A.T.I.P.E.C., Congr.* (Lucerne), **1957**, 207.
6. Kohn, L. S., and A. H. Horner, General Electric Co., (U.S.A.), U.S. Pat. 2,962,410 (1960).
7. Gutowsky, H. S., R. L. Rutledge, M. Tamres, and S. Searles, *J. Am. Chem. Soc.*, **76**, 4242 (1954).
8. Shell Chemical Co. Pty. Ltd., Bulletin on Silicone Modified Epikote Resins (1959).
9. Shell Chemical Co. Pty. Ltd., Bulletin on Acid Anhydrides as Curing Agents for Epikote Resins.
10. Tanaka, Y., and H. Kakiuchi, *J. Polymer Sci.*, **2**, 3405 (1964).
11. Fisch, W., and W. Hofmann, *J. Polymer Sci.*, **12**, 497 (1954).
12. Cooke, H. G., Jr., and J. E. Masters, Devoe and Raynolds Co., Inc., U.S. Pat. 2,947,711 (1960); Belanger, Wm. J., J. E. Masters, and D. D. Hicks, Devoe and Raynolds Co. Inc., U.S. Pat. 2,947,712 (1960); Belanger, Wm. J., and J. E. Masters, Devoe and Raynolds Co. Inc., U.S. Pat. 2,947,717 (1960); Belanger, Wm. J., Devoe and Raynolds Co. Inc., U.S. Pat. 2,947,726 (1960); Cooke, H. G., Jr., Devoe and Raynolds Co. Inc., U.S. Pat. 2,947,725 (1960); Pirozhnaya, L. N., *Plasticheskie Massy*, **1961**, 56.
13. Weiss, H. K., *Ind. Eng. Chem.*, **49**, 1089 (1957).
14. Tanaka, Y., and H. Kakiuchi, *J. Appl. Polymer Sci.*, **7**, 1063 (1963).
15. Fischer, R. F., *J. Polymer Sci.*, **44**, 155 (1960).
16. Kelly, D. P., G. J. H. Melrose, and D. H. Solomon, *J. Appl. Polymer Sci.*, **7**, 1991 (1963).
17. Shell Oil Company, U. S. Pat. 3,507,819.
18. Ullberg, Z. R., Z. T. Il'ina, V. P. Vasilenko, and E. M. Natanson, *Ukr. Khim. Zh.*, **36** 968 (1970).
19. Levine, L., Dow Chemical Company, U. S. Pat. 3,548,002 (1970).
20. Graver, R. B., *J. Paint Technol.*, **42**, 37 (1970).
21. Greenlee, S. O., G. J. Crocker, and C. L. Weidner, *J. Paint Technol.*, **42**, 31 (1970).
22. Alvey, F. B., *J. Paint Technol.*, **42**, 48 (1970).

CHAPTER 8

POLYURETHANES AND RELATED POLYMERS DERIVED FROM ISOCYANATES

Compounds that contain the isocyanate group, $-N{=}C{=}O$, are highly reactive both with themselves, and with compounds possessing a labile hydrogen atom (that is, one that can be replaced by sodium metal) (Gen. Refs. 1 & 2). In the surface-coatings industry, isocyanates have been utilized in a variety of polymer systems but, in all cases, the chemical reactions involved are similar to those of the simple monofunctional isocyanates. Additional reactions that are a function of other reactive components may also contribute to the formation of the polymeric coating. Therefore, the chemical reactions of the simple isocyanates are discussed first, and then the ways in which these reactions have been applied to formulation of surface coatings.

REACTIONS OF THE ISOCYANATE GROUP

The reactions that involve the isocyanate group and that are of particular importance to the formulation and preparation of surface-coating polymers are shown in Eqs. 8.1 to 8.9.

Reaction with alcohols.

$$R{-}N{=}C{=}O + R'{-}OH \longrightarrow R{-}NH{-}\underset{\underset{\displaystyle (urethane)}{OR'}}{\overset{\displaystyle C{=}O}{|}} \tag{8.1}$$

Reaction with amines.

$$R{-}N{=}C{=}O + R'{-}NH_2 \longrightarrow R{-}NH{-}\underset{\underset{\underset{\displaystyle (substituted\ urea)}{R'}}{\overset{\displaystyle |}{NH}}}{\overset{\displaystyle C{=}O}{|}} \tag{8.2}$$

Reaction with water.

$$R{-}N{=}C{=}O + H_2O \longrightarrow R{-}NH{-}\underset{\underset{\displaystyle (carbamic\ acid)}{OH}}{\overset{\displaystyle C{=}O}{|}} \tag{8.3}$$

$$R{-}NH{-}\underset{OH}{\overset{\displaystyle C{=}O}{|}} \longrightarrow R{-}NH_2 + CO_2 \uparrow \quad (amine)$$

211

Reaction with carboxylic acids.

$$R-N=C=O + R'-COOH \longrightarrow \left[\begin{array}{c} R-NH-C=O \\ | \\ O \\ | \\ CO \\ | \\ R' \end{array} \right]$$

$$\downarrow$$

$$R-NH-C=O + CO_2 \uparrow$$
$$\qquad\quad | $$
$$\qquad\quad R' $$

$$(8.4)$$

Reaction with ureas and urethanes.

$$R-N=C=O + R'-NH-CO-NH-R''$$
$$\downarrow$$
$$R-NH-C=O$$
$$\qquad\quad |$$
$$\qquad\quad N-R''$$
$$\qquad\quad |$$
$$\qquad\quad CO$$
$$\qquad\quad |$$
$$\qquad\quad NH \quad (biuret)$$
$$\qquad\quad |$$
$$\qquad\quad R'$$

$$R-N=C=O + R'-NH-CO-OR''$$
$$\qquad\qquad\qquad (urethane)$$
$$\downarrow$$
$$R-NH-C=O$$
$$\qquad\quad |$$
$$\qquad\quad N-R'$$
$$\qquad\quad |$$
$$\qquad\quad CO \qquad (allophanate)$$
$$\qquad\quad |$$
$$\qquad\quad OR''$$

$$(8.5)$$

Reaction with oximes.[1]

$$R-N=C=O + HO-N=C-R'$$
$$\qquad\qquad\qquad\qquad\qquad |$$
$$\qquad\qquad\qquad\qquad\qquad R''$$
$$\downarrow$$
$$\qquad\quad O$$
$$\qquad\quad \|$$
$$R-NH-C-O-N=C-R'$$
$$\qquad\qquad\qquad\qquad |$$
$$\qquad\qquad\qquad\qquad R''$$

$$(8.6)$$

Reaction with silylamines.[2]

$$R-N=C=O + (CH_3)_3Si-NR'_2$$
$$\downarrow$$

$$\begin{array}{ccc} R-N-C=O & & R-N=C-NR'_2 \\ \quad | \quad\; | & or & \qquad\quad | \\ (CH_3)_3Si \;\; NR'_2 & & OSi(CH_3)_3 \end{array}$$

$$(8.7)$$

Reaction with phenols, enols, etc.

$$R-N=C=O + \underset{\text{phenol}}{\text{(OH)}} \underset{\text{heat}}{\rightleftharpoons} R-NH-\underset{\underset{\text{(phenyl)}}{O}}{\overset{O}{C}} \tag{8.8}$$

Self-polymerization.[3]

$$2\ R-N=C=O \longrightarrow \text{(uretidione)} \tag{}$$

$$3\ R-N=C=O \longrightarrow \text{(cyanurate)} \tag{8.9}$$

$$R-N=C=O \xrightarrow[\substack{\text{irradiation in}\\ \text{solid state or}\\ \text{anionic}\\ \text{polymerization}}]{} \left[\ N-C\ \right]_n$$

The relative rates at which the reactions shown in Eqs. 8.1 to 8.9 occur depend on a number of factors (which will be discussed later), but a general order of decreasing reaction rates for common reagents toward isocyanates is as follows.[4]

1. Primary amines
2. Primary alcohols
3. Water
4. Ureas
5. Secondary and tertiary alcohols
6. Urethanes
7. Carboxylic acids
8. Carboxylic acid amides

The structure of the isocyanate has a marked influence on the reaction rate, and substituents that make the —NCO group relatively positive, or susceptible to nucleophilic attack, enhance the reaction. Baker and Holdsworth[5] have demonstrated this effect for the reaction of isocyanates with alcohols under triethylamine catalysis; their results are shown in Table 8.1.[5]

TABLE 8.1

THE INFLUENCE OF STRUCTURE ON THE REACTIVITY OF AN ISOCYANATE OF A GENERAL FORMULA RNCO[5]

Substituent R	Cyclohexyl	p-Methoxy phenyl	p-Methyl phenyl	Phenyl	p-Nitro
Reactivity constant					
$10^2 K_2$	0.021	9.9	12.3	22.6	ca 3000
Relative reactivity	1	471	590	1762	145,000

Similar effects are observed in diisocyanates, since the presence of a second isocyanate in an aryl ring can be expected to enhance the reactivity of the first group. The magnitude of this effect will depend on the relative positions of the two groups in the aromatic ring, and this is illustrated by the results of Kogon[6] (Table 8.2), who compared the isocyanates in their reaction with ethanol (no catalyst present).

TABLE 8.2

THE INFLUENCE OF A SECOND ISOCYANATE GROUP ON THE REACTION RATE WITH ETHANOL[6]

Isocyanate	Velocity Constant $K \times 10^4$ (liter mol^{-1} sec^{-1})	Relative Rate (PhNCO = 1)
Phenyl isocyanate	2.50	1
2:4 Tolylene diisocyanate	10.70	4.0
2:6 Tolylene diisocyanate	2.46	0.98

The relative reactivity of the *para* to *ortho* —NCO in 2:4-tolylene diisocyanate is between 7 and 8 to 1,[6,7] and this fact is made use of in coating compositions. Diisocyanates with an even greater difference in activity between the two groups are 4:4′-diisocyanato-hexahydrobisphenyl[8] and

3:5=diethyl 4:4′-diisocyanato-diphenylmethane.[9] In the latter case, the half-life of the two isocyanate groups is in the ratio of 49:1, and it has been suggested that the two ethyl groups reduce the activity of the 4-isocyanate group by steric hindrance.[9] The above results apply to reactions involving the isocyanate with alcohols, and care is necessary when applying these findings to other compounds. Cooper and Pearson[10] have compared the reactivity of various isocyanates with different hydrogen-donating compounds (Table 8.3),[10] and it is of interest to note that the 2:4-tolylene diisocyanate reacts more readily than the 2:6 isomer with water, whereas the converse is true in reaction with diphenyl urea.

TABLE 8.3

VELOCITY CONSTANTS FOR THE REACTION OF ISOCYANATES WITH HYDROGEN DONORS AT 100°C[10]

Isocyanate	Hydroxyl[a]	Water	Urea[b]	Amine[c]	Urethane[d]
p-Phenylene diisocyanate	36	7.8	13.0	17.0	1.8 at 130°C
2-Chloro 1:4-phenylene di-isocyanate	38	3.6	12.0	23.0	—
2:4-Tolylene diisocyanate	21	5.8	2.2	36	0.7 at 130°C
2:6-Tolylene diisocyanate	7.4	4.2	6.3	6.9	—
Hexamethylene diisocyanate	8.3	0.5	1.1	2.4	2×10^{-5} at 130°

The velocity constants are $\times 10^4$ liter mole^{-1} sec^{-1} except where otherwise indicated.

[a] Polyethylene adipate.

[b] Diphenyl urea.

[c] 3:3′-Dichlorobenzidine.

[d] p-Phenylene dibutyl urethane.

The structure of the compound containing the labile hydrogen atom also has a bearing on the rate of reaction and on the stability of the product formed. The relative reaction velocities of alcohols have been given as primary 1.0, secondary 0.3, and tertiary 0.005,[11] but the differences may be further accentuated by the conditions under which the reaction takes place. Of particular importance is the presence of catalysts (see page 217).

Blocked isocyanates are compounds that contain no free isocyanate group and are relatively inactive at room temperature. At elevated temperatures, however, the compounds are unstable and behave like—and undergo the typical reactions of—an isocyanate group. Typical blocking agents are phenols and thiols, tertiary alcohols, secondary aromatic amines, and 1:3-dicarbonyl compounds.[12] The temperature at which the blocked isocyanates dissociate or become available for reaction is of vital importance in deciding the conditions under which surface-coating compositions of this type should be cured. Unblocking temperatures for an 80/20 mixture of 2:4 and 2:6-tolylene diisocyanate are shown in Table 8.4.[13]

TABLE 8.4

UNBLOCKING TEMPERATURES OF BLOCKED DIISOCYANATES BASED ON TOLUENE DIISOCYANATE[13]

Blocking Agent	Unblocking Temperature (°C)			
	Infrared Spectrum		Carbon Dioxide Test[a]	
	In Paraffin Oil	In Methoxy-polyethylene Glycol	Uncatalyzed	Catalyzed
Ethanol	—	—	160	152
2-Methyl-2-propanol	No NCO up to 174	No NCO up to 174	150	125
m-Cresol	105	70	85	80
o-Nitrophenol	—	—	74	48
p-Chlorophenol	—	—	55	40
Guaiacol	92	60	87	87
Resorcinol	—	—	75	75
Phloroglucinol	—	—	110	106
1-Dodecanethiol	81	74	116	110
Benzenethiol	84	51	95	95
Ethyl acetoacetate	No NCO up to 154	110	97	97
Diethyl malonate	No NCO up to 100	79	85	85
ε-Caprolactam	125	108	132	140
Ethyl carbamate	—	No NCO up to 154	125	120
Boric acid	—	—	70	70

[a] The carbon dioxide test was based on the reaction of the liberated isocyanate with water (see Eq. 8.3).

Griffin and Willwerth[13] have concluded that blocked isocyanates based on phenols, on alkyl and aryl thiols, and on β-dicarbonyl compounds, dissociate to the free isocyanate when heated, although other evidence suggests that interchange reactions with alcohol are more likely (Gen. Ref. 1).

$$RNHCOOR' + R''OH \longrightarrow RNHCOOR'' + ROH$$

The dissociation or reaction temperature of a blocked isocyanate is very dependent on the other components present. For example, the poly(alkylene glycols) reduce the temperature, and this is relevant to the curing of blocked isocyanates with polyethers (see page 226).

The reaction conditions markedly influence the relative reaction rates of isocyanate groups. Whereas at room temperature the *para* group in 2:4-tolylene diisocyanate is 7 to 8 times more active than the *ortho* group, at 170°C the groups have equal reactivity (Gen. Ref. 2). Temperature can also influence the products of the reaction, as shown by the reaction of ethyl alcohol with phenyl isocyanate.[6]

The use of catalysts has a profound effect on the rate of addition of compounds to the isocyanate group. Metal ions, tertiary amines, and organo-metallic compounds are the common catalyst systems used. The comparative activity of typical examples of these compounds in the reaction of phenyl isocyanate with butanol is shown by the results of Hostettler and Cox,[14] listed in Table 8.5. The stronger catalytic action of the tin compounds is apparent.

Catalysts may vary in their effectiveness with the structure of the isocyanate, and with the compound supplying the labile hydrogen atom. In the reaction between tolylene diisocyanate and a secondary alcohol containing poly(oxypropylene) triol, Britain and Gemeinhardt[15] found the descending order of catalytic activity to be Bi, Pb, Sn, triethylene diamine, strong bases, Ti, Fe, Sb, U, Cd, Co, Th, Al, Hg, Zn, Ni, trialkylamines, Ce, Mo, V, Cu, Mn, Zr, trialkylphosphines. When m-xylylene diisocyanate or hexamethylene diisocyanate were used, the catalytic order changed; for

TABLE 8.5

CATALYTIC ACTIVITY IN THE ISOCYANATE/HYDROXYL REACTION[14]

Reactants:	Phenyl Isocyanate and Butanol	
Catalyst:	Catalyst, Mole Percent	Relative Activity at 1.0 Mole Percent
None	—	1.0
N-Methyl morpholine	1.0	4
Triethylamine	1.0	
N,N,N,N-Tetramethyl-1,3-butane diamine	1.0	27
1,4-diazabicyclo-(2:2:2) octane	1.0	120
Stannous chloride	0.10	2,200
Stannic chloride	0.10	2,600
Ferric acetyl acetonate	0.01	3,100
Tetra phenyl tin	1.0	9
Tetra n-butyl tin	1.0	160
Tri-n-butyl tin acetate	0.001	31,000
Di-n-butyl tin dilaurate	0.001	56,000
Dimethyl tin dichloride	0.001	78,000

hexamethylene diisocyanate, it became Bi, Fe, Sn, Pb, Ti, Sb, strong bases, Co, Zn, triethylenediamine, trialkylamine. The following mechanism has been proposed for the way in which a metal compound catalyzes the isocyanate/hydroxyl reaction. It has been suggested that the differences in catalytic activity observed with tolylene diisocyanate and aliphatic diisocyanates are related to steric hindrance.[15]

The structure of the alcohol also has a marked influence on the effectiveness of a catalyst in the isocyanate/alcohol reaction. Smith[16] has observed that triethylamine increased the rate constant for the 1-butanol/phenyl isocyanate reaction to approximately six times that for the 2-butanol/phenyl isocyanate reaction. Similarly, Wissman et al.[17] have found that dibutyl-tin dilaurate substantially increased the rate of reaction of polyethers containing

$$R-N=C=O$$

$$\downarrow MX_2$$

$$\left[\begin{array}{c} R-N=C=O^+ \\ \mid \\ {}^-MX_2 \end{array}\right] \longleftrightarrow \left[\begin{array}{c} R-N=\overset{+}{C}-O \\ \mid \\ {}^-MX_2 \end{array}\right]$$

$$H-O-R' \downarrow$$

$$\left[\begin{array}{c} R-\overset{+}{N}=C-O \\ \mid \ \ \mid \ \ \mid \\ H \ \ O \ \ {}^-MX_2 \\ \mid \\ R' \end{array}\right] \longleftarrow \left[\begin{array}{c} R-N=\overset{+}{C}-O \\ \mid \ \ \ \mid \\ H-\overset{+}{O}-MX_2^{-2} \\ \mid \\ R' \end{array}\right]$$

$$\updownarrow$$

$$\left[\begin{array}{c} R-N-C=O^+ \\ \mid \ \ \mid \ \ \mid \\ H \ \ O \ \ {}^-MX_2 \\ \mid \\ R' \end{array}\right] \longrightarrow \begin{array}{c} R-N-C=O + MX_2 \\ \mid \ \ \mid \\ H \ \ O \\ \mid \\ R' \end{array}$$

primary hydroxyl groups with tolylene diisocyanate, when compared to similar polyethers with terminal secondary hydroxyl groups. Ether groups influence the activity of nearby hydroxyl groups (an observation of considerable importance because many of the polyols used in commercial formulations are polyethers), although the type of catalyst controls the magnitude of this effect.[18] For example, in the series of methoxy-propanols shown below,

CH_2OH	CH_2OCH_3	CH_2OH
$\mid$	$\mid$	$\mid$
$CHOCH_3$	$CHOH$	CH_2
$\mid$	$\mid$	$\mid$
CH_3	CH_3	CH_2OCH_3
2-methoxy- 1-propanol	1-methoxy- 2-propanol	3-methoxy- 1-propanol
(1)	**(2)**	**(3)**

lead naphthenate increased the reaction rate of compounds **1** and **2** (adjacent ether and alcohol groups) to a greater extent than it increased the reaction rate of compound **3**. On the other hand, dibutyl-tin dilaurate preferentially catalyzes the reaction of the primary alcohol group (compounds **1** and **3**), and the presence of the ether groups adjacent to the hydroxyl gives a slower reaction rate than is observed for the compound in which the groups are separated. Satisfactory explanations for these effects have not been put forward.

The use of catalysts to promote reactions of the isocyanate groups with compounds other than alcohols may result in a different order of catalytic effectiveness, as illustrated in Table 8.6[8] for the action of zinc salts.[8] Therefore, it is necessary to evaluate catalysts under the conditions of use, and extrapolation of results should be done with due care.

Catalysts for polyurethane formation also promote the dimerization and trimerization of the isocyanate; the reaction is catalyzed vigorously by trialkylphosphines and more mildly by tertiary amines (Gen. Ref. 1). Even

TABLE 8.6

INFLUENCE OF ZINC CATALYSTS ON THE REACTION RATES OF ISO-CYANATES[8]

Additive	$K \times 10^3$ (liter mole^{-1} sec^{-1})				
	Hydroxyl 100°C	Water 100°C	Amine 130°C	Urethane 130°C	Urea 100°C
None	3.6	0.8	4.8	3.8	5.1
Zinc 15^{-3} mole/liter	5.3	2.3	23.0	—	19.0

standing at room temperature is sufficient for some aromatic isocyanates to form a dimer (Eq. 8.9, page 213). The dimers may be regarded as "blocked" isocyanates, since they are stable at room temperature but dissociate on heating, and then react. Trialkylphosphines and some other catalysts also promote trimerization of the isocyanate (Eq. 8.9, page 213). This is important in understanding some of the properties of polyurethanes, since the trimer is a relatively stable structure that leads to chain branching.

GENERAL REFERENCES

1. Saunders, J. H., and K. C. Frisch, *High Polymers* **16**, *Polyurethanes, Part 1*, Interscience Publishers Inc., New York, 1962.
2. Mort, F., *J. Oil Colour Chemists' Assoc.*, **45**, 95 (1962).
3. Frisch, K. C., and L. P. Rumao, *J. Macromol. Sci.,—Revs. Macromol. Chem.*, **C5**(1), 103-150 (1970).

REFERENCES

1. Campbell, T. W., V. S. Foldi, and R. G. Parrish, *J. Appl. Polymer Sci.*, **2**, 81 (1959).
2. Klebe, J. F., J. B. Bush, and J. E. Lyons, *J. Am. Chem. Soc.*, **86**, 4400 (1964).
3. Sobue, H., Y. Tabata, M. Hiraoka, and K. Oshima, *J. Polymer Sci.*, Part C, No. 4, 943 (1964).
4. Spitzer, W. C., *Offic. Dig. Federation Soc. Paint Technol.*, **36**, (**475**), 52 (1964).
5. Baker, J. W., and J. B. Holdsworth, *J. Chem. Soc. (London)*, **1947**, 713.
6. Kogon, I. C., *J. Am. Chem. Soc.*, **78**, 4911 (1956).
7. Gudgeon, H., and R. J. W. Reynolds, *J. Oil Colour Chemists' Assoc.*, **42**, 677 (1959).

8. BIOS, Final Report, 719, "Interview with Prof. O. Bayer," via Gen. Ref. 2.

9. Case, L. C., *J. Appl. Polymer Sci.*, **8**, 533, 935 (1964).

10. Cooper, W., R. Pearson, and S. Darke, *Ind. Chemist*, **36**, 121 (1960).

11. Davis, T. L., and J. McC. Farnum, *J. Am. Chem. Soc.*, **56**, 883 (1934).

12. Seeger, N. V., and T. G. Mastin, Goodyear Tire and Rubber Co., U.S. Pat. 2,801,990 (1957); U.S. Pat. 2,683,729 (1954); U.S. Pat. 2,683,728 (1954); U.S. Pat. 2,683,727 (1954).

13. Griffin, G. R., and L. J. Willwerth, *Ind. Eng. Chem., Prod. Res. Develop.* **1**, 265 (1962).

14. Hostettler, F., and E. F. Cox, *Ind. Eng. Chem.*, **52**, 609 (1960).

15. Britain, J. W., and P. G. Gemeinhardt, *J. Appl. Polymer Sci.*, **4**, 207 (1960).

16. Smith, H. A., *J. Appl. Polymer Sci.*, **7**, 85 (1963).

17. Wissman, H. G., L. Rand, and K. C. Frisch, *J. Appl. Polymer Sci.*, **8**, 2971 (1964).

18. Rand, L., B. Thir, S. L. Reegen, and K. C. Frisch, *J. Appl. Polymer Sci.*, **9**, 1787 (1965).

UTILIZATION OF ISOCYANATES IN SURFACE-COATING POLYMERS

Many of the reactions involving the isocyanate group occur readily at room temperature and, consequently, the formulation of polymer systems suitable for use as surface coatings has presented interesting chemical problems. The common coating compositions derived from isocyanates can be divided into two broad classes—namely, the one-component (pack) system and the two-component (pack) system.

In addition to the compositions where the isocyanate is a major component of the polymer system, there are a number of other uses for isocyanates. Mono- and difunctional isocyanates have been suggested as modifiers for other film formers, pigments, fillers, and surfaces. In coating compositions that dry by free radical mechanisms (unsaturated polyesters and alkyd resins), the hydroxyl groups retard the film-forming reactions; monofunctional isocyanates have been used to reduce the hydroxyl value by the formation of a urethane and to improve the drying properties of the polymer system.[1] An extension of this principle is to use an excess of a diisocyanate to remove the hydroxyl groups, and to give a polymer with free isocyanate groups that are claimed to improve adhesion to many substrates.[2]

The improved adhesion probably results from interaction of the isocyanate group with hydroxyl groups present in the surface that is being coated. This principle is also applied in the formulation of primer coats for use under a polyurethane, where the selection of a polymer that contains free hydroxyl groups aids in the development of good adhesion [3] by the chemical bonding of the topcoat to the undercoat. Many of the common pigments and fillers used in surface coatings have surface hydroxyl groups and, under appropriate conditions, these may be reacted with an isocyanate. This treatment changes the surface from hydrophilic to organophilic and is claimed to improve the wetting and dispersing characteristics of the pigment. The use of unsaturated isocyanates [4]—such as vinyl or allyl isocyanates—further extends this process, since it is possible to subsequently grow a polymer chain onto the unsaturated portion of the original isocyanate.

Pigment/unsaturated isocyanate reaction

$$\boxed{\text{Pigment}}-\text{OH} + \text{CH}_2{=}\text{CH}-\text{NCO} \longrightarrow \boxed{\text{Pigment}}-\text{OCONH}-\text{CH}{=}\text{CH}_2$$

$$\downarrow \quad \text{CH}_2{=}\underset{Y}{\overset{X}{C}}$$

$$\boxed{\text{Pigment}}-\text{OCO}-\text{NH}-\boxed{\text{Polymer}}$$

Pigment/diisocyanate reaction

$$\boxed{\text{Pigment}}-\text{OH} + \text{(tolylene diisocyanate)} \longrightarrow \boxed{\text{Pigment}}-\text{O}-\text{CO}-\text{NH}-\cdots-\text{NCO}$$

$$\downarrow \boxed{\text{Polymer}}-\text{OH}$$

$$\boxed{\text{Pigment}}-\text{O}-\text{CO}-\text{NH}-\cdots-\text{NH}-\text{C}{=}\text{O}-\boxed{\text{Polymer}}$$

Alternatively, the use of an excess of a diisocyanate converts the pigment surface to one containing free isocyanate groups; these can be combined with a polymer system that contains free hydroxyl groups.

Thermoplastic block copolymers with urethane and acrylic blocks have been prepared by capping an isocyanate prepolymer with cumene hydroperoxide and using the peroxycarbamate so formed as the initiator for the polymerization of styrene, methyl methacrylate, or acrylonitrile.[5]

$$OCN-[Polymer]-NCO + 2R'OOH \longrightarrow R'OO.\overset{\overset{\displaystyle O}{\|}}{C}-NH-[Polymer]-NH.\overset{\overset{\displaystyle O}{\|}}{C}-OOR'$$

$$\downarrow \Delta$$

$$2R'O^{\cdot} + {}^{\cdot}OOCHN-[Polymer]-NHCOO^{\cdot}$$

$$\xrightarrow{\text{vinyl monomer}}$$

$$[Poly(vinyl)]OOCHN-[Polymer]-NH.COO-[Poly(vinyl)]$$

See the General References that follow for a discussion on the more technological aspects of urethane coatings.

GENERAL REFERENCES

1. Saunders, J. H., and K. C. Frisch, *High Polymers* **16**, *Part II Polyurethanes*, Interscience Publishers Inc., New York, 1964.
2. Patton, T. C., *Paint Ind. Mag., No. 9*, 15-(1959).
3. Taft, D. D., and A. F. Mohar, *J. Paint Technol.*, **42**, 615 (1970).
4. Blomeyer, F., *J. Oil Colour Chemists' Assoc.*, **54**, 141 (1971).
5. Bieneman, R. A., *J. Paint Technol.*, **43** (555), 92 (1971).

REFERENCES

1. Solomon, D. H., Unpublished observations.
2. Nischk, G., E. Müller, and L. Goerden, Farbenfabriken Bayer Akt.-Ges., Ger. Pat. 940,018 (1956).
3. Vogel, H. A., Pittsburgh Plate Glass Co., U.S. Pat. 2,910,381 (1959).
4. Grotenhuis, T. A., U.S. Pat. 3,156,576 (1964).
5. Zaganiaris, E., and A. Tobolsky, *J. Appl. Polymer Sci.*, **14**, 1997 (1970).

One-Component (Pack) Coatings. The most common coating compositions that utilize the reactions of an isocyanate group, and yet give stable solutions are the following:

1. Urethane oils or alkyds.
2. Moisture-cured isocyanates-terminated adducts.
3. "Blocked" isocyanates.

Types 1 and 2 form films at room temperature, whereas the blocked isocyanates require elevated temperatures for interaction between the polymer components.

The term urethane oil is somewhat ambiguous. It may be used to indicate the reaction product of a diisocyanate with the small amount of mono- and diglycerides in a naturally occurring oil, or to describe the polymer resulting from the interaction of a monoglyceride and a diisocyanate. The polymer that results from the monoglyceride diisocyanate reaction is comparable to an alkyd resin in which the phthalic anhydride has been replaced by the diisocyanate and, consequently, the term "uralkyd" is commonly used for

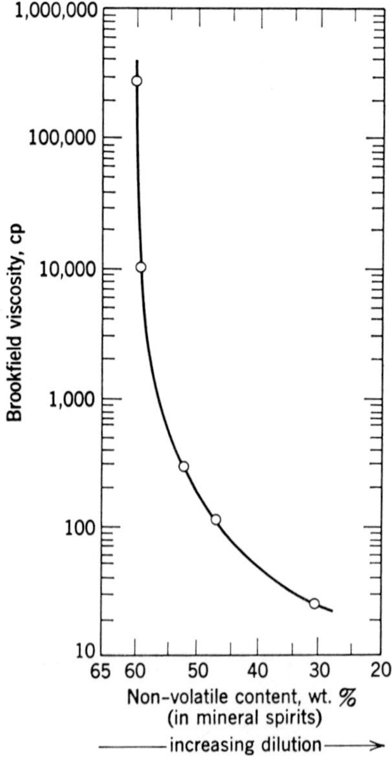

Figure 8.1.   Solids viscosity relationship for typical uralkyd.[2]

such polymer systems. Uralkyds are prepared by first subjecting the glyceride oil to an alcoholysis reaction (as in alkyd manufacture), but it is preferable to avoid the use of alcoholysis catalysts that also promote urethane formation. Calcium oxide or sodium hydroxide are often used for the alcoholysis reaction, and these catalysts can be neutralized before the addition of the diisocyanate. The most commonly used polyfunctional isocyanate is tolylene diisocyanate, and this is added to the cooled "monoglyceride" mixture (see page 86). After the initial reaction is complete, the temperature may be raised (for example to 80°C), and a catalyst may be added to complete the polyurethane formation. In practice, a slight excess of hydroxyl groups is used so that there are no free isocyanate groups in the final polymer. The film-forming ability of uralkyds is related to the fatty acids present in the original glyceride oil, and involves autoxidative polymerization in a similar manner to an alkyd resin. The uralkyd is characterized by giving hardei, tougher, and more flexible films than an alkyd of comparable oil length. Consequently, uralkyds are used as floor varnishes, metal primers, and marine varnishes.[1] Formulation changes similar to those made in alkyd resins can be applied to uralkyds, and the influence of these changes on properties has been discussed by Jolly.[1] Polymer systems in which both phthalic anhydride and tolylene diisocyanate are used to link up the "monoglyceride" molecules are claimed to give improved solubility and wetting characteristics, and to reduce the tendency of a uralkyd to exhibit "false body." The relationship between viscosity and solids of a linseed oil/polyol/tolylene diisocyanate uralkyd is shown in Figure 8.1.[2]

Isocyanate adducts or prepolymers are prepared by the reaction of excess diisocyanate with a polyol. Usually, tolylene diisocyanate is used and the choice of the polyol will govern the properties of the polymer system. Polyethers, polyesters, and castor oil are the common sources of the hydroxy compounds. Isocyanate adducts, therefore, may contain ether, ester, and/or urethane groups in the polymer chain.

$$n\text{HO}\!-\!\text{R}\!-\!\text{OH} + (n+1)\text{OCN}\!-\!\text{R}'\!-\!\text{NCO}$$

$$\downarrow$$

$$\text{OCN}\!-\!\!\left(\!\text{R}'\!-\!\text{NHCOOR}\!-\!\text{OCONH}\!\right)_{\!n}\!\!\text{R}\!-\!\text{NCO}$$

$$(\text{R} = \text{polyester, polyether,}$$
$$\text{or a simple alkyl group})$$

The alcohol component can also be more highly functional and can give an adduct with a greater cross-linking potential. When the adduct is applied as a film, some of the isocyanate groups react with moisture in the air, or substrate, and form amine groups and carbon dioxide. As pointed out on page 213, the amine-isocyanate reaction is rapid and, hence, the amine groups are quickly converted to urea cross-links by reaction with remaining isocyanate groups. The liberated carbon dioxide diffuses out of the film.

$$\boxed{Polymer}-NCO + H_2O \longrightarrow \boxed{Polymer}-NH_2 + CO_2 \uparrow$$

$$\boxed{Polymer}-NH_2 + OCN-\boxed{Polymer}$$

$$\downarrow$$

$$\boxed{Polymer}-NH-CO-NH-\boxed{Polymer}$$

Blocked isocyanate systems consist of a polyester or polyether that contains terminal hydroxyl groups and a suitable masked polyfunctional isocyanate. The masking agent is usually phenol or acetoacetic ester. Wide variations in properties of blocked isocyanate systems are possible by modification of the individual polymer components. Since isocyanates are toxic substances, the isocyanate component is usually a low molecular weight polymer with terminal isocyanate groups masked by phenol. This technique avoids the liberation of the monomeric isocyanate during the curing reaction. A typical masked isocyanate is produced from 1 mole trimethylol propane, 3 mole tolylene diisocyanate, and 3 mole phenol.[3]

The temperature that is necessary to promote reaction in a blocked isocyanate system depends on a number of factors; these include the nature of the blocking or masking group, the reactivity of the hydroxyl end groups in the polyester or polyether, and the presence of catalysts. As a general guide, curing schedules of 30 minutes at 150°C are a useful starting point in the evaluation of masked isocyanate coatings. The unblocking of isocyanates in urethane coatings has been studied by Evaporative Rate Analysis and this technique has enabled a more accurate assessment of the unblocking temperature than was given by previous methods.[4]

REFERENCES

1. Jolly, A. C., *J. Oil Colour Chemists' Assoc.*, **47**, 919 (1964).
2. Remington, W. J., *Factors in the Design of Oil-Base Urethane Coatings*, PB-10, Feb. 1960, published by E. I. du Pont de Nemours & Co.
3. Mielke, K. H., W. Bunge, and E. Windemuth, Farbenfabriken Bayer Akt.-Ges., Ger. Pat. 1,061,013 (1959).
4. Hill, H. E., C. S. Pietra and D. J. Damico, *J. Paint Technol.*, **43** (553), 55 (1971).

Two-Component (Pack) Systems. The use of two-component polymer systems removes many of the technical difficulties associated with the formulation

of coating compositions based on isocyanates. The two systems used in practice are the ones below.

1. An isocyanate prepolymer-polyol.
2. A catalyzed isocyanate prepolymer.

The isocyanate-polyol system cures by the reaction shown in Eq. 8.1 on page 211. The composition of the polyol may be modified over a wide range to provide the necessary balance of properties required for specific uses. However, the polyols can be broadly classified into two classes—polyesters and polyethers. The polyesters are usually prepared from a linear dibasic

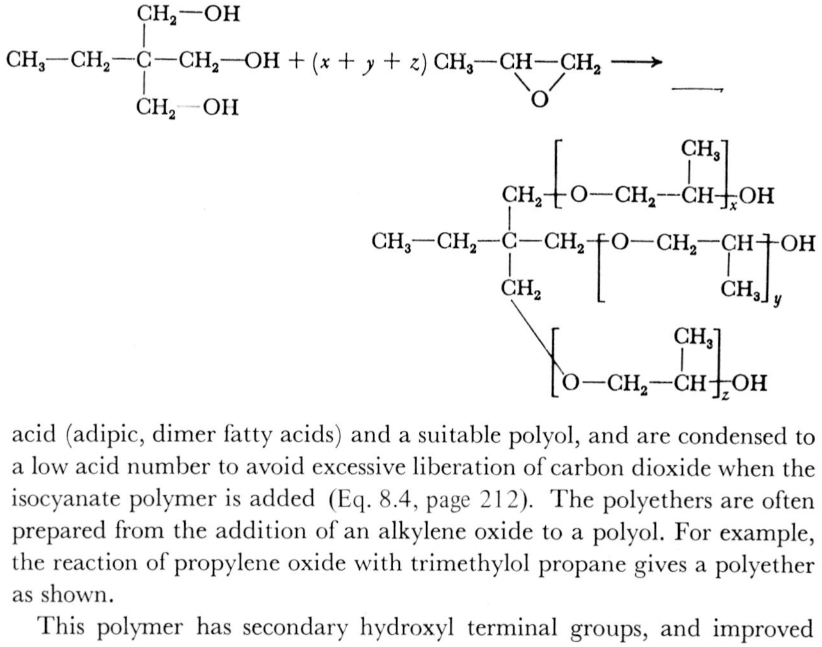

$$CH_3-CH_2-\underset{\underset{CH_2-OH}{|}}{\overset{\overset{CH_2-OH}{|}}{C}}-CH_2-OH + (x + y + z)\ CH_3-CH-CH_2 \longrightarrow$$

$$CH_3-CH_2-\underset{\underset{CH_2}{|}}{\overset{\overset{CH_2\left[O-CH_2-\underset{\underset{}{}}{CH}\right]_x OH}{|}}{C}}-CH_2\left[O-CH_2-\underset{\underset{CH_3}{|}}{CH}\right]_y OH$$

$$\left[O-CH_2-\underset{\underset{}{}}{\overset{\overset{CH_3}{|}}{CH}}\right]_z OH$$

acid (adipic, dimer fatty acids) and a suitable polyol, and are condensed to a low acid number to avoid excessive liberation of carbon dioxide when the isocyanate polymer is added (Eq. 8.4, page 212). The polyethers are often prepared from the addition of an alkylene oxide to a polyol. For example, the reaction of propylene oxide with trimethylol propane gives a polyether as shown.

This polymer has secondary hydroxyl terminal groups, and improved reaction with the isocyanate results if primary groups are present. Reaction of the above polymer with ethylene oxide is a method of forming polyethers with primary terminal hydroxyl groups.

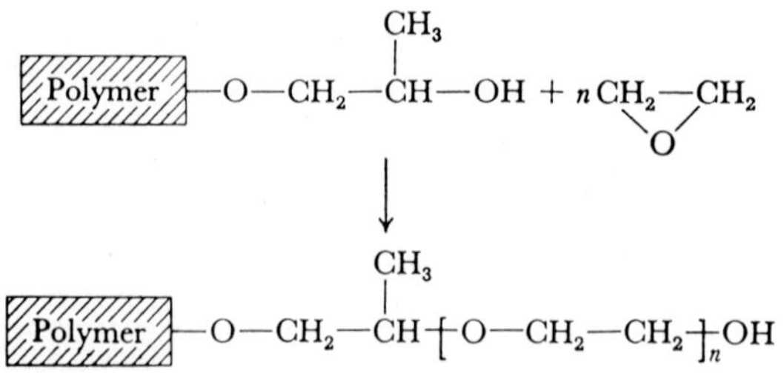

$$\boxed{Polymer}-O-CH_2-\underset{\underset{}{}}{\overset{\overset{CH_3}{|}}{CH}}-OH + n\ CH_2-CH_2$$

$$\downarrow$$

$$\boxed{Polymer}-O-CH_2-\underset{\underset{}{}}{\overset{\overset{CH_3}{|}}{CH}}\left[O-CH_2-CH_2\right]_n OH$$

A range of polyethers is available in which the size of the polypropylene oxide and the polyethylene oxide blocks is varied.

An isocyanate prepolymer is chosen mainly to reduce toxicity hazards associated with the volatile low molecular weight monomeric isocyanates, (see trimethylol propane-tolylene diisocyanate reaction product, page 227). Alternatively, diisocyanates with low vapor pressures may be used, and a commercial product of this type is diphenyl methane-4:4-diisocyanate.[1]

$$OCN - \langle\!\bigcirc\!\rangle - CH_2 - \langle\!\bigcirc\!\rangle - NCO$$

Even in the cured polyurethane coating some isocyanate groups are present, and care is necessary in operations where cutting or sanding of the film results in the formation of a dust.[2]

Isocyanate prepolymers are reported to be effective additives for systems which normally dry by solvent evaporation, e. g. nitrocellulose, acrylic resins, cellulose acetate-butyrate, and vinyl chloride-acetate copolymers.[7] In these systems the hydroxyl groups of the lacquer type polymer form urethane crosslinks with the isocyanate groups of the prepolymer. The resultant films have solvent, chemical and water resistance properties which are better than those of the unmodified polymers.

The catalyzed isocyanate polymer systems used as a two-component composition rely on the addition of a catalyst to bring about reaction of the isocyanate group with moisture in the atmosphere (see page 211 for reactions involved), or with hydroxyl groups that form part of the "catalyst." The conventional autoxidation catalysts, such as lead or cobalt naphthenate, promote the reaction of isocyanate groups with hydroxyl groups, or the self-condensation of the isocyanate to allophanates or trimers (see Eqs. 8.5 and 8.9, page 212 and 213).

The isocyanate-hydroxyl reaction is particularly suited to the formulation of high solids coatings, since by careful selection of the polyester or polyether, and of the isocyanate prepolymer, solvent-free systems are available.[1] The successful formulation of these high solids compositions requires the polymers to be water free; atmospheric moisture must also be removed in order to prevent carbon dioxide liberation (Eq. 8.3, page 211) and the formation of bubbles in the film. Molecular sieves, which are aluminum silicates with long tubular canals within their structures, have been successfully used to take up the water in the polymer systems and in the atmosphere.[1] When pigmented coatings are required, the removal of water is even more important, since most pigments have water molecules adsorbed on their surfaces. Molecular sieves are also used to remove this water;[3] an alternative technique is to grind the pigments in the isocyanate prepolymer, and remove the water by reaction with some of the free isocyanate groups. Molecular sieves can also be used to improve the stability of polyurethane coatings by the controlled release of curing agents, or catalysts, from the sieve. This release is brought about by displacement from the sieve by heating, or by the preferential absorption of water from the air.[2]

The development of a yellow coloration is one of the main problems associated with polyurethane coatings and, in general, this color is more pronounced with the one-pack than with the two-pack systems. The color development has been attributed to the presence of free amine groups that oxidize to give compounds with quinonoid-type structures.[4]

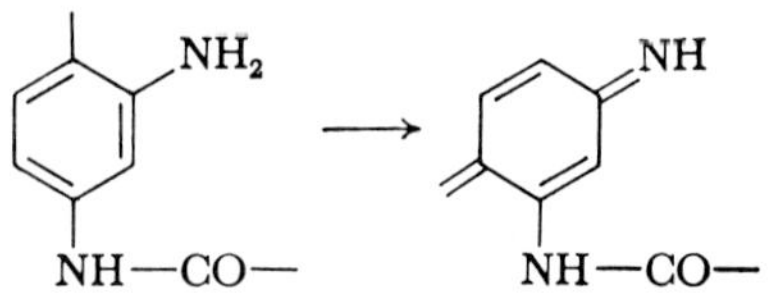

The curing mechanism discussed previously would support the presence of more amino groups in a moisture cured one-pack system and the consequent development of a more intense color. Color development is accelerated by elevated temperatures, and a mechanism to explain this change has been proposed[5] in which the urethane links dissociate to yield free tolylene diisocyanate that gives structures of the type shown opposite and other colored oxidation products.

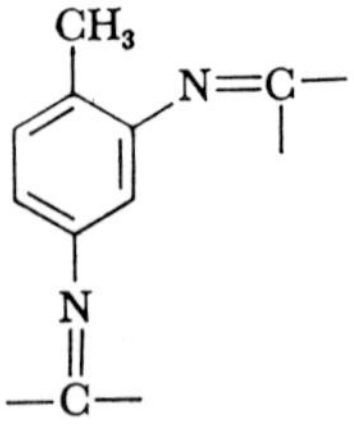

An additional source of yellowing of aromatic isocyanate based urethanes is reported to be the active methylene groups in urethane polymers derived from 4,4-diphenylmethane diisocyanate.[8]

Color development can be minimized by the use of *N*-substituted poly-urethanes, since the labile hydrogen of the carbamate group favors dissociation.[6]

$$R\text{—}NH\text{—}C\overset{\displaystyle O}{\underset{OR'}{\big\langle}} \;\overset{\text{Heat}}{\rightleftharpoons}\; R\text{—}N\text{=}C\text{=}O + R'OH$$

The treatment of urethanes with monoisocyanates, ketones, ethylene oxide or acetic anhydride reduces yellowing by limiting the number of amino or potential amino groups at chain ends.[9] The use of aliphatic or other isocyanates in which the —NCO group is not adjacent to an aromatic ring also reduces color development. Typical aliphatic isocyanates which minimize yellowing are hexamethylene diisocyanate and 4-4'-methylene bis(cyclohexyl isocyanate), commonly referred to as $H_{12}MDI$.[10] Stabilizers, such as antioxidants and ultraviolet screening agents, assist in the prevention of color development.[5,6] An evaluation of the methods used for accelerated testing of urethane coatings for yellowing has been reported[11] and a simple equation developed to express the yellowing of coatings.

REFERENCES

1. Gruber, H., VI, *F.A.T.I.P.E.C. Congr.*, **1962**, 53.
2. Freeborn, A. S., *J. Oil Colour Chemists' Assoc.*, **48**, 539 (1965).
3. O'Connor, F. M., and D. J. Waythomas, *Offic. Dig. Federation Soc. Paint Technol.*, **34**, 162 (1962).
4. Solomon, D. H., *Rev. Pure Appl. Chem.*, **13**, 171 (1963).
5. Beachell, H. C., and C. P. Ngoc Son, *J. Appl. Polymer Sci.*, **7**, 2217 (1963).
6. Beachell, H. C., and C. P. Ngoc Son, *J. Appl. Polymer Sci.*, **8**, 1089 (1964).
7. Mennicken, G., *J. Paint Technol.*, **43** (552), 83 (1971).
8. Shollenberger, C. S., and K. Dinsberg, *SPE Trans.*, **1** (1), 31 (1961).
   Shollenberger, C. S., H. Scott, and G. R. Moore, *Rubber World*, **137**, 549 (1958).
9. Wilson, C. L., U. S. Pat. 2,921,866.
10. Kaplan, M., and G. Wooster, *J. Paint Technol.*, **41**, 551 (1969).
11. Hill, H. E., *J. Paint Technol.*, **41**, 321 (1969).

CHAPTER 9

AMINE- AND PHENOL-FORMALDEHYDE RESINS

Compounds that contain an active hydrogen atom react with aldehydes and ketones by adding across the carbon-oxygen double bond.

$$R\!-\!X\!-\!H + R^1\!-\!\overset{\overset{\displaystyle O}{\|}}{C}\!-\!R^2 \underset{\longleftarrow}{\overset{OH^-}{\longrightarrow}} R^1\!-\!\underset{\underset{\displaystyle X\!-\!R}{|}}{\overset{\overset{\displaystyle OH}{|}}{C}}\!-\!R^2$$

This type of reaction has a number of important uses in the preparation of chemicals and resinous intermediates for use in surface coatings. Formaldehyde is by far the most common carbonyl compound used. This choice results from the ready availability—at an attractive price—of formaldehyde solutions (formalin, formcels), or polymers (paraformaldehyde, trioxane), and from the unique chemical activity of formaldehyde when compared to higher aldehydes or the ketones. The compounds that supply the active hydrogen atom include aldehydes, ketones, aromatic hydrocarbons, amines, amides, and phenols.

The condensation of aldehydes with formaldehyde under basic conditions is used to produce some of the synthetic polyols that are used in polyesters, alkyd resins, etc. Acetaldehyde forms trimethylol acetaldehyde, which is reduced by excess formaldehyde (crossed Cannizzaro reaction) to give pentaerythritol.

$$CH_3\!-\!CHO + 3HCHO \longrightarrow HOCH_2\!-\!\underset{\underset{\displaystyle CH_2OH}{|}}{\overset{\overset{\displaystyle CH_2OH}{|}}{C}}\!-\!CHO$$

$$\downarrow H\!-\!CHO$$

$$HOCH_2\!-\!\underset{\underset{\displaystyle CH_2OH}{|}}{\overset{\overset{\displaystyle CH_2OH}{|}}{C}}\!-\!CH_2OH$$

Similarly, propionaldehyde yields trimethylol ethane, and butyraldehyde forms trimethylol propane.

The reaction between formaldehyde and ketones can be illustrated by considering the example of cyclohexanone. With excess formaldehyde, and under basic conditions, tetramethylol cyclohexanone is formed.[1] This product can be reduced by a crossed Cannizzaro reaction to the penta-hydroxyl compound, which is of potential interest as a polyol for plasticizer or polymer manufacture.

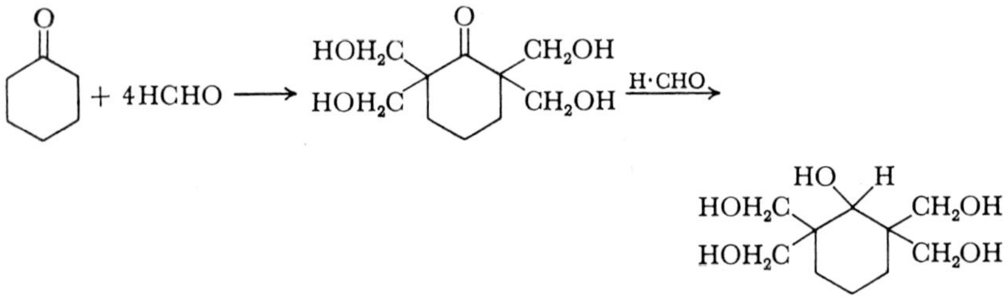

Other products are possible from the reaction of formaldehyde with cyclohexanone. When excess cyclohexanone and room temperature are used for the reaction, methylene cyclohexanone is formed[2] by the internal dehydration of the methylol compound.

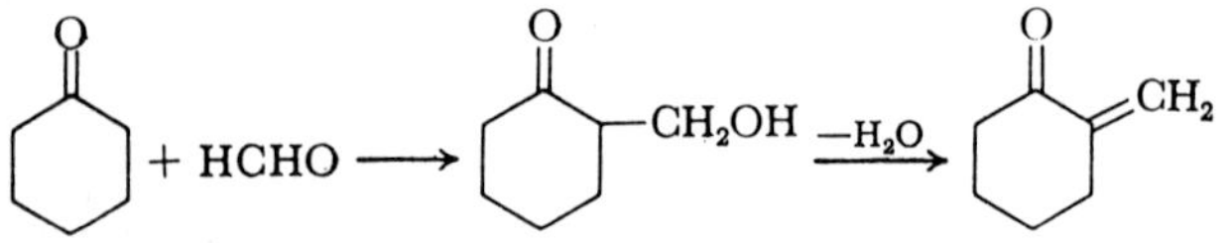

On the other hand, resinous materials are formed when 1.2 to 1.9 moles of formaldehyde per mole of cyclohexanone react with an alkaline catalyst[3] and, in this case, the polymer results from an intermolecular dehydration of the methylol group. These resins are usually water white, and are claimed to have good color stability. They are used in combination with cellulose esters (Gen. Ref. 1).

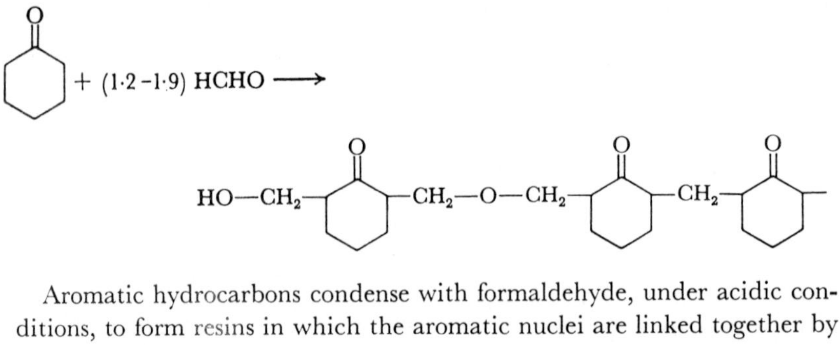

Aromatic hydrocarbons condense with formaldehyde, under acidic conditions, to form resins in which the aromatic nuclei are linked together by methylene bridges.

These reactions are favored by substituent groups that activate the *ortho* and *para* positions. The strong catalytic action of hydrogen chloride in promoting polymer formation is related to the formation of chlormethyl compounds (Gen. Ref. 1). These readily undergo intermolecular dehydrochlorination.

In the surface-coating industry, amines, amides, and phenols are the most important compounds for condensation and polymerization with formaldehyde.

These reactions are considered in more detail than those discussed above, but it should be realized that the same general principles apply to the manufacture of formaldehyde polymers with other compounds that are hydrogen atom donors.

A detailed review of the reactions of formaldehyde with compositions that contain an active hydrogen atom has been published (Gen. Ref. 2).

GENERAL REFERENCES

1. Walker, J. F., *Formaldehyde*, Reinhold Publishing Corp., New York, 1964.

2. Drumm, M. F., and J. Leblanc, *The Kinetics and Mechanisms of Step-Growth Polymerizations*, Ed. Solomon, D. H., Marcel Dekker Inc., New York 1972.

REFERENCES

1. Mannich, C., and W. Brose, *Chem. Ber.*, **56**, 833 (1923).
2. Dimroth, K., K. Resin, and H. Zetzsch, *Chem. Ber.*, **73**, 1399 (1940).
3. Hurst, D. A., and W. J. Miller, Allied Chem. & Dye Corp. U.S. Pat. 2,540,885–6 (1951).

AMINE- AND AMIDE-FORMALDEHYDE RESINS

Amines and amides form resinous products with formaldehyde by the polymerization of the methylol compounds, formed as the initial products of the reaction. The steps in polymer formation are each susceptible to catalysis and, as a general rule, the rate of the methylolation reaction increases with the pH. The polymerization step is favored by acid conditions and by heat. The specific conditions needed to prepare a polymer suitable

$$R—X—NH_2 + H—CHO$$
$$(X = C{=}O \text{ or } CH_2) \downarrow OH^-$$
$$R—X—NH—CH_2—OH$$
$$\downarrow H^+$$
$$\text{Polymer}$$

for use as a surface-coating resin will depend on the relative rates of the methylolation and polymerization reactions. This aspect is discussed further under the individual compounds.

Amide-Formaldehyde Resins. The amide structure, $R—CO—NH_2$, would suggest that two hydrogen atoms are available for addition across the carbonyl group of formaldehyde. However, under the conditions used in resin manufacture, only one hydrogen is usually available. Consequently, simple amides are essentially monofunctional in their reaction with formaldehyde although, often, more than one mole of formaldehyde for each mole of amide is used merely to increase the rate of reaction; they are, therefore, more often used as modifiers for other amines or amides. Compounds with more than one amide group are susceptible to polymer formation; adipamide and other aliphatic diamides are used in resin production. Alternatively, the multifunctional amide can be an acrylic or vinyl polymer containing acrylamide or methacrylamide residues. Because of the importance of this type of polymer, the reactions are considered in more detail, although the principles underlying the methylolation, polymerization, and etherification apply equally well to the other amides. The base polymer usually contains up to 15% of the amide (acrylamide, methacrylamide, etc.), the remaining monomer composition being dictated by the requirements that the coating has to satisfy. The reaction with formaldehyde can be carried out under acidic,[1] or basic,[2] conditions. However, basic conditions give a more controllable reaction product, since the competing polymerization of the methylol groups is minimized. It is preferable to use a volatile base, such as a tertiary amine, that can subsequently be removed by distillation. The resistance properties of the coating are then not degraded by the presence of a water-soluble base. Another important advantage of the basic method is apparent where acid groups are also present in the acrylic or vinyl backbone; it is probable that the solution will be buffered, as it will contain a weak base and the salt of the weak base. As a result, the control of pH should be more easily accomplished than in non-acid systems, where it is common for pH drifts of 2 to 3 units to occur. Such drifts lead to poor control of the polymer; the degree of methylolation and polymerization is likely to vary. In a number of cases, the methylolated acrylamide copolymers have shown poor compatibility with other film formers,[3] or when self-cured, as shown by the low gloss of the composition. The etherification of the methylol groups by an alcohol, usually butanol, reduces this problem. The butylation of methylol groups is catalyzed by acid, present either as part of the copolymer or subsequently added. The degree of etherification can be controlled by the time and temperature of the reaction, the pH, and the amount of butanol present. Low temperature, a low pH, and a large excess of butanol, all favor etherification over polymerization.

Urea-formaldehyde resins are of major importance to the surface-coating industry. Although they do not form satisfactory films by themselves, they are widely used as cross-linking agents for resins that contain hydroxyl, carboxyl, or amide groups (see Chapters 3 and 10). Urea may be regarded as possessing both amide and amine functions or, alternatively, as an amine-imide.

Urea forms a monomethylol (**1**) and a dimethylol (**2**) derivative; the trimethylol derivative has been claimed to be present in strong aqueous

$$\underbrace{NH_2\!-\!CO\!-\!NH_2}\qquad \underbrace{NH_2\!-\!\overset{\displaystyle OH}{\overset{|}{C}}\!=\!NH}$$

(amine) (amide) (amine) (imide)

solutions that contain a large excess of formaldehyde,[4,5] but negligible amounts of tetramethylol urea are formed. These observations are in conformity with urea having a functionality of approximately 3 in its reactions with formaldehyde.

$$\begin{array}{ll}
NH\!-\!CH_2OH & NH\!-\!CH_2OH \\
| & | \\
CO & CO \\
| & | \\
NH_2 & NH\!-\!CH_2OH \\
\qquad(1) & \qquad(2)
\end{array}$$

The methylolation of urea proceeds most rapidly under alkaline conditions. At low pH's, the concurrent polymerization of the methylol ureas is very much faster than methylolation; consequently, polymeric material, which is usually insoluble in the reaction mixture or common solvents, is formed. This precipitate has a low formaldehyde/urea ratio.

$$\begin{array}{c}
\begin{array}{ccc}
NH_2 & & NH\!-\!CH_2OH \\
| & & | \\
CO & +\ HCHO\ \longrightarrow & CO \\
| & & | \\
NH_2 & & NH_2
\end{array} \\[2em]
\downarrow \\[1em]
\begin{array}{cccc}
-N\!-\!CH_2\!-\!N\!-\!CH_2\!-\!N\!-\!CH_2\!-\!N\!- \\
|\qquad\quad |\qquad\quad |\qquad\quad | \\
CO\qquad CO\qquad CO\qquad CO \\
|\qquad\quad |\qquad\quad |\qquad\quad | \\
NH_2\qquad NH_2\qquad NH\qquad NH_2 \\
|\\
CH_2\\
|
\end{array}
\end{array}$$

Therefore, it is essential to use a pH of > 7 so as to give a high degree of methylolation, before acidifying for the polymerization stage. The structure of a urea-formaldehyde resin is still uncertain, but Marvel et. al.[6] have suggested that cyclic structures are formed as a result of trimerization of monomeric methylene urea or methylol methylene urea.

Urea resins usually have a formaldehyde/urea ration of 2.0 (approximately)/1—that is, $X = CH_2OH$ in the formula shown on the next page; although in achieving a combined formaldehyde content equivalent to two moles, it is necessary to use up to 2.4 moles of formaldehyde in the resin formulation.

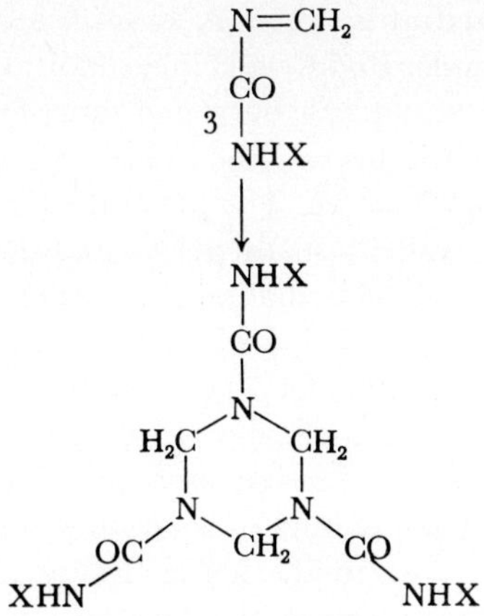

$$\begin{array}{c}
\mathrm{N}\!=\!\mathrm{CH_2} \\
| \\
\mathrm{CO} \\
3 \quad | \\
\mathrm{NHX}
\end{array}$$

$$\downarrow$$

$$\begin{array}{c}
\mathrm{NHX} \\
| \\
\mathrm{CO} \\
|
\end{array}$$

$$(\mathrm{X} = \mathrm{H} \text{ or } \mathrm{CH_2OH})$$

The resins prepared from urea and formaldehyde are water soluble at low molecular weights, and alcohol soluble at slightly higher degrees of condensation; they are used as adhesives, laminating resins, and textile modifiers. The action of heat and/or acid catalysts converts them into insoluble products. Resins of this type are not suitable for use in surface-coating formulations for several reasons: they are insoluble in the common solvents; they tend to react with themselves rather than intercondense with other polymer systems; and they are comparatively unstable.

Modification of the urea/formaldehyde resin with an alcohol overcomes the above problems, by converting some of the methylol groups to methylol ethers. The alcohol that is used to modify the resin must fulfill the following conditions: It must be a solvent for the methylol ureas; it should have a reasonably long alkyl chain to enhance hydrocarbon solubility; the alkyl ether must be sufficiently reactive to condense with other polymer systems; for ease of production, the alcohol should form an azeotrope with water which readily separates into two layers; and the alcohol should react readily to form the methylol ether. These conditions are satisfied by *n*-butanol, which is the alcohol most widely used in preparing resins for use in surface coatings.

The formaldehyde source chosen will depend on the economic balance between the price of the formaldehyde and the yield from the resin kettle. Paraformaldehyde is usually more expensive than formalin (37% aqueous solution), but it gives higher kettle yields. Butyl "Formcel"* (solution of formaldehyde in butanol) is another possible formaldehyde source. When

* Trade name, Celanese Corp. of America.

paraformaldehyde is used, it is necessary to form a concentrated aqueous solution by dissolving under slightly alkaline conditions.

The preparation of a typical butylated urea-formaldehyde resin is discussed in more detail to illustrate the influence of processing conditions on the chemical reactions involved. The pH of the formaldehyde solution is adjusted to the specified value (usually pH 8), and the urea is added to the solution in the resin kettle. The formation of dimethylol urea will take place at room temperature over a period of 24 hours (approximately), or by heating the mixture under reflux for 1 hour. Butanol is then added and the mixture acidified. The mixture is heated under reflux at a controlled temperature cycle; water is removed at a specified rate. The water is removed by condensing the distillate, separating the aqueous layer, and returning the organic phase to the kettle. Many processes use vacuum stripping to aid the final removal of water at a low temperature. The addition of aromatic hydrocarbons (xylene, toluene) with the butanol lowers the temperature of water removal, since the ternary azeotrope boils at a lower temperature than the butanol-water binary mixture. The amount of butylation will be governed by the reaction conditions; low pH values, a large excess of butanol, and low temperatures all favor increased butylation. These conditions minimize the competing self-polymerization of the methylol groups, as shown in Eqs. 9.1 and 9.2.

$$2 \,{-}NH{-}CH_2OH + 2 \, C_4H_9OH \longrightarrow 2 \,{-}NH{-}CH_2{-}O{-}C_4H_9 + 2 \, H_2O \quad (9.1)$$

$$2 \,{-}NH{-}CH_2OH \longrightarrow {-}NH{-}CH_2{-}O{-}CH_2{-}NH{-} + H_2O \quad (9.2)$$

The most difficult reaction condition to control is the pH, which usually drifts over the course of the reaction. Unless the pH drift is the same for each batch, variation in resin properties will occur. The pH change can be related to a number of factors.

1. Formaldehyde undergoes the Cannizzaro reaction to produce formic acid.

$$2 \, HCHO \longrightarrow HCOOH + CH_3OH$$

2. The presence of trace amounts of impurities that hydrolyze to acid components is possible, particularly in some sources of formalin, for example, methyl formate present as 1 part per 1000 can alter pH drift by hydrolysis to formic acid.

$$HCOOCH_3 \xrightarrow{+H_2O} HCOOH + CH_3OH$$

3. Amine stabilizers and iron salts influence the reaction, and pH changes.

Variation in the processing conditions of a urea/formaldehyde resin shows up in a number of ways. The water of reaction will increase or decrease from that normally expected; when excessive butylation occurs at the expense of polymerization, more water is formed, since self-condensation of the methylol groups gives $\frac{1}{2}$ mole, and etherification 1 mole of water per methylol group. The tolerance of the final resin for mineral spirits or petroleum ether will increase with the degree of butylation, since the molecular weight and hydrophilicity of the resins are altered. The solids content of the resin solution will be higher than specified when excess butylation occurs, because of the increase in weight due to combined butanol. Lastly, the viscosity of the resin will decrease with increasing butylation as a result of the butyl ether acting as a partial chain stopper and giving a lower molecular weight resin. Low pH's also favor the formation of butyl formal, which is readily detected by its characteristic sweet odor.

$$\text{HCHO} + 2\,\text{C}_4\text{H}_9\text{OH} \longrightarrow \text{CH}_2\!\!\begin{array}{l}\nearrow \text{OC}_4\text{H}_9 \\ \searrow \text{OC}_4\text{H}_9\end{array} + \text{H}_2\text{O}$$

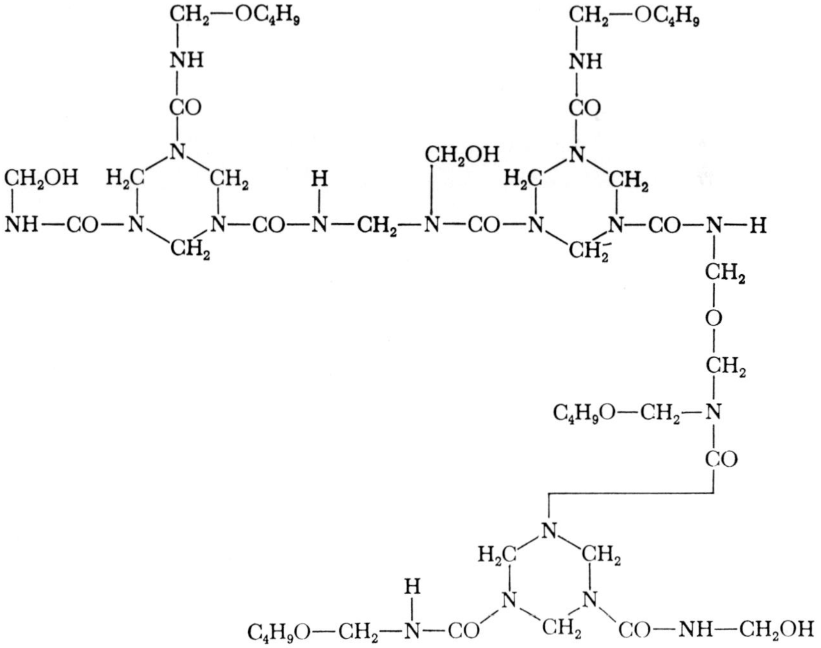

It should be noted that measurement of the solids content on this type of resin solution does not give an absolute figure. The value depends on the conditions used for the analysis, since the resin usually intercondenses and eliminates butanol, formaldehyde, and water during the drying. However, under standardized conditions, reproducible results can be obtained, and the method is suitable as a production control. The degree of butylation can be measured by infrared spectroscopy, or by titration with a standardized petroleum solvent. The structure of a typical resin, based on Marvel's theory of a cyclic trimer, is shown opposite.

Urea resins used in surface coatings are often described as either slow or fast curing, the slow curing being the more highly etherified resins. The curing rate refers to the hardness developed by a film of the resin in combination with another film former (often an alkyd resin), and this parameter should not be confused with the amount of chemical combination between the polymer entities. It is most likely that the faster curing resin gives less co-reaction and more self-polymerization; this is often reflected in the long-term film properties, particularly outdoor weathering, where very fast curing resins give inferior properties. The fast curing is a reflection of the difficulty of etherifying a methylol group with the hydroxyl group of a large polymer molecule (such as an alkyd resin), and the film hardness is developed as a result of rapid self-polymerization of the urea resin.

$$
2\ \boxed{\text{Polymer}} \longrightarrow \boxed{\text{Polymer}}\ +\ HCHO\ +\ H_2O
$$

$$
\begin{array}{ccc}
 & \text{CO} & \quad\ \text{CO} \\
 & | & \quad\ | \\
 & \text{NH} & \quad\ \text{NH} \\
 & | & \quad\ | \\
 & \text{CH}_2\text{OH} & \quad\ \text{CH}_2 \\
 & & \quad\ | \\
 & & \quad\ \text{NH} \\
 & & \quad\ | \\
 & & \quad\ \text{CO} \\
 & & \quad\ | \\
 & & \boxed{\text{Polymer}}
\end{array}
$$

On the other hand, the butyl ether does not polymerize readily but does undergo ether interchange. This is a slower reaction than the self-polymerization of the methylol groups and can lead to co-condensation of the urea resin with the other polymer.

Alcohols other than butanol are used where modified resins, with specific properties, are required. In general, the secondary and tertiary alcohols form ethers more slowly than the normal alcohols; the longer the carbon chain length of the alcohol, the slower is ether formation. Since the higher alcohols are not solvents for the methylol ureas, attempts to prepare resins

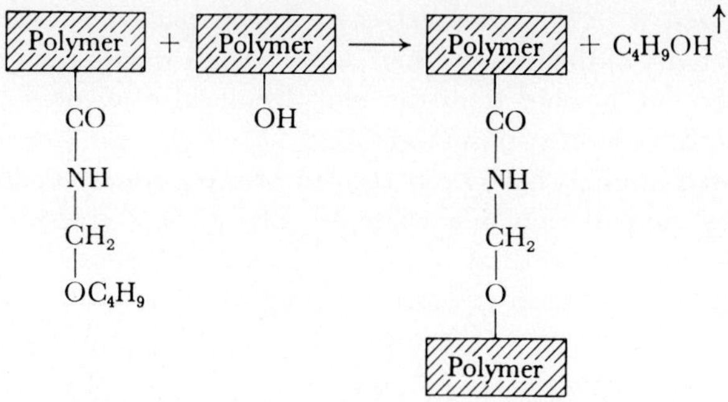

$$\boxed{Polymer} + \boxed{Polymer} \longrightarrow \boxed{Polymer} + C_4H_9OH\uparrow$$

$$
\begin{array}{ccc}
| & | & | \\
CO & OH & CO \\
| & & | \\
NH & & NH \\
| & & | \\
CH_2 & & CH_2 \\
| & & | \\
OC_4H_9 & & O \\
& & | \\
& & \boxed{Polymer}
\end{array}
$$

by the normal process described above lead to a highly polymerized, but only slightly etherified, product. However, the higher alcohols can be introduced into the resin by ether interchange with the butyl resin.

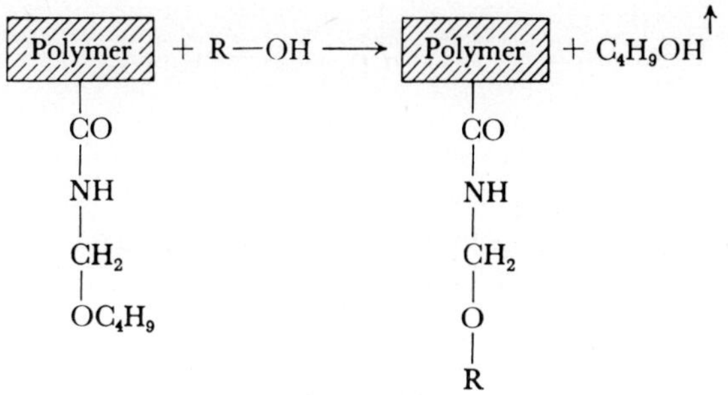

$$\boxed{Polymer} + R\!-\!OH \longrightarrow \boxed{Polymer} + C_4H_9OH\uparrow$$

$$
\begin{array}{cc}
| & | \\
CO & CO \\
| & | \\
NH & NH \\
| & | \\
CH_2 & CH_2 \\
| & | \\
OC_4H_9 & O \\
& | \\
& R
\end{array}
$$

$$(R = > C_4)$$

The hydrocarbon tolerance of the resin increases with the chain length of the alcohol used for etherification; but the curing rate and the ease of

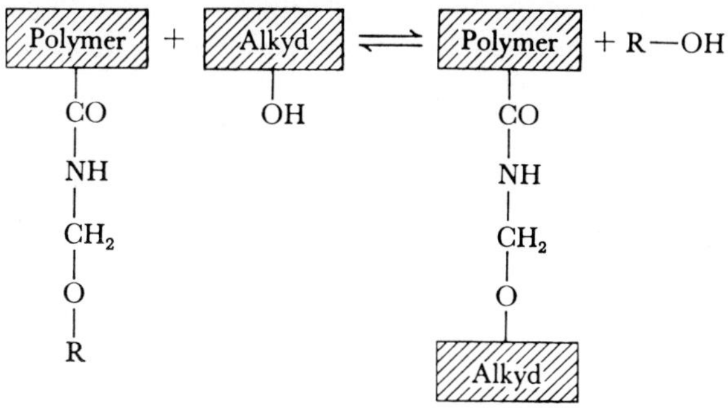

$$\boxed{Polymer} + \boxed{Alkyd} \rightleftharpoons \boxed{Polymer} + R\!-\!OH$$

$$
\begin{array}{ccc}
| & | & | \\
CO & OH & CO \\
| & & | \\
NH & & NH \\
| & & | \\
CH_2 & & CH_2 \\
| & & | \\
O & & O \\
| & & | \\
R & & \boxed{Alkyd}
\end{array}
$$

$$(R = > C_4)$$

co-condensation with an alkyd decrease. This is partly as a result of the decreased activity of the higher ethers, and is partly due to the difficulty of removing the free alcohol from the film. Removal of alcohol drives the reaction, shown opposite, to the right.

Unsaturated alcohols have been used to prepare ethers capable of film formation by autoxidation. For example, allyl alcohol gives an allylated urea-formaldehyde resin[8] that can cure by oxidative attack at the active methylene of the allyl ether group.

$$
\boxed{\text{Polymer}} \qquad\qquad + \ CH_2{=}CH{-}CH_2{-}OH
$$
$$
\overset{|}{CO}{-}NH{-}CH_2{-}OR
$$
$$
\downarrow
$$
$$
\boxed{\text{Polymer}} \qquad\qquad \text{point of oxidative attack}
$$
$$
\overset{|}{CO}{-}NH{-}CH_2{-}O{-}\overset{\downarrow}{CH_2}{-}CH{=}CH_2 \ + \ ROH\uparrow
$$

Ether interchange is an important reaction where resins are prepared from formalin, since many commercial formalins contain up to 10% methanol as a stabilizer. It is likely that in resin manufacture, the methyl ethers are formed first, and these interchange with butanol as shown below.

$$
\boxed{\text{Polymer}} \ + \ C_4H_9OH \longrightarrow \boxed{\text{Polymer}} \qquad\qquad + \ CH_3OH\uparrow
$$
$$
\overset{|}{CO}{-}NH{-}CH_2{-}OCH_3 \qquad\qquad\qquad \overset{|}{CO}{-}NH{-}CH_2{-}OC_4H_9
$$

This situation possibly arises because of the tendency of formaldehyde to exist as the hemiformal in solution and, in fact, a decreased rate of reaction between urea and formaldehyde in the presence of methanol has been attributed to hemiformal formation.[7]

$$
HCHO \ + \ CH_3OH \ \rightleftharpoons \ CH_3{-}O{-}CH_2OH
$$
$$
\Big\downarrow \begin{array}{l} NH_2 \\ | \\ CO \\ | \\ NH_2 \end{array}
$$
$$
NH{-}CH_2{-}O{-}CH_3
$$
$$
\overset{|}{CO} \qquad\qquad + \ H_2O
$$
$$
\overset{|}{NH_2}
$$

The suggestion that methanol plays a part in the chemical reactions leading to the butylated resins, explains why it is often difficult to match exactly a resin produced from formalin with one prepared from other formaldehyde sources. Some paraformaldehyde formulations include small amounts of methanol to assist in ether formation, and this makes the process more akin chemically to the use of formalin.

REFERENCES

1. Christenson, R. M., and D. P. Hart, *Offic. Dig. Federation Soc. Paint Technol.*, **33**, 684 (1961).
2. Kelly, D. P., G. J. H. Melrose, and D. H. Solomon, *J. Appl. Polymer Sci.*, **7**, 1991 (1963).
3. Melrose, G. J. H., and D. H. Solomon, *J. Appl. Polymer Sci.*, **8**, 1777 (1964).
4. De Jong, J. I., and J. De Jonge, *Rec. Trav. Chim.*, **71**, 643 (1952); **72**, 88, 139, 202, 207, 213, 1027 (1953).
5. Ito, Y., *J. Chem. Soc., Japan, Ind. Chem. Sect.*, **64**, 382 (1961).
6. Marvel, C. S., J. R. Elliott, F. E. Boettner, and H. Yuska, *J. Am. Chem. Soc.*, **68**, 1681 (1946).
7. Walker, J. F., *Formaldehyde*, Reinhold Publishing Co., New York, 1964, p. 378.
8. Ada, R., and H. Senda, *Chem. High Polymers* (*Japan*), **8**, 154 (1951); via *Chem. Abstr.*, **47**, 342 (1953).

Amine-Formaldehyde Resins. The reaction of formaldehyde with the simple aliphatic diamines is not of great importance for resin manufacture. However, in a number of coating formulations that contain formaldehyde resins, volatile amines are used as stabilizers. In these systems, the ability of the amine to react with formaldehyde governs the choice of stabilizer. Primary and secondary amines can cause a "depolymerization" of formaldehyde resins where free methylol groups are present; this results from the initial dissociation of the methylol compounds, as shown in the equations below.

$$\boxed{\text{Polymer}}-NH-CH_2-OH \rightleftharpoons$$

$$\boxed{\text{Polymer}}-NH_2 + H-CHO \xrightarrow{+ R-NH_2} R-NH-CH_2-OH$$

Consequently, tertiary amines are usually preferred as stabilizers in systems of this type.

Aromatic amines react at the amine group, but under acidic conditions give resinous products in which substitution in the aromatic nucleus has also occurred.[1] Attack of the aromatic ring is analogous to the formation of phenol-formaldehyde resins, and it enables polymers to be formed from tertiary aromatic amines. In general, these polymers have only limited use in coating compositions.

Melamine, which is a cyclic trimer of cyanamide, is widely used to produce resins with formaldehyde. Polymers of this type are used in high performance automotive and industrial enamels. The melamine resins have better heat stability, durability, hardness, and drying speed than the urea/formaldehyde resins.[2] The chemistry of melamine-formaldehyde resin formation and reaction is basically similar to that described for the urea resins, but with a number of significant differences. Melamine has six replaceable hydrogens and is hexafunctional in its reactions with formaldehyde. This results in a much faster rate of cross-linking than is possible with urea, where the functionality is approximately 3. Another important difference between melamine and urea is that with the melamine the methylolation and polymerization reactions occur at comparable rates under acid conditions. Consequently, melamine-formaldehyde resins can be prepared by an all-acid process or, if preferred, by an alkaline methylolation step, followed by an acid polymerization and butylation stage. The ratio of formaldehyde to melamine influences the polymerization rate; at the high ratios (6/1) used in most resins, polymerization is much slower than at lower formaldehyde contents (4/1).[3] The properties of melamine-formaldehyde resins depend on the same factors as the urea resins—namely, the ratio of formaldehyde/melamine and the degree of polymerization and of butylation. Rapid removal of the water of reaction results in a low molecular weight highly butylated resin, whereas slow removal yields a highly condensed polymer with a low degree of butylation.[4] As a general guide to the types of resin in use, a typical resin would have a formaldehyde/melamine ratio of 6/1, a molecular weight of 800 to 1500, and approximately one half of the methylol groups etherified (fast cure resin). The melamine molecules would be linked by either methylene, methyleneoxy, or dimethyleneoxy groups. A detailed study of the effect of the processing conditions used to prepare a melamine formaldehyde resin and the performance of the resin in paint compositions is reported in reference 17. The melamine resin undergoes cross-linking reactions with other polymers that contain hydroxyl, carboxyl, or amide groups (see Chapters 3 and 10). It should be clearly understood that the use of external acid catalysts to speed up the cure results in self-polymerization of the methylol groups and poorer film properties.[5] More highly etherified melamine resins can be prepared by the use of lower temperatures, an increased butanol charge, and a lower pH; these are the slower curing resins used in some appliance enamels.

Modification of melamine condensates with higher alcohols is possible. These higher alcohols alter the properties in the same manner as noted with urea resins.

Typical Structure of a Butylated Melamine-Formaldehyde Resin

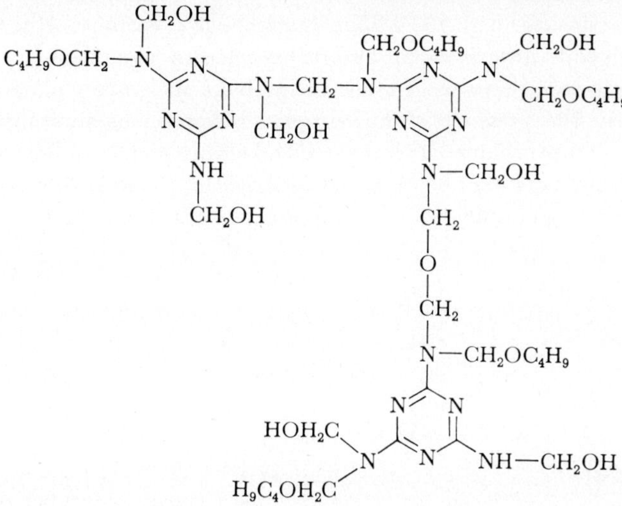

Some attempts have been made to form esters of the methylol melamines and to incorporate the melamine resin into a polyester structure. Fatty acid esters of the methylol melamines have been prepared by direct esterification,[6] but some self-polymerization of the methylol groups takes place.

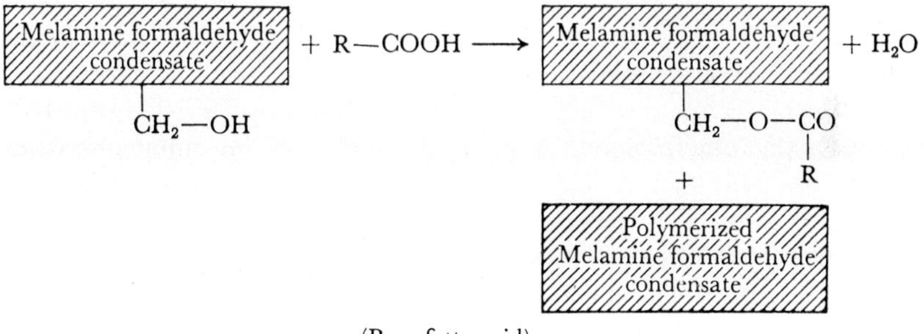

(R = fatty acid)

A more desirable process is interchange between an alkyl ether and the fatty acid or acid-containing polymer.[7]

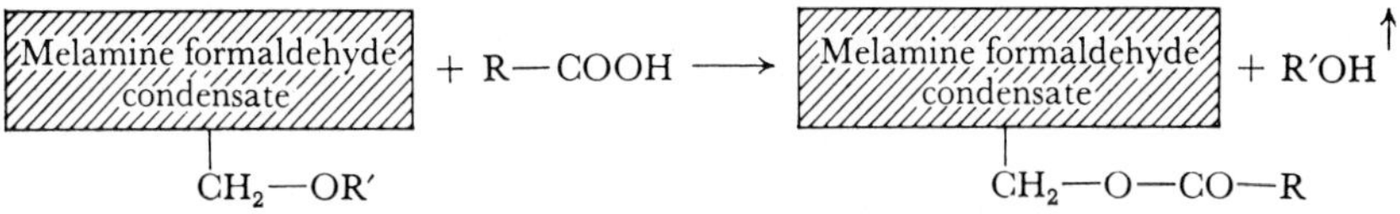

Co-condensates of melamine resins with alkyds can be prepared by this method,[8] and these co-condensates offer potential advantages over a mixture of the two resins (see page 103). Some interchange of the butyl ether groups with the alkyd hydroxyls would also be expected to take place.

The unsaturated fatty acid esters of methylol melamines undergo film formation in the presence of conventional oxidation catalysts or driers. Similar results are obtained if unsaturated ethers are prepared. For example, 2-butene-1-ol ethers may be prepared as shown in the following equation, and these undergo autoxidation in the presence of cobalt driers.[9]

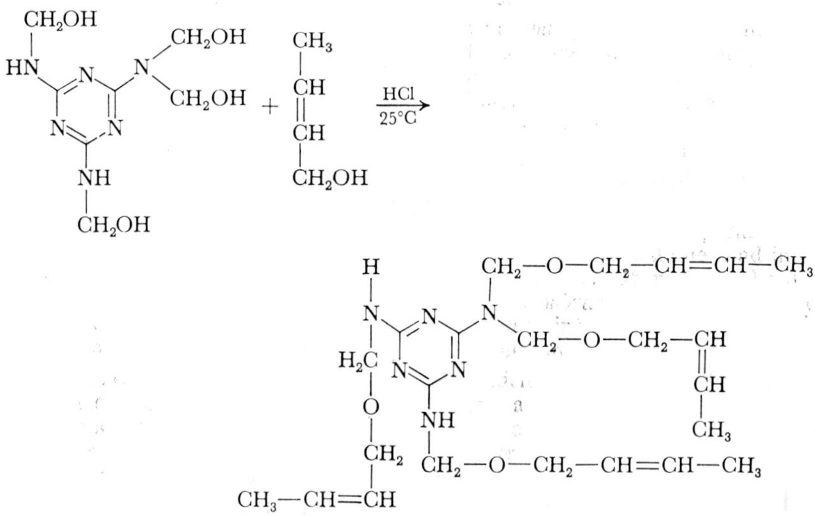

$$
\begin{array}{c}
\text{melamine-methylol} \; + \; \underset{\substack{| \\ \text{CH}_2\text{OH}}}{\overset{\substack{\text{CH}_3 \\ | \\ \text{CH} \\ \| \\ \text{CH} \\ |}}{}} \;\; \xrightarrow[25°\text{C}]{\text{HCl}}
\end{array}
$$

Similarly, allyl ethers have been prepared[10] and used in conjunction with unsaturated polyesters. Other unsaturated methylol melamines have been prepared by ether interchange of unsaturated esters that carry a hydroxyl substituent, for example, hydroxyethyl methacrylate as shown overleaf.[11]

Esters of methylolmelamines with maleic anhydride are another means of converting the melamine derivatives into entities which can copolymerize with vinyl or acrylic monomers. A typical reaction sequence is shown on next page:[14]

Melamine-formaldehyde resins are becoming increasingly important as the cross-linking agents for water dilutable surface-coating compositions. The early water-soluble melamine resins were developed primarily for use in the textile and adhesive industries. They were prepared by dissolving melamine in formalin, under alkaline conditions, with a melamine/formaldehyde ratio of approximately 4/1, then heating the mixture to effect a small degree of polymerization. The resin structure was derived from a

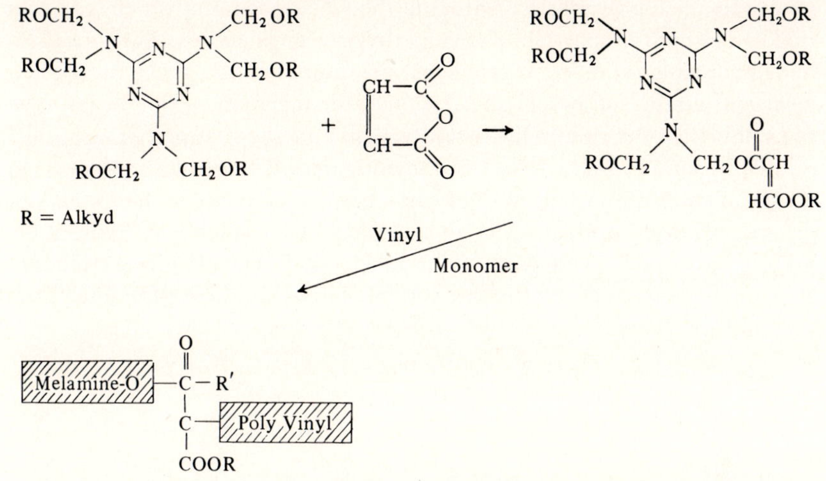

R = Alkyd

Vinyl Monomer

R' = Initiator residue

tetramethylol melamine, with some etherification (approximately 1 in 8 methylol groups) from the methanol present in the formalin.

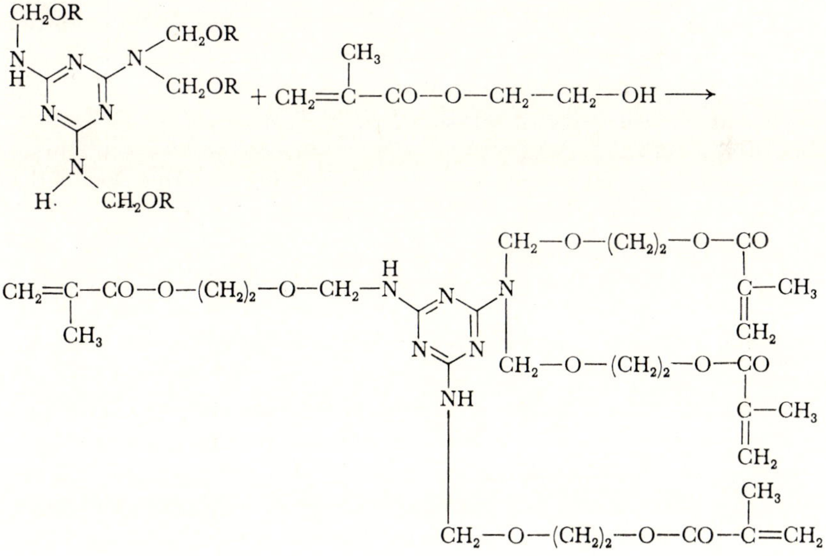

These resins gave very fast curing rates when used in conjunction with other water-dilutable polymers as the basis for surface-coating type formulations. However, the highly reactive nature of the melamine resin, which is a function of the high proportion of free methylol groups and the low formaldehyde/melamine ratio, led to problems of storage stability; precipitates often formed as a result of self-polymerization of the melamine resin. To improve the stability of the resin solutions, alternative methods of preparing water-soluble melamine resins have been investigated. The general approach to the problem of improving stability has been to convert the free methylol groups to ethers, and in order to maintain water solubility, hydrophilic groups have often been part of the alkyl ether. Polyols give melamine/formaldehyde monoethers that are water soluble, but these cure slowly; the

Typical Structure of a Water-Soluble Melamine Resin as Used for Textiles or Adhesives

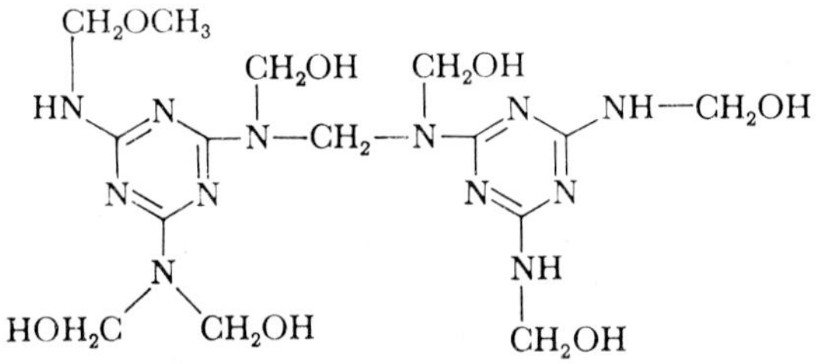

$$CH_2OCH_3$$
$$CH_2OH \quad CH_2OH$$

free hydroxyl groups of the polyols are not as reactive as the methylol groups. Also, reaction by ether interchange is slow, since the liberated glycol is not very volatile. A typical resin structure, where ethylene glycol has been used as the polyol, is shown below.

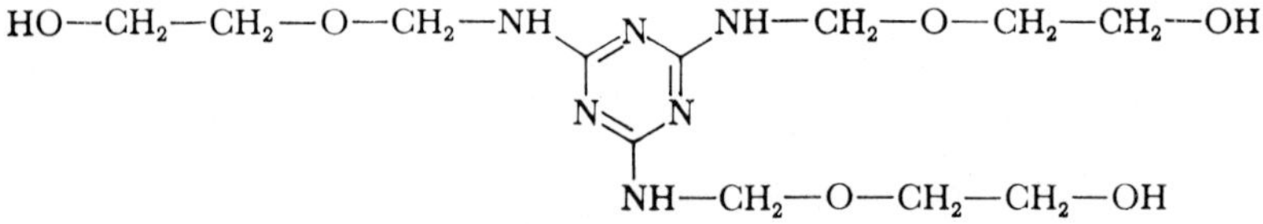

$$HO-CH_2-CH_2-O-CH_2-NH \quad \quad NH-CH_2-O-CH_2-CH_2-OH$$

$$NH-CH_2-O-CH_2-CH_2-OH$$

Similarly, ether-alcohols have been used to give stable water soluble melamine resins;[12] however, these resins are also relatively unreactive. Melamine resins that are water soluble, stable in solution, and yet highly reactive in film-forming reactions, have been successfully prepared by making three important changes in the traditional textile/adhesive type of formulation. First, the ratio of formaldehyde to melamine has been increased up to 6/1, so that no active amino hydrogen atoms are left in the resin structure. Second, the hexamethylol melamine has been etherified with methanol, which—because of its short alkyl chain—still permits the resin to be water soluble (in the presence of small amounts of an alcohol). Third, the molecular weight of the resin has been controlled to as low as possible by choosing reaction conditions that favor etherification over polymerization. Actually, one commercial product is hexamethoxymethyl melamine, a crystalline compound with the structure shown below.[13]

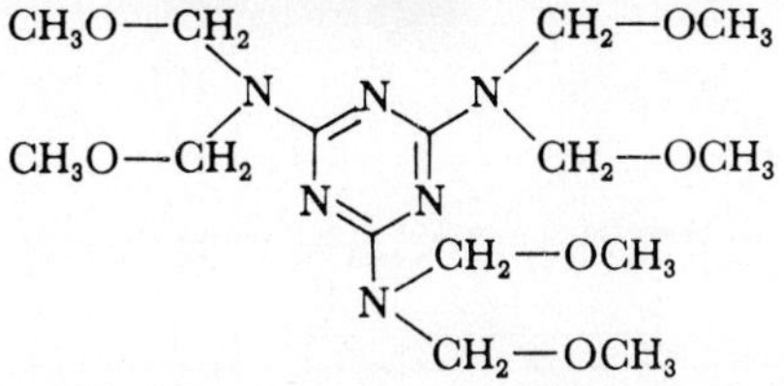

$$
\begin{array}{c}
CH_3O{-}CH_2 \qquad\qquad CH_2{-}OCH_3 \\
\diagdown N \quad N \quad N \diagup \\
CH_3O{-}CH_2 \qquad\qquad CH_2{-}OCH_3 \\
N \diagdown \diagup N \\
CH_2{-}OCH_3 \\
N \\
CH_2{-}OCH_3
\end{array}
$$

The synthesis of monomeric etherified hexamethylol melamine and their characterization by Gel Permeation Chromatography and by Proton Magnetic Resonance Spectroscopy have been reported recently.[15] Other commercially available resins of this type are derived from hexamethoxymethyl melamine by partial polymerization, which converts the melamine derivative from a solid to a viscous liquid. The partially polymerized hexamethyoxy methyl melamine (HMM) is a better crosslinking agent for hydroxy acrylic polymers, than the monomeric (HMM) provided the methylol groups are still etherified.[16] This is a reflection of the increased functionality of the partially polymerized product. This is illustrated below. The use of partially polymerized HMM has been suggested as a method of reducing the amino resin content of a coating composition and hence reducing the cost.

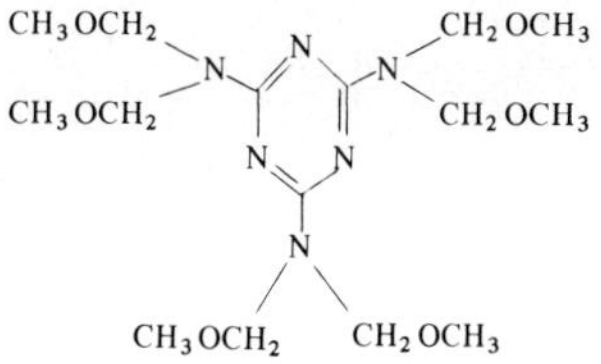

HMM
(Potential functionality 6)

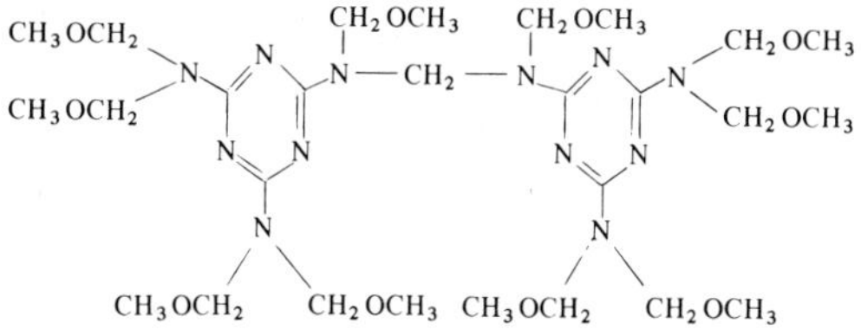

$$\left(\begin{array}{c}\text{Dimer of HMM}\\ \text{Potential functionality 10}\end{array}\right)$$

NB. The potential functionality is unlikely to be realized in practical cross-linking reactions. The figures are quoted to illustrate the increase in functionality which results from dimerization. The enhanced solubility of the methoxymethyl melamines, when compared to the methylol compounds, is possibly a result of the breaking of the strong intramolecular hydrogen bonding that exists in the free methylol compounds.

Formaldehyde condensation polymers with other triazines are also of interest for surface coatings. Cross-linked films of these polymers with alkyd resins have higher gloss and considerably increased resistance to alkali and soap solutions, when compared to the melamine resins. The exterior durability of the triazine resins is usually poorer than that obtained with melamine.[2] Typical triazines are shown below.

(substituted melamine)   (guanamine)

The methods of preparation and control of resin manufacture are similar to those used with melamine resins.

REFERENCES

1. Walker, J. F., *Formaldehyde*, Reinhold Publishing Co., New York, 1964, p. 370.
2. Rohm and Haas Co., Pamphlet on Uformite Resins, 1959.
3. Norris, W. C., and J. C. Bacon, *Offic. Dig. Federation Soc. Paint Technol.*, **20**, 785 (1948).
4. Grimshaw, F. P., *J. Oil Colour Chemists' Assoc.*, **40**, 1060 (1957).
5. Goppel, J. M., P. Bruin, and J. J. Zonsveld, VI *F.A.T.I.P.E.C. Congr.*, **1962**, 31.

6. Hodgins, T. S., A. G. Hovey, S. Hewett, W. R. Barrett, and C. J. Meeske, *Ind. Eng. Chem.*, **33**, 769 (1941). Protex Products Chimiques, Fr. Pat., 1,547,691. (1969)

7. Soc. pour l'ind. chim. a Bale., Brit. Pat. 611,012 (1948); via *Chem. Abstr.*, **43**, 4050 (1949); Swiss Pat. 231,424 (1944); via *Chem. Abstr.*, **43**, 2448 (1949).

8. Coveney, L. W., and S. L. M. Saunders, Pinchin, Johnson and Associates Ltd., Brit. Pat. 665,473 (1952).

9. Widmer, G., C I B A Ltd., U.S. Pat. 2,953,536 (1960).

10. Zuppinger, P. and G. Widmer, C I B A Ltd., U.S. Pat. 2,885,382 (1959).

11. Magrane, J. K., and R. E. Layman, American Cyanamid Co., U.S. Pat. 3,020,255 (1962).

12. Celanese Corp. of America, Brit. Pat. 823,246 (1959).

13. American Cyanamid Co., Booklet on Cymel 300.

14. American Cyanamid Co., Aust. Pat. Appl. 20,123/67.

15. Anderson, D. G., D. A. Netzel and D. J. Tessari, *J. Appl. Polymer Sci.*, **14**, 3021 (1970).

16. OHe, O., T. Tanaka, T. Sato and Y. Motoyama, *J. Appl. Polymer Sci.*, **12**, 213 (1968).

17. Evers, J. J. M., X F. A. T. I. P. E. C. Congr., **1970**, 135.

PHENOL-FORMALDEHYDE RESINS

Phenol undergoes ring substitution when reacted with formaldehyde as a result of activation of the *ortho* and *para* positions by the phenolic OH. Consequently, phenol has a potential functionality of three in its reactions with formaldehyde. Substituents in the phenol ring influence the rate of the methylolation step; in general, any substituent group that increases the electron density of the ring positions enhances the reaction; electron-withdrawing groups ($-NO_2$, $-Cl$, etc.) greatly reduce the likelihood of attack by formaldehyde (Gen. Refs. 1 to 4). The position of the substituent group relative to the phenolic group will have a bearing on the course of the reaction with formaldehyde; if the substituent is in the *ortho* or *para* position, it will reduce the functionality of the phenol, and this is a method that is widely used in commercial formulations for controlling the phenol/formaldehyde reaction. Under most conditions, the *m*-positions to the phenolic group do not become substituted by formaldehyde but, if substituent groups block all the available *ortho* and *para* positions, *meta*-substitution is possible. However, for appreciable *meta*-substitution to occur, the substitutent groups must be electron releasing, and the reaction must be conducted under acidic conditions. This ensures that the contributions of the mesomeric forms of the phenoxide ion to the electron density of the aromatic ring are at a minimum. The reaction involving phenols with blocked *ortho* and *para* positions is more akin to that of an aromatic hydrocarbon and formaldehyde, and should be regarded as such. For example, mesitol, 2:4:6-trimethylphenol, reacts with formaldehyde under acidic conditions,[1] as shown

(mesitol)

H—CHO

(3:3′-dihydroxy-2:2′, 4:4′, 6:6′-hexamethyl-diphenyl methane)

+

(3:5-bis(3-hydroxy 2:4:6 trimethylbenzyl) 2:4:6 trimethylphenol)

on this page, to give compounds in which two and three aromatic nuclei are linked together.

The ratio of formaldehyde/phenol and the reaction conditions have a pronounced effect on the relative rates of the methylolation and polymerization stages of the reaction. In general terms, alkaline conditions and high formaldehyde/phenol ratios favor methylolation; a low pH and formaldehyde/phenol ratios of less than one give a polymerized product. The influence of reaction conditions on the phenol-formaldehyde reaction is directly applied in the formulation of the two basic types of resins used in surface coatings. The resoles or thermosetting type are prepared by the condensation of excess formaldehyde with phenol under alkaline conditions; these resins are sometimes referred to as one-stage type, since they are

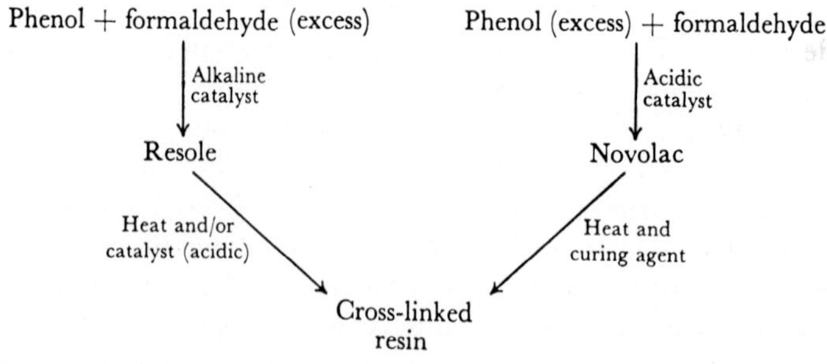

Figure 9.1. Types of phenolic resins and conditions used for their preparation.

capable of forming a cross-linked structure without any additional curing agent. The novolacs are thermoplastic and are prepared under acidic conditions with excess phenol; they require the addition of a curing agent and, therefore, are sometimes referred to as a two-stage type of phenolic resin. The differences between the two types are illustrated in Figure 9.1.

Resoles. The reaction of phenol with formaldehyde under alkaline conditions gives methylol phenols as the initial reaction products. The methylol groups will be located at either the *ortho* or *para* positions in the benzene ring; relative rate studies by Martin (Gen. Ref. 1) have established that the *para* position of phenol is slightly more reactive toward formaldehyde but, since there are two *ortho* positions available, the *ortho* methylol compound is formed at a faster overall rate.

Later in the reaction, some 2:4-dimethylol phenol forms; the concentration of the *ortho*, or 2-methylol phenol, decreases as a result of its greater reactivity in subsequent polymerization reactions. The introduction of one methylol group into the phenolic nucleus increases the likelihood of methylolation in the molecule, when compared to the parent phenol.[2] This observation is of great importance in attempting to predict a phenol-formaldehyde resin structure; it means that a mixture of methylol phenols is formed, some of which have more methylol groups than others. For example, Goldblum[3] has shown that trimethylol phenol is formed when much less than the required three moles of formaldehyde are reacted with phenol in aqueous alkaline solution (Table 9.1). Conversely, some methylol phenols with much

TABLE 9.1

EFFECT OF PHENOL/FORMALDEHYDE RATIO ON YIELD OF TRIMETHYLOL-
PHENOL[3]

Mole ratio phenol/formaldehyde	1/0.5	1/1	1/1.5	1/2	1/2.5
Percent trimethylolphenol	3.38	9.24	22.16	31.00	40.70

less than the original phenol/formaldehyde ratio must also be present; this is shown in an idealized manner by the following equation.

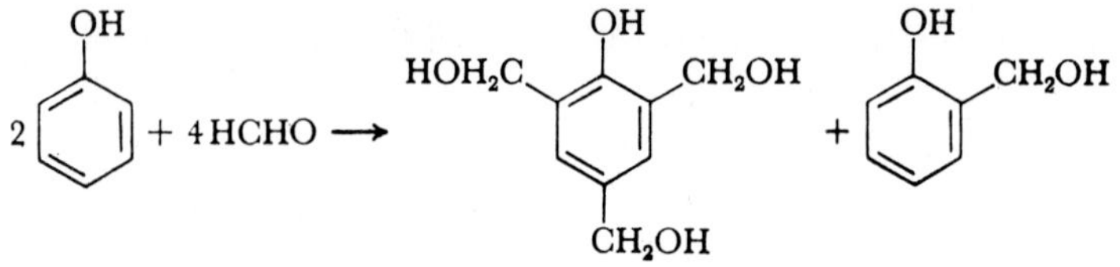

$$2\,C_6H_5OH + 4\,HCHO \longrightarrow \text{(trimethylol phenol)} + \text{(o-methylol phenol)}$$

The phenol alcohols or methylol phenols are relatively stable at low temperatures (less than 20°C), in a dilute reaction medium, and in the absence of strong acid catalysts (Gen. Ref. 1) but under the conditions

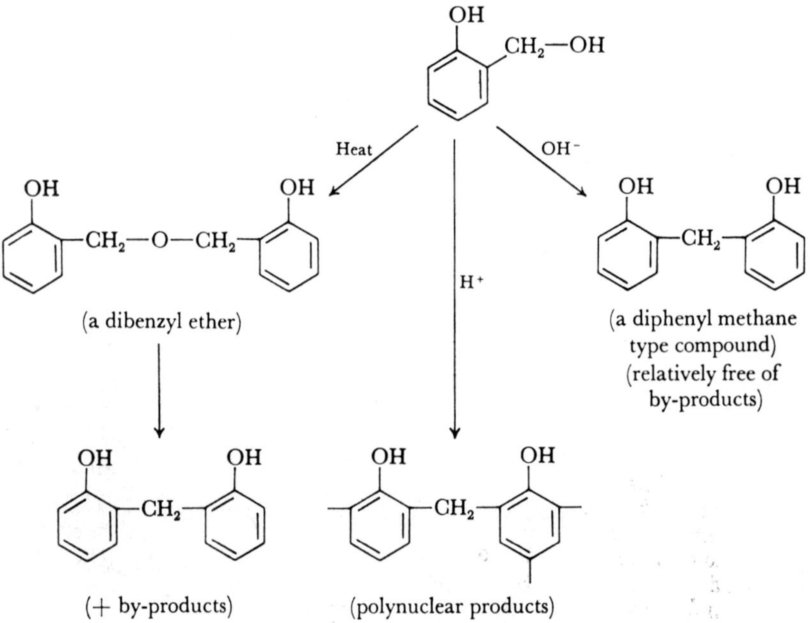

Figure 9.2.   Influence of reaction conditions on the polymerization of methylol phenols.

normally encountered in resin manufacture, polymerization takes place. A number of polymeric structures can develop, and the precise course of the polymerization varies with the reaction conditions as shown in Figure 9.2.

In the processing of resole-type phenolic resins, the polymerization or dehydration stage can be conducted under the original alkaline conditions, or after the neutralization of the methylol phenol mixture. The structure of the resole will be influenced by the conditions chosen, as shown in Figure 9.2. Therefore, a resole resin (represented as in Figure 9.3) is a complex

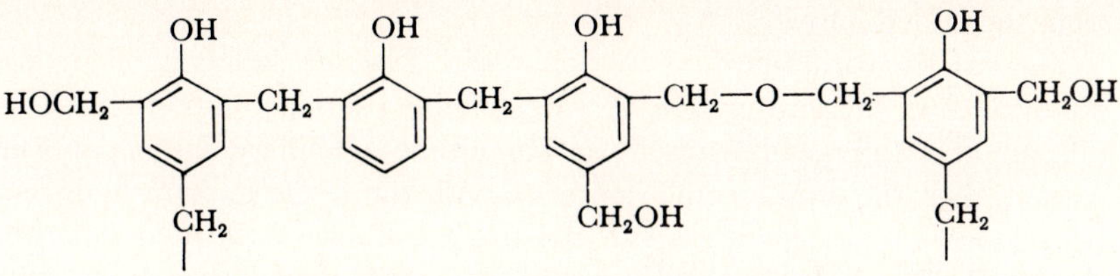

Figure 9.3. Probable structure of a resole-type phenolic resin.

mixture of mono- and polynuclear molecules joined via methylene or dimethyleneoxy linkages. Free methylol groups are present; these allow the resin to undergo self-curing by the action of acids and/or heat, and to react with oils and with unsaturated compounds (Page 77). Reaction with other polymer systems, which contain hydroxyl, amino, or carboxyl substituents, is essentially the same as for urea- and melamine-formaldehyde resins (see Eqs. 10.1 to 10.12). The probable course of the thermal hardening of a resole is through the quinone methide, as shown in the equation below.

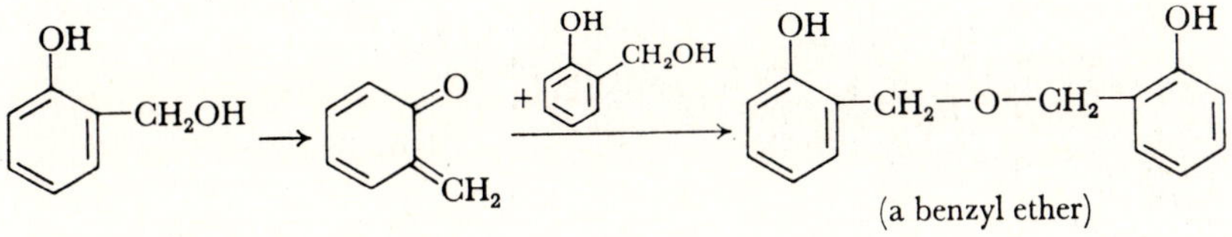

(a benzyl ether)

Resole resins prepared in the presence of amines, or ammonia, differ from those in which the stronger bases of the alkaline earth type are used. The amine apparently enters into the polymer structure and forms bis- and tris-(hydroxybenzyl) amine type derivatives.[4]

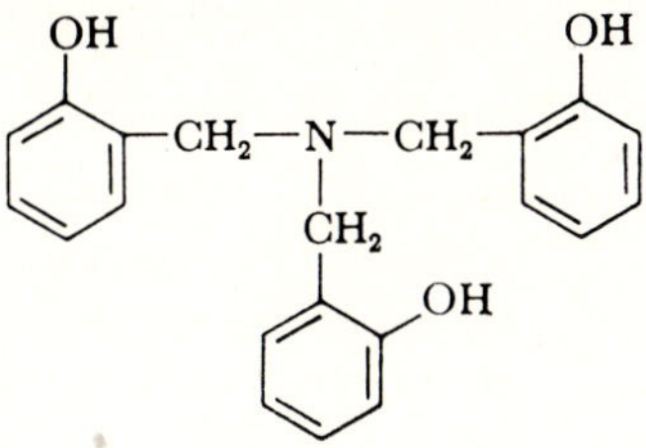

The amine has, therefore, increased the overall functionality of the system. These reactions are of commercial significance for two main reasons: amines are not generally used to prepare water-soluble resins, since the increased functionality results in a rapid loss of water solubility; and the curing of novolacs relies partly on the increased functionality which results from the added amine.

The reaction of amines or ammonia with phenolic resins is of further importance in the formulation of water-soluble coating compositions. If a mixture of a water-soluble resole, and an amine-solubilized acidic polymer is used, then the amine could condense with the resole to form hydroxybenzyl amine type structures. Also, the amine can reverse the methylolation step, as discussed previously for melamine-formaldehyde resins (page 101). These problems are avoided by the use of a tertiary amine to solubilize the acidic polymer.

Novolacs. When phenols are condensed with formaldehyde under acidic conditions, the polymerization of the methylol phenols is a much faster reaction than the initial methylolation. Consequently, where less than equimolar amounts of formaldehyde to phenol are used, a polymer structure is built up consisting of phenolic nuclei, linked together by methylene bridges, and with no free methylol groups.[5] This is a typical novolac structure, and is shown in Figure 9.4.

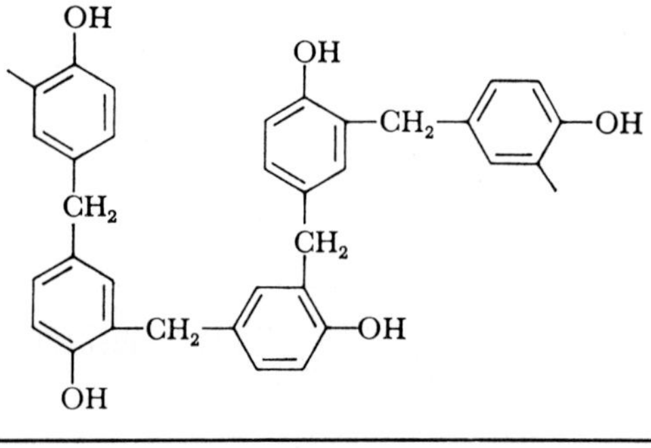

Figure 9.4.   Typical structure of a novolac phenolic resin.

The novolac requires the addition of formaldehyde to form a crosslinked polymer structure; by far the most common means of adding the formaldehyde is as hexamine (Gen. Ref. 1). The ammonia that is liberated when the hexamine dissociates also assists in the development of the three-dimensional resin structure by the formation of (hydroxylbenzyl) amine-type structures (page 257).

Recently, the use of cyclic formals and a strong acid as the curing agents for novolac resins has been reported.[6] This system is claimed to offer advantages over normal phenolics; it allows low or contact pressure techniques to be used in molding operations.[6] The mechanism of the curing reaction with a cyclic formal is not simply the release of formaldehyde, and then the condensation of the phenolic resin, although this is one of the reactions involved. The scheme shown below has been proposed to account for the novolac-cyclic formal reaction.[6]

$$\xrightarrow{H^+} \quad H{-}CHO + {}^+\overset{|}{C}{-}\overset{|}{C}{-}\overset{|}{C}{-}OH$$

Oil-Soluble Phenolic Resins. The resole and novolac resins derived from phenol tend to be brittle, and they require plasticization to produce satisfactory surface coatings.

One of the most attractive plasticizers for a phenolic resin is a vegetable oil; the combined system would be expected to be hard, and solvent and moisture resistant (from the phenolic), and to possess air-drying and film-forming properties (from the oil). However, many of the early phenolic resins were not compatible with oils unless they were first reacted with rosin or a rosin ester. Subsequently, it was noted that the presence of a large oleophilic group in the parent phenol enhanced oil solubility. Since this type of resin does not require reaction with rosin, it is termed "100% phenolic." The commonly used phenols have the substituent in the *para* position, and typical examples are *para*-phenylphenol, *para-tert*-butylphenol, and *para-tert* amyl phenol. Besides modifying the solubility of the resin by providing a large oleophilic group, the *para* substituent also reduces the phenols' functionality, and minimizes the likelihood of cross-linked structures being formed.

Two main types of "100% phenolic resin" are available, depending on whether the resin is related to the novolac or resole structures. The novolac type is prepared under acidic conditions from a low formaldehyde/phenol ratio; this resin is thermoplastic and does not react chemically to any appreciable extent with the oil during resin processing. However under suitable conditions reaction is claimed to occur during film formation; it is suggested that the hydroperoxides formed on the oil molecule interact with the phenolic resin.[13] The resole type of resin, prepared with a high formaldehyde/phenol ratio and an alkaline catalyst, combines chemically with the oil during resin processing.

The resole-type phenolic resins interact with the unsaturated centers in the oil molecule and with the ester links. The *ortho*-methylol phenol structure adds to the double bond to form chroman-type compounds in a similar manner to the reaction with rosin (see page 74). In the particular case of methyl oleate and 2-methyl-4-*tert* butyl-6-methylol-phenol, the chromans (**3**) and (**4**), are formed as a result of addition to either end of the double bond.[7]

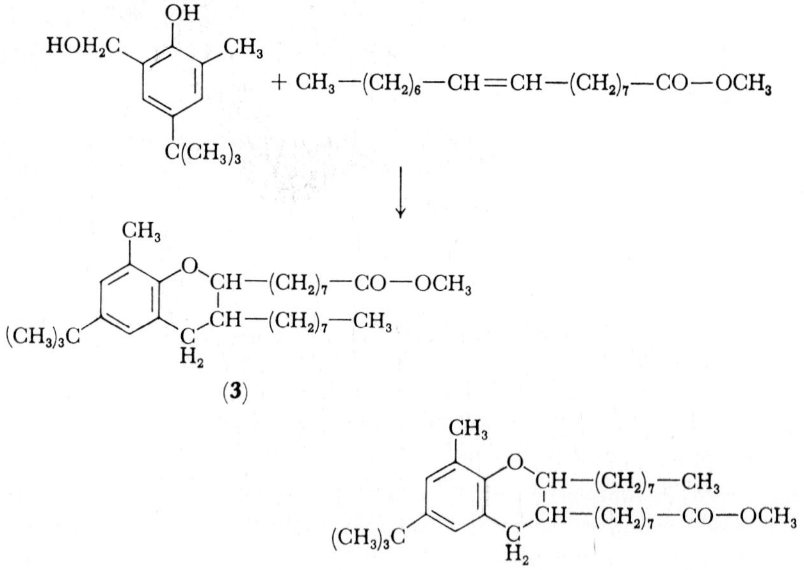
OH
HOH₂C
CH₃
+ CH₃—(CH₂)₆—CH=CH—(CH₂)₇—CO—OCH₃
C(CH₃)₃
CH₃
O
CH—(CH₂)₇—CO—OCH₃
CH—(CH₂)₇—CH₃
(CH₃)₃C
C
H₂
(3)
CH₃
O
CH—(CH₂)₇—CH₃
CH—(CH₂)₇—CO—OCH₃
(CH₃)₃C
C
H₂

(**4**)

The above reaction probably proceeds by way of a quinone methide.[8]

Phenolic resins with a *para*-methylol phenol structure cannot form a chroman ring because of steric factors, but they possibly react with the α-methylene groups.

Ester-interchange type reactions between the oil and phenolic resins are possible, but it is unlikely that these are of great significance in gaining oil solubility (Gen. Ref. 1). However, phenolic resin esters of fatty acids may

be prepared by a number of other techniques. Direct esterification of the methylol groups with a fatty acid is possible[9] but, in the presence of strong acid catalysts, self-polymerization of the methylol groups would also be expected. Fatty acid esters of phenolic resins can be prepared by first converting the phenolic hydroxyl and the methylol group to a conventional alcohol. This is done by reaction with an epoxide,[10] usually ethylene or propylene oxide.

These fatty acid esters are used as plasticizers for poly(vinyl chlorides); they have also been copolymerized, at the unsaturated centers of the fatty acid, with vinyl monomers.[11]

Oil solubility of phenolic resins is improved when some of the methylol groups are etherified with an alcohol, such as butanol.[12] These products are used as film formers or plasticizers for poly(vinyl chloride).[12]

GENERAL REFERENCES

1. Martin, R. W., *The Chemistry of Phenolic Resins,* John Wiley & Sons, New York, 1956.

2. Gould, D. F., *Phenolic Resins*, Reinhold Publishing Corp., New York, 1959.
3. Ellis, C., *The Chemistry of Synthetic Resins*, Reinhold Publishing Corp., New York, 1935.
4. Walker, J. F., *Formaldehyde*, Reinhold Publishing Corp., New York, 1964, p. 304.

REFERENCES

1. Finn, S. R., and J. W. G. Musty, *J. Soc. Chem. Ind.*, **69**, Suppl. No. 1, S3 (1950); via *Chem. Abstr.*, **45**, 7073 (1951).
2. Sprung, M. M., *J. Am. Chem. Soc.*, **63**, 334 (1941).
3. Goldblum, K. B., personal communication to R. W. Martin, via Gen. Ref. 1.
4. Zigeunar, G., and O. Gabriel, *Monatsh. Chem.*, **81**, 952 (1950).
5. Martin, R. W., Unpublished observations, via Gen. Ref. 1.
6. Heslinga, A., and A. Schors, *J. Appl. Polymer Sci.*, **8**, 1921 (1964).
7. Sprengling, G. R., *J. Am. Chem. Soc.*, **74**, 2937 (1952).
8. Hultzsch, K., *J. Prakt. Chem.*, **158**, 275 (1941), via Gen. Ref. 1; Cunneen, J. I., E. H. Farmer, and H. P. Koch, *J. Chem. Soc. (London)*, **1943**, 472.
9. Sprengling, G. R., personal cummunication to R. W. Martin, via Gen. Ref. 1.
10. The Dow Chemical Co., Brit. Pat. 755,459–60 (1957); Case, E. N., Sinclair Refining Co., U.S. Pat. 2,973,340 (1961); Martin, R. W., General Electric Co., U.S. Pat. 2,606,934 (1952); U.S. Pat. 2,606,935 (1952), via *Chem. Abstr.*, **47**, 3880 (1953).
11. Hanle, J. E., A. M. Tringali, and H. Yuska, Interchemical Corp., U.S. Pat. 2,880,187 (1959).
12. Martin, R. W., General Electric Co., U.S. Pat. 2,606,929 (1952); via *Chem. Abstr.*, **47**, 1980 (1952).
13. Chen, L. W. and J. Kumanotani, *J. Appl. Polymer Sci.*, **9**, 2785 (1965).

CHAPTER 10

THERMOSETTING VINYL AND ACRYLIC COPOLYMERS

The widespread acceptance of the thermoplastic vinyl and acrylic finishes has stimulated the development of thermosetting compositions of this type. The general characteristics of thermoplastic and thermosetting surface coatings have been pointed out previously (see page 2) but, in the particular case of the vinyl and acrylic polymers, the thermosetting compositions have the following possible advantages.

1. Higher solids at application viscosity in much cheaper solvents.
2. Better gloss and general appearance after baking.
3. Improved chemical, solvent, acid, and alkali resistance.
4. Less softening at higher temperatures and, as a consequence, better blocking resistance.

On the other hand, some of the thermosetting acrylics are of the two-pack type or have limited pot-life where the paint is constantly circulated; therefore, these types are unsuitable for mass production painting of automobiles. However, other thermosetting acrylics similar to alkyd/amine formaldehyde compositions can be stabilized to a degree that makes them acceptable for use in circulating systems. The temperature required for the stoving of thermosetting acrylics was once considered a limitation to their use but, in the automotive industry at least, this is no longer so. The reflow temperatures currently being used on thermoplastic acrylics (see page 155) are in excess of the stoving conditions necessary for the thermosetting reactions. Also, the thermosetting acrylics cure under comparable conditions to those used for alkyd/melamine resin blends.

The trade commonly uses the term "thermosetting acrylic" to cover a wide variety of polymer systems, some of which are based predominantly on vinyl copolymers. It has been suggested that the term "thermosetting acrylic" can be justified on the grounds that the functional groups, which are responsible for the cross-linking reactions, are derivatives of acrylic or methacrylic acid.[1] This definition satisfactorily includes, for example, a styrene/acrylic acid (92:8) copolymer (which is predominantly a vinyl-type copolymer), but excludes a styrene/maleic acid (92:8) copolymer. In the

263

present context, this distinction is unwarranted, since we are primarily concerned with the chemistry of the film-forming reactions. Consequently, the term "thermosetting acrylic" will be taken to include vinyl and/or acrylic copolymers that contain a reactive functional group; this definition avoids the necessity of making a distinction between copolymers containing similar functional groups derived by different means; for example, a hydroxyl group may be introduced into the polymer by the hydrolysis of a vinyl acetate residue, from copolymerized allyl alcohol, or from hydroxy-ethyl methacrylate and, in each case, the copolymer will undergo similar chemical reactions during film formation.

The chemical reactions used to cure thermosetting acrylic compositions are summarized by the Eqs. 10.1 to 10.12.

Chemical Reactions Used in the Cross-linking of Thermosetting Acrylics

$$[\text{Polymer}]-CH-CH_2 \;+\; [\text{Polymer}]-NH_2$$

(epoxide)   (amine)

$$\downarrow$$

$$[\text{Polymer}]-CH-CH_2-NH-[\text{Polymer}]$$
$$\qquad\qquad\quad\ \ OH$$

(10.1)

$$[\text{Polymer}]-CH-CH_2 \;+\; [\text{Polymer}]$$
$$\qquad\qquad\ \ O \qquad\qquad\qquad CO\ \ CO$$
$$\qquad\qquad\qquad\qquad\qquad\qquad\qquad\ \ O$$

(epoxide)   (acid anhydride)

$$\downarrow H^+$$

$$[\text{Polymer}]-CH-CH_2-O-CO-[\text{Polymer}]$$
$$\qquad\qquad\quad\ \ OH \qquad\qquad\qquad\qquad OC^+$$

(10.2)

$$\fbox{Polymer}-\!\!\underset{\underset{O}{\diagdown\diagup}}{CH}\!\!-CH_2 + \fbox{Polymer}-COOH$$

(epoxide)   (carboxyl)

$$\downarrow$$

$$\fbox{Polymer}-\underset{\underset{OH}{|}}{CH}-\underset{\underset{OCO-\fbox{Polymer}}{|}}{CH_2}$$

(10.3)

$$\fbox{Polymer}-\!\!\underset{\underset{O}{\diagdown\diagup}}{CH}\!\!-CH_2 + \fbox{Polymer}\!\!-\!\!\overset{\overset{OH}{|}}{\bigcirc}\!\!-\fbox{Polymer}$$

(epoxide)   (phenolic hydroxyl)

$$\downarrow$$

$$\fbox{Polymer}-\underset{\underset{OH}{|}}{CH}-CH_2-O-\bigcirc\!\!\!\genfrac{}{}{0pt}{}{\fbox{Polymer}}{\fbox{Polymer}}$$

(10.4)

$$\fbox{Polymer}-\!\!\underset{\underset{O}{\diagdown\diagup}}{CH}\!\!-CH_2 + \fbox{Polymer}-OH$$

(epoxide)   (hydroxyl)

$$\downarrow$$

$$\fbox{Polymer}-\underset{\underset{OH}{|}}{CH}-CH_2-O-\fbox{Polymer}$$

(10.5)

$$\fbox{Polymer}-\!\!\underset{\underset{O}{\diagdown\diagup}}{CH}\!\!-CH_2 + \fbox{Polymer}-NH-CH_2OR$$

(epoxide)   (*N*-methylol
or *N*-methylol ether)

$$\downarrow$$

$$\fbox{Polymer}-\underset{\underset{OR}{|}}{CH}-CH_2-O-CH_2-NH-\fbox{Polymer}$$

(10.6)

$$[\text{Polymer}]-NH-CH_2OR + [\text{Polymer}]-COOH$$

(*N*-methylol   (carboxyl)
or *N*-methylol ether)

$$\downarrow$$

$$[\text{Polymer}]-NH-CH_2-O-CO-[\text{Polymer}] + ROH\uparrow$$

(10.7)

$$[\text{Polymer}]-NH-CH_2OR + [\text{Polymer}]-OH$$

(*N*-methylol   (hydroxyl)
or *N*-methylol ether)

$$\downarrow$$

$$[\text{Polymer}]-NH-CH_2-O-[\text{Polymer}] + ROH\uparrow$$

(10.8)

$$[\text{Polymer}]-NH-CH_2OR + [\text{Polymer}]-NH-CH_2OR$$

(*N*-methylol   (*N*-methylol
or *N*-methylol ether)   or *N*-methylol ether)

$$\downarrow$$

$$[\text{Polymer}]-NH-CH_2-O-CH_2-NH-[\text{Polymer}]$$

$$\downarrow$$

$$[\text{Polymer}]-NH-CH_2-NH-[\text{Polymer}] + HCHO\uparrow$$

(10.9)

$$[\text{Polymer}]-COOH + [\text{Polymer}]-NCO$$

(carboxyl)   (isocyanate)

$$\downarrow$$

$$[\text{Polymer}]-CO-NH-[\text{Polymer}] + CO_2\uparrow$$

(10.10)

$$\boxed{Polymer}\!-\!OH \;+\; \boxed{Polymer}\!-\!NCO$$

$$\text{(hydroxyl)} \qquad\qquad \text{(isocyanate)}$$

$$\downarrow \qquad\qquad\qquad\qquad (10.11)$$

$$\boxed{Polymer}\!-\!O\!-\!CO\!-\!NH\!-\!\boxed{Polymer}$$

$$\boxed{Polymer}\!-\!NH_2 \;+\; \boxed{Polymer}\!-\!NH\!-\!CH_2OR$$

$$\text{(amine)} \qquad\qquad \text{(N-methylol}$$
$$\text{or N-methylol ether)} \qquad (10.12)$$

$$\downarrow$$

$$\boxed{Polymer}\!-\!NH\!-\!CH_2\!-\!NH\!-\!\boxed{Polymer} \;+\; ROH \nearrow$$

The functional groups shown in Eqs. 10.1 to 10.12 may both be present in the one polymer molecule, each in a separate acrylic copolymer, or one group in an acrylic copolymer and the other in a polymeric cross-linking agent. In general, both groups are not introduced into the same molecule, since the conditions under which the acrylic monomers are polymerized (for example, 4 hours at 130°C) are often equivalent to, or more forcing than, those required for cross-linking (for example, 1 hour at 130°C). Consequently, gelation can occur readily during the acrylic polymer formation. Two acrylic polymers are used where the specification warrants such an approach, but the majority of thermosetting acrylic formulations have a cross-linking agent that is not an acrylic polymer and is usually of a comparatively low molecular weight. This latter approach is capable of giving higher solids at application viscosity, because the overall average molecular weight of the polymer system is lower than in the systems containing only acrylic entities.

Typical monomers that contain the functional groups shown in Eqs. 10.1 to 10.12 are listed in Table 10.1.

In some instances, it may be necessary to prepare the acrylic copolymer by chemical treatment of a more stable, or more economically available, copolymer. Examples of this approach are to be found in acrylic copolymers, where acid groups are formed by the opening of a copolymerized anhydride, where hydroxyl groups result from esterification of an acid copolymer with an epoxide, or where an ester group is hydrolyzed to an alcohol or acid.

Aspects such as this are considered further when the individual polymer system is discussed.

TABLE 10.1

FUNCTIONAL MONOMERS USED IN THERMOSETTING ACRYLIC COPOLYMERS

Functional Group	Monomer
Epoxide	Glycidyl acrylate Glycidyl methacrylate Allyl glycidyl ether
Amine	Dimethylaminoethyl-methacrylate Vinyl pyridine *tert*-Butylaminoethyl-methacrylate
Anhydride	Maleic anhydride Itaconic anhydride
Hydroxyl	Allyl alcohol, monoallyl ethers of polyols Hydroxyethyl methacrylate Hydroxypropyl methacrylate Hydroxypropyl acrylate
Amide	Acrylamide, methacrylamide Maleamide
Methylol amide and alkyl ethers	*N*-methylol acrylamide, *N*-methylolmethacrylamide, etc. and corresponding ethers
Isocyanate	Vinyl isocyanate, allyl isocyanate

THE PREPARATION AND COMPOSITION OF THE ACRYLIC COPOLYMERS

Most thermosetting acrylic compositions are applied as solutions and, since the process of dissolving solid polymers is often difficult and slow, the polymer solutions are prepared directly from the monomer by polymerization in solution. The solvents used in the polymerization are chosen so that their solvent power and boiling range are acceptable to the final coating composition requirements. The polymerization is initiated by the use of free radical initiators, and follows the general mechanism discussed previously (see page 7). To have a satisfactory solids content at application viscosity (40 to 50%), the molecular weight of the polymer is controlled to 20,000 to 30,000 by the use of relatively high initiator concentrations (2% on the monomer) and temperatures (100 to 140°C), and by adding chain-transfer agents to the polymerization mixture. The solvent chosen for the polymerization must not react with the functional group of either the acrylic or the cross-linking polymer system and is, therefore, usually a hydrocarbon, such as xylene. In a few cases, the thermosetting acrylic is used in a dispersion or latex form; these polymers are prepared by an aqueous or nonaqueous emulsion polymerization.

The monomer composition, other than the functional group monomer, is related to the cost limitations and the properties demanded of the coating composition. Where alkali and detergent resistance are of prime importance, a hydrocarbonlike monomer (styrene, vinyl-toluene) is preferable to an ester (methyl methacrylate); where exterior durability is required, styrene is unsatisfactory (see page 175), and compositions based predominantly on methyl methacrylate are chosen. In thermosetting systems, it is usual for the plasticizer to be chemically incorporated into the final three-dimensional polymer network, since external plasticizers are often exuded from the film; this can cause low gloss and poor adhesion. The plasticizing entity can be part of the cross-linking agent or of the acrylic copolymer, and the same general principles as for thermoplastic acrylics apply; the acrylates and long chain methacrylates are frequently used as internal plasticizers for thermo-setting acrylics. The plasticizing monomer may form a normal random copolymer with the other monomer units, or conditions may be chosen so that a graft copolymer structure develops.[2]

The choice of the monomer composition for an exterior acrylic enamel has been recently placed on a systematic basis by Graham et al.;[3] it is now possible to predict compositions that will give the maximum durability and the required mechanical properties at the lowest cost. Graham's work was conducted on acrylic copolymers containing acrylamide, cross-linked by a urea-formaldehyde resin, but the results and underlying theory should be applicable to all thermosetting acrylics (see page 163 for further details). Briefly, Graham showed that copolymers of a hard and soft monomer had maximum exterior durability at a specific composition; blends of the individual copolymers that showed maximum durability, or terpolymers corresponding in compositions to this blend, also showed maximum durability. Therefore, from a study of simple copolymer compositions, it is possible to select a formulation that gives the most desirable balance of properties.

The concentration of functional groups in the acrylic polymer may range from 3 to 25%; the actual level will depend on the molecular weight of the polymer, on the functionality of the cross-linking agent, and on the availability of the reactive groups. At higher polymer molecular weights, fewer functional groups are required to give a satisfactory cured film, but a balance between solids at application viscosity (which decrease as the molecular weights increases) and curing rate must be made. It is generally desirable to keep the functional groups to a minimum, since they are derived from relatively expensive monomers. The distribution of functional groups along the polymer chain greatly influences the overall number of groups required, particularly where the functional monomer does not enter the growing chain at a rate equivalent to the other monomers. By allowing for the reactivity ratios of the monomers, and by gradually adding the more reactive monomer throughout the polymerization, more distributed polymers may be prepared (Figure 10.1). Nondistributed polymers require a higher level of functional groups[4] because it is necessary to ensure that individual molecules have a minimum number of reactive groups. Consequently, a high overall level of functional monomer is used and some molecules have an excessive number of functional groups (Figure 10.1a);

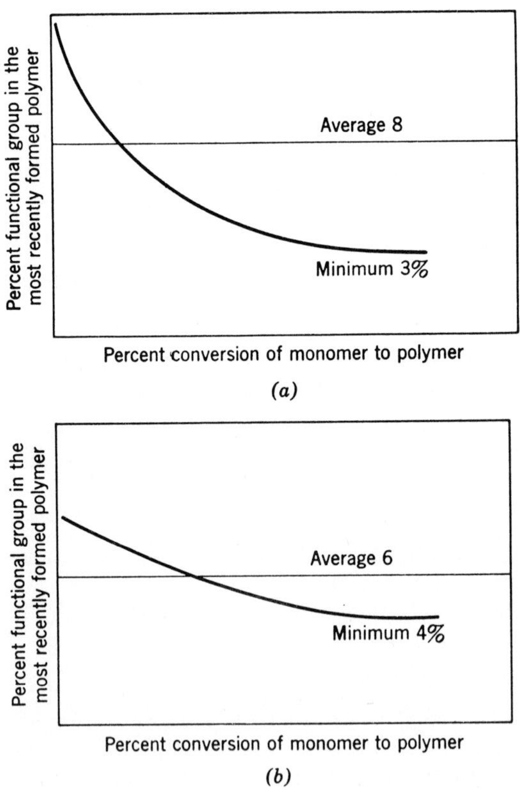

Figure 10.1.   (a) Non-distributed functional groups.   (b) Distributed functional groups.

these molecules contribute to poor storage stability. With a distributed polymer of lower overall level of functional groups, the minimum functional group content can be higher than in the nondistributed composition (Figure 10.1*b*).

By analogy with thermoplastic polymer solutions, uniform distribution of plasticizing groups would be expected to result in more effective plasticization. Therefore, in selecting the monomer composition, due regard should be paid to the likely distribution of the monomer units in the final polymer. Where necessary, controlled addition of the more reactive monomer is used as a method of forming a more evenly distributed copolymer.

THE CROSS-LINKING REACTIONS OF ACRYLIC POLYMERS

The particular functional group and cross-linking agent chosen for a thermosetting coating composition will depend on: the final film properties required, the conditions of the curing reaction, cost, etc. Each of the systems discussed below has some properties that are superior to the other compositions, but the choice between the various alternatives is often restricted by patents. The extent to which the cross-linking reactions have taken place may be measured by solvent resistance, durability, mechanical properties[5] or conductivity[6] of the films, or by following the reactions by physical methods such as infrared spectroscopy[2] or gas chromatography.

Epoxide-Amine Reaction. The reaction between an epoxide group of an acrylic copolymer and an amine (Eq. 10.1) takes place readily at room temperature. Therefore, this is a route to thermosetting acrylics when elevated temperatures cannot be used. The presence of the epoxide group on an acrylic backbone results in a number of significant differences from the conventional bisphenol-A type epoxy resins; the acrylics are more resistant to degradation by ultraviolet light, and are soluble in nonpolar solvents.[7] However, the choice of cross-linking amine is more restricted with the acrylic polymers, since many of the common amines are exuded from the cured films. Menthane diamine is the most satisfactory cross-linking amine when the epoxy group is derived from glycidyl methacrylate, and this amine gives good compatibility and pot-life with the acrylic copolymers.[7] In general, the aliphatic amines give more rapid cross-linking at room temperature than aromatic amines.[7]

The alternative method of utilizing the epoxide-amine reaction is to prepare an acrylic copolymer containing a primary or secondary amine group and to use a conventional epoxide resin as the cross-linking agent. *tert*-Butylaminoethyl methacrylate has been suggested as a suitable amine source;[5,8] alternatively, the amine group can be formed by reaction of an epoxide group in an acrylic copolymer with a primary amine or ammonia.

$$-CH\!-\!\!-\!CH_2 + RNH_2 \longrightarrow -CH\!-\!CH_2$$

Ester-Forming Reactions. The reaction of an anhydride with an epoxide group appears to have been utilized to cross-link polymers only where the anhydride is part of the acrylic polymer. This is understandable in view of the difficulty of synthesizing suitable cross-linking agents containing an anhydride group and the comparative ease with which acrylic-maleic anhydride copolymers can be prepared. In addition, maleic anhydride is the lowest cost functional monomer available if both of its reactive groups (double bond and the anhydride) are utilized.[9] A previous limitation on the use of maleic anhydride was the lack of suitable methods for preparing non-equimolar copolymers, particularly those with a high styrene content. These are now available, and copolymers of a desired composition can be synthesized by routine methods[10] that involve careful control of the monomer feed composition. The cross-linking reaction shown in Eq. 10.2 requires a catalyst and elevated temperatures in order to proceed at an acceptable rate; typical curing conditions are $\frac{1}{2}$ hour at 170°C with 0.5% amine catalyst.[9,10] The base catalyst probably functions by providing a mechanism for the initial opening of the anhydride ring.[9]

The anhydride copolymers react with polyols under conditions similar to those used for epoxides, although the reaction is more facile. It should be noted that when an anhydride copolymer is reacted with a bisphenol-A resin, the first reaction is probably between the anhydride and hydroxyl groups; the carboxyl group so formed subsequently attacks the epoxide ring as discussed below.

The acid-epoxide reaction (Eq. 10.3, page 253) was used in one of the early thermosetting acrylic systems, and is the basis of a number of commercial appliance and general industrial enamels. The field performance of these compositions has contributed greatly to the current reputation of thermosetting acrylics. The epoxide group may be incorporated into the acrylic copolymer,[7,11] but this approach is limited by the relatively expensive and hazardous nature of the epoxy monomers used. By far the most common acid-epoxide systems use an acid acrylic copolymer and a bisphenol-A type epoxy resin; the general formulating principles can be illustrated by considering the specific case of an appliance enamel, where alkali and detergent resistance are of prime importance. A typical copolymer composition is styrene/ethyl acrylate/acrylic acid = 72/20/8 by weight, where the styrene is the hard monomer and, being a hydrocarbon, is not susceptible to hydrolysis; the ethyl acrylate is the internal plasticizer, and the acrylic acid the functional group for the subsequent cross-linking reaction. If this copolymer is prepared by adding the monomer mixture to refluxing solvent, the acid is poorly distributed (Figure 10.1a); it enters the growing chain more readily than the other two monomers. Regulated addition of the acrylic acid gives a more uniform distribution (Figure 10.1b) and much more desirable properties.[12] By evaluating copolymers that varied in molecular weight and acid content, it was found that adequate cross-linking was obtained by having approximately 25 carboxyl groups per molecule. To obtain good spraying characteristics, the molecular weight had to be lower than 30,000 which meant an acid content of not less than 6%.[12] For an appliance enamel, the cross-linking agent must be alkali resistant; consequently, bisphenol-A type epoxide resins are used. The single ester link formed between the epoxy resin and the acrylic polymer is the only part of the backbone polymer network susceptible to hydrolysis; other ester groups are in side chains. Theoretically, each epoxide group is capable of reaction with one acid group (Eq. 10.3), but it has been found that not all the carboxyl groups are available for reaction because of steric restrictions. Consequently, less than equivalent quantities of the epoxide resin are used, since this is the more expensive resin component.[12] The epoxide-acid reaction requires stoving schedules of approximately $\frac{1}{2}$ hour at 150°C in the presence of a base catalyst.[13] Typical examples of the catalysts used are shown below.[9]

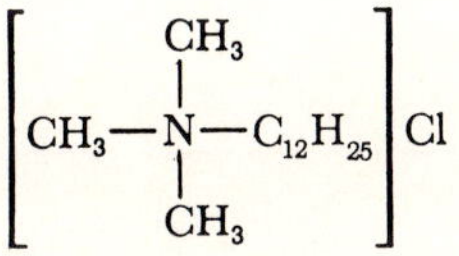

$$\left[\begin{array}{c} CH_3 \\ | \\ CH_3-N-C_{12}H_{25} \\ | \\ CH_3 \end{array} \right] Cl$$

(dodecyl trimethyl
ammonium chloride)

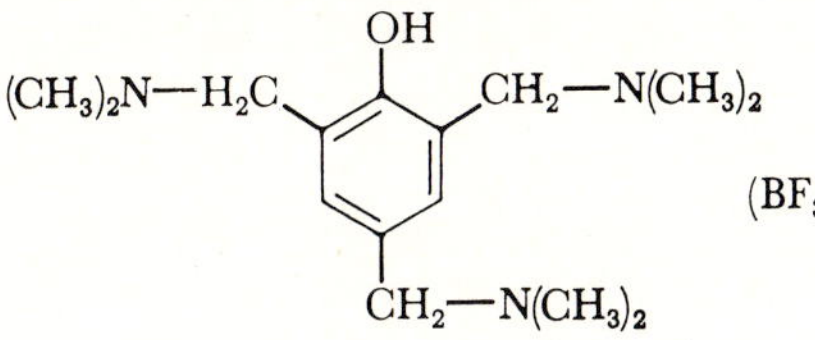

$$(CH_3)_2N-H_2C \underset{\underset{CH_2-N(CH_3)_2}{}}{\overset{OH}{\bigcirc}} CH_2-N(CH_3)_2$$

[Tri(dimethylaminomethyl)phenol]
(DMP-30)

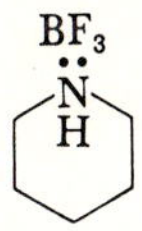

(BF$_3$-piperidine)

Melamine-formaldehyde resins are sufficiently basic to be used as the catalyst, and these have the additional advantage of entering into the cross-linking reaction (Eqs. 10.6 and 10.7). Where possible, the base catalyst is added to the other components just prior to use; the acid-epoxide system has limited pot-life when mixed with catalysts, and is normally sold and regarded as a two-pack composition. Typical properties of an appliance enamel, based on the composition discussed above, are shown in Table 10.2.[12] An alternative means of catalyzing the acid-epoxide reaction is to use small quantities (0.1 to 0.2%) of vinylpyridine in the copolymer.[14]

TABLE 10.2

PROPERTIES OF A THERMOSETTING APPLIANCE ENAMEL[12]

Copolymer	Vinyltoluene/ethyl acrylate/acrylic acid		
	72	20	8
Diepoxide	Epon[a] 828		
Gloss	93 at 60°		
Adhesion to Bonderite[b] 1000	Excellent		
Adhesion to alodized Al	Excellent		
Flexibility	Fair		
Grease resistance (15 days at 60°C in 1:1 cottonseed oil:oleic acid)	Unaffected		
Detergent resistance (2% "Tide" at 80°C)	One coat (1.5 mils), 200 hours Two coats (2.2 mils), 1000 hours		
Heat resistance	6 hours at 315°F to cause noticeable change		
Pencil hardness	3H		

[a] Epon is a registered trademark of Shell Chemical Co.
[b] Bonderite is a registered trademark of Parker Rust-Proof Co.

Numerous variations of the formulation discussed above are possible, and some of the more technically and commercially interesting are considered below. In an attempt to overcome the poor can stability of a catalyzed acid-epoxide system, the acid acrylic copolymer has been reacted, in solution, with the epoxy resin. A ratio of two moles of epoxide to one of acid was used so that the product shown in Eq. 10.13 would result.[15]

In a film, this polymer would cure by the hydroxyl-epoxide reaction. Problems associated with overspray cratering, which are particularly prevalent where more than one type of coating composition is being applied, have been minimized by replacing the styrene by vinyl toluene[12] or by using 2-ethylhexyl acrylate as the plasticizer. However, the long chain acrylates resulted in compositions with poor stain and grease resistance, and it is necessary to compromise between the resistance properties and the tendency to crater. Where properties other than detergent and alkali resistance are required, the backbone polymer has been altered accordingly; vinyl

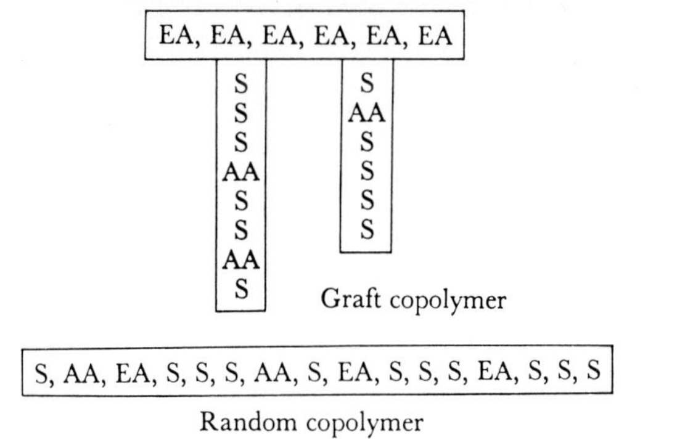

$$(10.13)$$

acetate has been used in coatings for light metals, steel, and wood, since it is a cheap monomer;[16] methyl methacrylate[17,18] has been used where coatings with good exterior durability are required.

Alternative methods of plasticizing the acrylic polymer have been studied; grafting the styrene-acrylic acid chains onto a preformed poly(ethyl acrylate) gives a cross-linked film (with an epoxy resin) with greater flexibility than is obtained from a random copolymer of the same composition. Conversely, at equivalent degrees of flexibility, the graft copolymer uses less ethyl acrylate and is cheaper than the random copolymers. The difference between the two types of copolymer is illustrated in Figure 10.2. Rubber-like polymers (Buton 100, marketed by Esso Co.) offer an even cheaper means of plasticization by using the grafting technique.[19]

Figure 10.2.   Schematic representation of difference between random and graft styrene/ethyl acrylate/acrylic acid copolymers.

Epoxide resin esters, which have been prepared by selective esterification of the hydroxyl groups by long chain acids, are another approach to the plasticization of acrylic acid-epoxide polymer systems. The methods of synthesizing such epoxide esters have been discussed on page 202.

The conventional source of carboxyl groups in thermosetting acrylic enamels has been acrylic and/or methacrylic acids, but recent advances in the control of the copolymerization of maleic anhydride (see page 177) indicate that this could be a more economical source of acid groups. Treatment of a solution of the anhydride copolymer with an alcohol readily yields the half-ester.[9] In addition to creating a carboxyl group, this reaction gives a residue equivalent to an acrylate; consequently, the polymer is plasticized by this treatment.[9,20] The maleic acid half-ester can be formed prior to polymerization, and this technique can be of assistance in overcoming solubility problems[21] sometimes associated with the preparation of maleic anhydride copolymers.

The introduction of ester groups by the opening of the anhydride ring has a further possible advantage in that the plasticizing group is distributed along the chain and is always adjacent to the carboxyl function. Alternatively, the anhydride copolymer can be decarboxylated to an acid copolymer by heating in the presence of moisture. The anhydride group can also be converted to a carboxyl by reaction with an amine to give the acid-amide, as shown below.[9]

Finally, diepoxide compounds, other than epoxy-type resins, have been used to cross-link acid copolymers, particularly where it is desirable to avoid the degradation which ultraviolet light causes to bisphenol-A coatings.[22] Typical diepoxides include vinylcyclohexene diepoxide, diglycidyl azelate[17] and diglycidyl sebacate.[16] Acrylic copolymers that contain epoxy groups have also been used.

REFERENCES

1. Gerhart, H. L., *Offic. Dig. Federation Soc. Paint Technol.*, **33**, 680 (1961).
2. Kelly, D. P., G. J. H. Melrose, and D. H. Solomon, *J. Appl. Polymer Sci.*, **7**, 1991 (1963).
3. Graham, N. B., F. R. Crowne, and D. E. MacAlpine, *Chem. Eng. News*, **1965**, 67; *Offic. Dig. Federation Soc. Paint Technol.*, **37**, 1228 (1965).
4. Solomon, D. H., and J. J. Hopwood, Australian Pat. Applic. 48,478/64.
5. Brown, W. H., and T. J. Miranda, *Offic. Dig. Federation Soc. Paint Technol.*, **36**, (**475**), 92 (1964).
6. Chadwick, G. F., Am. Chem. Soc., Div. Org. Coatings Plastic Chem., Abstr. of Papers, 145th Meeting, New York, **1963**, 11-R.
7. Simms, J. A., *J. Appl. Polymer Sci.*, **5**, 58 (1961).
8. Rohm and Haas Co., Special product bulletin, 84.
9. Dennis, K. S., C. E. Lyons, and D. R. Spiekerman, *Offic. Dig. Federation Soc. Paint Technol.*, **37**, 378 (1965).
10. Moore, E. R., R. L. Zimmerman, and D. R. Spiekerman, *Offic. Dig. Federation Soc. Paint Technol.*, **37**, 410 (1965). *J. Oil Colour Chemists' Assoc.*, **52**, 830 (1969).
11. Fryling, C. F., and L. F. Gusiak, Koppers Co. Inc., U.S. Pat. 3,058,947 (1962).
12. Murdock, J. D., and G. H. Segall, *Offic. Dig. Federation Soc. Paint Technol.*, **33**, 709 (1961).
13. Shechter, L., and J. Wynstra, *Ind. Eng. Chem.*, **48**, 86 (1956).
14. Allenby, D. C. W., Canadian Industries Ltd., Brit. Pat. 731,946 (1955).
15. Segall, G. H., and J. F. C. Dixon, Canadian Industries Ltd., and Du Pont Co. of Canada Ltd. Brit. Pat. 730,869 (1955).
16. Abbotson, W., D. H. Coffey, R. Hurd, J. M. Phillipson, and E. J. Vickers, Imperial Chemical Industries Ltd., Brit. Pat. 810,940 (1959).
17. Fang, J. C., E. I. du Pont de Nemours & Co., U.S. Pat. 2,866,767 (1958).
18. Leutner, F. S., and G. P. Roeser, American-Marietta Co., Can. Pat. 680,815 (1964).
19. Murdock, J. D., Canadian Industries Ltd., Can. Pat. 688,029 (1964).
20. Vasta, J. A., E. I. du Pont de Nemours & Co., U.S. Pat. 2,967,162 (1961).
21. Barrett, G. R., Monsanto Chemical Co., Brit. Pat. 673,849 (1952).
22. Mercurio, A., *Offic. Dig. Federation Soc. Paint Technol.*, **36** (**475**), 135 (1964).

Reactions Involving N-Methylol or N-Methylol Ether Groups. In considering the reactions of polymers that contain an —NH—CH$_2$OH group, it is essential to realize that this group differs considerably from a simple alcohol; it is far more acidic and reactive. Similarly, the methylol ether is a less stable, and more reactive, entity than a conventional dialkyl ether. It will be recalled that the cross-linking reactions which take place between an alkyd resin and a melamine- or urea-formaldehyde condensate involve N-methylol groups or their ethers, and the reaction can be summarized as follows:

Polymer compositions of the alkyd/melamine-formaldehyde type can be satisfactorily stabilized for use in large scale paint-circulating systems, and they have been used widely as the "super enamels" on many automobiles. The same cross-linking reactions have been applied to acrylic systems designed for use in automotive enamels and for general industrial use. These polymer systems may be regarded as being derived from the alkyd/melamine-formaldehyde condensate system by the replacement of one or other, or even both, of these polymer species by an acrylic copolymer containing similar functional groups.

In the preparation of an alkyd resin, it is necessary to have both hydroxyl and carboxyl groups present, since these are involved in the polymer-forming reactions. This situation does not arise with acrylic resins where it is a relatively easy matter to prepare copolymers that contain acid, hydroxyl, or both acid and hydroxyl groups. Consequently, the control of the type and number of functional groups, and of the molecular weight is much simpler with acrylics than with alkyd resins. The replacement of the alkyd resin by acrylic copolymers has consequently resulted in the development of a number of polymer systems, which vary in the hydroxyl-carboxyl ratio, molecular weight, etc. (Gen. Ref. 1 & 2).

Acrylic polymers that contain acid groups can be prepared by any of the methods described above. Reaction of these polymers with urea, melamine, and benzoguanamine-formaldehyde condensates (Eq. 10.7, page 265) has been used to give thermosetting acrylic enamels suitable for appliance or general industrial enamels.[1] Evidence supporting the reaction illustrated by Eq. 10.7 is the development of a strong ester band in the infrared spectrum of a cured film;[2-4] the reaction, which is acid catalyzed, possibly proceeds as follows:[3,4]

$$[\text{Amine condensate}]-\text{N}-\text{CH}_2-\text{OCH}_3 + \text{H}^+ \longrightarrow [\text{Amine condensate}]-\text{N}-\text{CH}_2-\overset{+}{\underset{\text{H}}{\text{O}}}\text{CH}_3$$

$$\downarrow$$

$$[\text{Amine condensate}]-\text{N}-\overset{+}{\text{CH}}_2 + \text{CH}_3\text{OH}$$

$$+ [\text{Acrylic}]-\text{COOH}$$

$$[\text{Amine condensate}]-\text{N}-\text{CH}_2-\text{O}-\text{CO}-[\text{Acrylic}] + \text{H}^+$$

The normal curing conditions for these systems are in the vicinity of $\frac{1}{2}$ hour at 150°C; at this temperature the acid-methylol reaction is slow, and considerable self-condensation of the formaldehyde condensate takes place.[4] Further changes in film properties occur on continued heating, and it appears that not all of the functional groups have reacted under the conditions normally used in technological practice to obtain cured films. The functionality of the cross-linking resins also varies from the value expected, probably because the molecules lose mobility once cross-linking begins. In a system using hexakis (methoxymethyl) melamine **1** as the formaldehyde condensate, 1 mole of **1** reacted with only 1.5 moles of acid in the copolymer; the full functionality of 6 was not realized.[3]

$$\begin{array}{c}
(\text{CH}_2\text{OCH}_3)_2 \\
| \\
\text{N} \\
| \\
\end{array}$$

Hexakis(methoxymethyl)melamine

(1)

The presence of hydroxyl groups in an acrylic polymer enables the reaction with formaldehyde condensates to proceed much more readily than with carboxyl groups.[4] The hydroxy-containing acrylic polymers are readily prepared by the use of hydroxy-acrylic or vinyl monomers (see Table 10.1 on page 267), or by chemical treatment of suitable polymers. Hydrolysis of vinyl acetate copolymers is commonly used with systems in which vinyl chloride is the predominant monomer (vinyls), and this is a cheap method of providing hydroxyl sites.

$$-CH_2-CH-CH_2-CH-$$
$$\ \ \ \ \ \ \ \ \ \ |\ \ \ \ \ \ \ \ \ \ \ \ \ \ \ |$$
$$\ \ \ \ \ \ \ \ \ \ Cl\ \ \ \ \ \ \ \ \ \ \ \ OAc$$

$$\downarrow$$

$$-CH_2-CH-CH_2-CH-$$
$$\ \ \ \ \ \ \ \ \ \ |\ \ \ \ \ \ \ \ \ \ \ \ \ \ \ |$$
$$\ \ \ \ \ \ \ \ \ \ Cl\ \ \ \ \ \ \ \ \ \ \ \ OH$$

The hydroxyl group can also be incorporated into an acrylic polymer by the esterification of copolymerized acid with a monofunctional epoxy compound under such conditions that further esterification of the resultant hydroxyl group with unreacted acid does not occur. Usually, temperatures of 50°C or less and base catalysts are used to promote the reaction.[2,5-7]

$$\boxed{Polymer}-COOH + R-CH-CH_2$$
$$\ \backslash O /$$

$$\downarrow \text{Base}$$

$$\boxed{Polymer}-CO-O-CH_2-\overset{R}{\underset{|}{C}H}-OH$$

An example of direct commercial interest is the reaction of a polymer containing a maleic acid half-ester with an epoxide; this represents a most

$$\boxed{Polymer}\begin{matrix} -CO \\ \ \ \ \ \ \ \ \ \ \ \ O + ROH \\ -CO \end{matrix}$$

$$\downarrow \text{Step 1}$$

$$\boxed{Polymer}\begin{matrix} -COOR \\ \\ -COOH \end{matrix}$$

$$\downarrow \begin{matrix} + R-CH-CH_2 \\ \text{Step 2}\ \ \ \backslash O / \end{matrix}$$

$$\boxed{Polymer}\begin{matrix} -COOR \ \ \ \ \ \ \ \ \ \ \ \ \ R \\ | \\ -CO-O-CH_2-CH-OH \end{matrix}$$

economical method of plasticizing the polymer and providing a reactive hydroxyl site.

Step 2 occurs much more readily than the reaction between an epoxide and anhydride group; therefore, it is usual to carry out the reactions in the order shown. Epoxide groups in acrylic polymers can be readily reacted with amines to produce hydroxy-amine groups.[2] By using secondary amines, the risk of gelation is avoided; a primary amine would yield a hydroxy-*sec*-amine capable of further reaction, while a tertiary amine will not form a hydroxyl compound by opening the epoxide ring.

$$\boxed{Polymer}-CH-CH_2 + R_2NH$$
$$\diagdown \diagup$$
$$O$$

$$\downarrow$$

$$\boxed{Polymer}-CH-CH_2$$
$$\qquad\qquad | \quad\ |$$
$$\qquad\qquad OH \quad NR_2$$

(R = H or alkyl)

These alternative methods of introducing hydroxyl groups into thermosetting acrylic polymers are of interest where they offer a cost, or technical, advantage over the process using the hydroxy monomer. In some systems, they can give a more even distribution of the hydroxyl groups than is obtained with the hydroxy monomer.

Reaction of the hydroxy polymer with an amide or amine-formaldehyde resin requires an acid catalyst;[2,4,6,7] it then proceeds by way of ether interchange, as shown below for a melamine-formaldehyde resin:

$$\boxed{\substack{Melamine\\formaldehyde}}-NHCH_2-OR + HO-\boxed{Acrylic}$$

$$\downarrow$$

$$\boxed{\substack{Melamine\\formaldehyde}}-NHCH_2-O-\boxed{Acrylic} + ROH\uparrow$$

Stoving schedules of $\frac{1}{2}$ hour at 125°C are sufficient for the curing reaction. Excessive acid catalysis and the presence of free *N*-methylol groups induces self-condensation of the melamine resin, whereas the use of an external acid catalyst can give rise to poor can stability.[4] It is usually preferable to include in the acrylic backbone polymer some acid groups to act as an internal catalyst and, to a lesser extent, to take part in the cross-linking reactions. This can be done by using a suitable acid monomer or by forming a half-ester of some of the hydroxyl groups with an anhydride; this has led to the development of polymers that contain both acid and hydroxyl groups.

Thermosetting acrylics with both acid and hydroxyl functional groups and cross-linked by amine-formaldehyde resins are covered in a number of patents.[8] Formulations based predominantly on the acrylates and methacrylates have given films that satisfactorily resist Florida exposure while styrene-rich systems have provided detergent resistant finishes.[7] Reflow thermosetting acrylic resins based on hydroxy-acid acrylic copolymers and melamine resins have been formulated. In these systems the coating is given an initial low bake (e.g. 70-100°C) to remove the solvent. At this stage the film is still soluble and repairs or respray can be carried out in much the same manner as for thermoplastic polymers. The film is then stoved at a higher temperature (e.g. 140°C) to bring about crosslinking. This process aims at combining the ease of repair associated with thermoplastic acrylics with the high solids at application viscosity and general resistance properties of the thermosetting acrylic. In order to control the interaction between the acrylic and melamine resin the acid content of the acrylic copolymer has been reduced.[18]

Organosol-type thermosetting acrylics, which utilize the acid or hydroxyl/melamine-formaldehyde resin interaction, have been described.[9] The example discussed here illustrates the manner in which the organosol systems can be applied to thermosetting reactions in general (see page 167 for a discussion on thermoplastic organosols and stabilizers for organosols). For the polymer particles of an organosol to fuse at an acceptably low temperature, it is necessary to use an external plasticizer, often in amounts in excess of those required in the final film. However, by using a plasticizer that will subsequently react with the polymer good fusion and final film properties are obtainable. For automotive enamels, a polymer composed predominantly of methyl methacrylate, and with hydroxyl and carboxyl functional groups, has been used; the plasticizer was a simple hydroxy ester, while the third component was a melamine resin. By adding the acid monomer early in the organosol preparation, it was possible to concentrate the acid toward the center of the drop; hence, no curing reactions occur until after particle fusion. A cross-linked film results in which the melamine resin ties the plasticizer and methacrylate polymer into the one structure. Other organosols have been described in which controlled cross-linking has been carried out in the polymer particles before film formation.[20]

The alternative method of using the alkyd/melamine type of cross-linking reactions in a solution acrylic system is to prepare acrylic copolymers in which the functional groups are similar to the melamine resin. Methylol and methylol ether groups can be introduced into an acrylic polymer by the use of the appropriate monomer;[10] the polymerization must be conducted under mild conditions to avoid self-condensation of the methylol groups and the formation of a gel-structure. It is preferable to prepare an amide copolymer, and subsequently methylolate, and etherify.[11,12] Amide polymer preparation is straightforward, provided suitable solvents are chosen; with acrylamide, butanol is commonly chosen as the solvent, since it can also be used in the subsequent etherification reactions of the methylol groups. The amide-acrylic polymers are capable of cross-linking with an amine resin,[13] but the reaction requires more stringent conditions than when a methylol amide is used. One novel use of the amide-aminoplast system is in the formation of mat finishes, where the low gloss results from a surface microwrinkle, rather than from high pigment contents. These mat finishes are claimed to have better mar resistance than conventional finishes. The surface wrinkle is controlled by the use of a latent catalyst (secondary or tertiary amine salt of sulfuric acid) which preferentially promotes surface cure.[13]

The methylolation, and subsequent etherification, of amide copolymers involves reactions analogous to those that take place in the formation of urea- or melamine-formaldehyde resins. The formaldehyde is conveniently added as paraform[2] or a formcel,[11] and the methylolation reaction is catalyzed either by base[2] or by acid.[11] As with other amine-formaldehyde reactions, the pH of the reaction mixture can drift as the methylolation proceeds; this problem is minimized when the basic method is used, particularly where the copolymer contains acid groups. It is probable that the solution will be buffered, since a weak base—and the salt of that base—are present.[2] The etherification of the methylol amide copolymer can be brought about by removing the base (usually a volatile tertiary amine) and

$$\boxed{Polymer}-CO-NH_2$$

$$\downarrow HCHO$$

$$\boxed{Polymer}-CO-NH-CH_2-OH$$

$$\downarrow ROH$$

$$\boxed{Polymer}-CO-NH-CH_2-OR + H_2O$$

acidifying the mixture. Water of reaction is removed and the formation of *N*-methylol ethers is followed by infrared spectroscopy, petroleum ether tolerance, or compatibility measurements.[2,11] Copolymerized acid has also been used to catalyze etherification; this method has the added advantage that the acid contributes to the curing rate, adhesion, and compatibility with other film formers.[11] A single stage process has been described in which the vinyl polymerization of the acrylamide/monomer mixture and the methylolation of the amide group take place simultaneously.[19]

It is usual to use 2 moles formaldehyde for each amide group in addition to a large excess of the etherifying alcohol. The alcohol is often used also as the solvent for the reaction. The excess formaldehyde compensates for losses due to side reactions, the most likely of which are oxidation to formic acid, a Cannizzaro-type reaction to yield methanol and formic acid, and the formation of a formal.

$$HCHO + 2\,ROH$$

$$\downarrow$$

$$CH_2 \begin{array}{c} \diagup OR \\ \diagdown OR \end{array} + H_2O$$

Formal

The film-forming reactions of the formaldehyde modified amide copolymers are similar to those that a melamine-formaldehyde resin undergoes. The resins may be self-condensed (Eq. 10.9, page 265) or cross-linked with another amine resin. For satisfactory coreaction with a melamine resin, the copolymer should contain predominantly methylol ethers, since the methylol derivatives give poor cure[6] and lower gloss films[12] as a result of self-condensation. The cross-linking reactions are acid catalyzed; copolymerized acid, external acid catalysts, or latent catalysts, such as the morpholine salt

TABLE 10.3

PROPERTIES OF ACRYLIC AND ALKYD-MELAMINE ENAMELS[16]

Test Procedure	Enamel 1[a]	Coconut Alkyd-Melamine Enamel	Dehydrated Castor Alkyd-Melamine Enamel	Enamel 2[b]
Gloss	90+	85+	80+	90+
Pencil hardness	2H	F	HB	3H
Mar resistance	Excellent	Good	Fair	Excellent
Adhesion	Excellent	Poor	Excellent	Excellent
Recoat adhesion	Excellent	Very poor	Excellent	Excellent
Impact resistance[c]	6–12	<2	24+	<2
Vegetable oil immersion[d]	H	2B–B	Less than 3B	2H
5% Salt spray[e]	Nil	1/16″	$\frac{1}{8}$″	Nil
Humidity (1000 hours)	No change	No change	Minute blisters	No change
Detergent resistance[f]	Few #6 (good)	Medium #8 (fair)	Dense #8 (poor)	Nil (excellent)
Stain resistance[g]				
Mustard	Very faint mark	Stain	Heavy stain	No change
Lipstick	No change	Stain	Heavy stain	No change
Ink	No change	Slight stain	Heavy stain	No change

[a] Enamel 1 was based on acrylamide/ethyl acrylate/styrene = 15/45/40.

[b] Enamel 2 was based on acrylamide/ethyl acrylate/styrene = 15/–/85.

The resins used to make Enamel 1 and Enamel 2 were methylolated, pigmented, and blended with EPON 1001 equivalent to 9% by weight on a solids basis.

[c] Impact resistance—measured in inch-pounds.

[d] Vegetable oil immersion—rated after 100 hours at 100°F as pencil hardness.

[e] Salt spray—rated after 250 hours in 5% NaCl fog at 100°F.

[f] Detergent resistance—rated after 100 hours immersion at 140°F.

[g] Stain resistance—rated after 100 hours contact.

of *p*-toluene sulfonic acid, have been used. Pigmented compositions require higher stoving temperatures[11] to give equivalent curing.

The methylol-amide copolymers react readily with polyols (Eq. 10.8, page 265), two common examples being alkyd resins[14] and partial esters of polyols such as a monoglyceride.[12] Reaction with an epoxy resin[15] also involves the hydroxyl group but, in addition, the methylol ether can add across the epoxide group (Eq. 10.6, page 264). This reaction, which requires acid catalysts, has been followed by infrared spectroscopy.[2] Formulations of this type are used as appliance enamels, since both polymer entities (that is, the acrylic and the epoxy) are devoid of ester links in the polymer backbone. Therefore, they resist hydrolysis. The properties of an acrylic enamel of this type are compared with typical alkyd-melamine formaldehyde enamels in Table 10.3.[16]

Reactions in Resin Mixtures. In commercial practice, resin mixtures are sometimes used to provide the required balance of properties in a coating. The chemistry of the cross-linking reactions in these formulations is a combination of the interactions considered above. For example, a blend of an acid-acrylic copolymer, a phenol-formaldehyde condensate, and a polyepoxide, with a butylated melamine resin as catalyst has been suggested as an appliance or automotive finish.[17] The reactions expected in this system are the following:

1. Reaction of the acid-acrylic polymer with the polyepoxide (Eq. 10.3, page 264). This will be catalyzed by the melamine resin.

2. The phenolic resin may undergo:

(a) Self-condensation, catalyzed by the acid group of the acrylic polymer.

(b) Reaction with the acid groups of the acrylic polymer.

$$\boxed{Polymer} \text{—}\overset{\displaystyle OH}{\underset{\displaystyle CH_2OR}{\bigcirc}}\text{—}CH_2OH \; + \; \boxed{Acrylic}\text{—}COOH \longrightarrow$$

$$ROH \; + \; \boxed{Polymer}\text{—}\overset{\displaystyle OH}{\underset{\displaystyle CH_2\text{—}O\text{—}CO\text{—}\boxed{Acrylic}}{\bigcirc}}\text{—}CH_2OH$$

(c) Etherification, and ether interchange, with the hydroxyl groups of the polyepoxide entity.

$$\boxed{Polymer}\text{—}\overset{\displaystyle OH}{\underset{\displaystyle CH_2OR}{\bigcirc}}\text{—}CH_2OH \; + \; CH_2\underset{O}{\overset{}{\diagdown}}CH\text{—}\boxed{Polymer}\text{—}\underset{OH}{CH}\text{—}CH_2\underset{O}{\diagup} \longrightarrow$$

$$\boxed{Polymer}\text{—}\overset{\displaystyle OH}{\underset{\displaystyle CH_2OR}{\bigcirc}}\text{—}CH_2\text{—}O\text{—}\boxed{Polymer}$$

(d) Addition across the epoxide group of the polyepoxide resin.

3. The melamine resin would catalyze the acid-epoxide reaction, and would chemically combine with the acrylic polymer (via the carboxyl), the phenol, and the epoxide resin.

GENERAL REFERENCES

1. King, R. J., *J. Oil Colour Chemists' Assoc.*, **52**, 1075 (1969).
2. Wampner, H. L., *J. Oil Colour Chemists' Assoc.*, **52**, 309 (1969).

REFERENCES

1. Barrett, G. R., Monsanto Chemical Co., Brit. Pat. 673,849 (1952); Strolle, C. H., E. I. du Pont de Nemours & Co., U.S. Pat. 3,068,183 (1962); Rohm & Haas Co., Brit. Pat. 843,139 (1960).
2. Kelly, D. P., G. J. H. Melrose, and D. H. Solomon, *J. Appl. Polymer Sci.*, **7**, 1991 (1963).
3. Saxon, R., and J. H. Daniel, *J. Appl. Polymer Sci.*, **8**, 325 (1964).
4. Saxon, R., and F. C. Lestienne, *J. Appl. Polymer Sci.*, **8**, 475 (1964).
5. Canadian Industries Ltd., Brit. Pat. 955,670 (1964).
6. Petropoulos, J. C., C. Frazier, and L. E. Cadwell, *Offic. Dig. Federation Soc. Paint Technol.*, **33**, 719 (1961).
7. Costanza, J. R., and E. E. Waters, *Offic. Dig. Federation Soc. Paint Technol.*, **37**, 424 (1965).
8. American Cyanamid Co., Can. Pat. 536,704; Gaylord, N. G., Interchemical Corp., U.S. Pat. 2,853,462 (1958), U.S. Pat. 2,853,463 (1958); Rohm and Haas Co., Brit. Pat. 939,211 (1963).
9. Cousens, R. H., D. W. J. Osmond, and D. H. Solomon, Imperial Chemical Industries Ltd., Brit. Pat. 980,633 (1965).
10. American Cyanamid Co., Technical Bulletin on *N*-Methylol Acrylamide; Graulich W., and K. E. Müller, Farbenfabriken Bayer Akt.-Ges., Can. Pat. 601,430 (1960).
11. Christenson, R. M., and D. P. Hart, *Offic. Dig. Federation Soc. Paint Technol.*, **33**, 684 (1961).
12. Melrose, G. J. H., and D. H. Solomon, *J. Appl. Polymer Sci.*, **8**, 1777 (1964).

13. Forrest, D. B., and J. F. Smelko, Canadian Industries Ltd., Brit. Pat. 929,973 (1963); Nelan, N., Canadian Industries Ltd., Australian Pat. 250,615 (1963).
14. Christenson, R. M., and H. G. Bittle, Pittsburgh Plate Glass Co., Can. Pat. 607,933 (1960).
15. Pittsburgh Plate Glass Co., Brit. Pat. 840,325 (1960).
16. Vogel, H. A., and H. G. Bittle, *Offic. Dig. Federation Soc. Paint Technol.*, **33**, 699 (1961).
17. Leutner, F. S., and G. P. Roeser, American Marietta Co., Can. Pat. 680,815 (1964).
18. Taylor, J. R., and H. Foster, *J. Oil Colour Chemists' Assoc.*, **51**, 975 (1968).
19. Sekmakas, K., and R. F. Stancl, *J. Paint Technol.*, **38**, 217 (1966).
20. Balm Paints Ltd., Australian Pat. Appln. 29088/67.

Miscellaneous Cross-linking Reactions. The reactions of an isocyanate group with a carboxyl, hydroxyl, or amine group (Eqs. 10.10 to 10.12, pages 265–266 have been applied to acrylic systems.[1] Either the isocyanate or the other reactive group can be made a part of the acrylic polymer by suitable choice of monomers (Table 10.1) and cross-linking agents. These reactions are discussed more fully in Chapter 8.

Photochemical cross-linking can be brought about with selected polymer systems in the presence of a photosensitizer.[2] A polyacrylate containing 2-methylanthraquinone as sensitizer, after irradiation at 365 mμ, becomes insoluble in solvents for the untreated polymer. The cross-linking apparently involves the tertiary hydrogen of the acrylate, since poly(methyl methacrylate) is not readily attacked.

$$\begin{array}{ccc}
\text{(possible point of attack)} & & \\
\qquad\quad H \nearrow & & CH_3 \\
\qquad\quad | & & | \\
-CH_2-C-CH_2- & & -CH_2-C-CH_2- \\
\qquad\quad | & & | \\
\qquad\quad COOR & & COOR \\
\text{(acrylate chain)} & & \text{(methacrylate chain)}
\end{array}$$

A number of acrylic polymers that undergo autoxidative polymerization at room or slightly elevated temperatures have been described.[3-6] The mechanism of the curing reaction is similar to that discussed for the autoxidation of drying oils; in fact, some of the acrylic polymers may be regarded as analogous to oils in which the glycerol has been replaced by a polymeric acrylic-hydroxy derivative. Copolymers of epoxide monomers, such as allyl glycidyl ether and glycidyl methacrylate, can be prepared by free radical polymerization, and the epoxide ring opened by an unsaturated fatty acid.[4] This gives an "acrylic" fatty acid ester which is analogous to, but much more highly functional than, a vegetable oil.

$$\text{Polymer} + R\text{---COOH}$$

$$\begin{array}{c} HC \\ \diagdown \\ \quad O \\ H_2C \diagup \end{array}$$

$$\downarrow$$

$$\text{Polymer}$$

$$CH\text{---}OH$$
$$CH_2\text{---}O\text{---}CO\text{---}R$$

$$(R = C_{17}H_{(35-2x)})$$

$$(x = \text{number of double bonds in fatty acid chain})$$

Alternatively, the epoxide copolymer can be converted to a dioxolane by reaction with acrolein under acidic conditions. The reaction of these systems with oxygen is discussed in Chapter 5.

$$\text{Polymer} + \begin{array}{c} CHO \\ | \\ CH \\ \| \\ CH_2 \end{array}$$

$$\begin{array}{c} HC \\ \diagdown \\ \quad O \\ H_2C \diagup \end{array}$$

$$\downarrow H^+$$

$$\text{Polymer}$$

$$\begin{array}{c} CH\text{---}O \\ \diagdown \\ \qquad CH\text{---}CH{=}CH_2 \\ \diagup \\ CH_2\text{---}O \end{array}$$

$$(\text{a dioxolane})$$

The reaction of acid copolymers with unsaturated epoxide containing monomers has been used to give acrylic polymers susceptible to autoxidation. Allyl glycidyl ether is generally preferred in this process, since the methylene group of this system is readily attacked by oxygen as discussed previously.

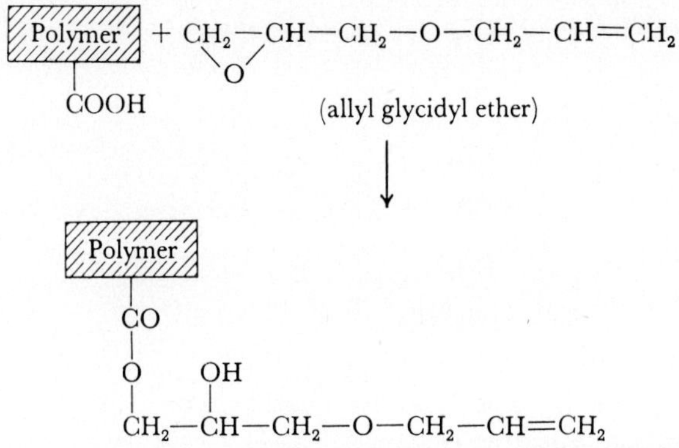

$$\boxed{Polymer} + CH_2-CH-CH_2-O-CH_2-CH=CH_2$$
$$\underset{COOH}{|} \qquad \underset{O}{\diagdown\diagup}$$

(allyl glycidyl ether)

$$\downarrow$$

$$\boxed{Polymer}$$
$$|$$
$$CO$$
$$|$$
$$O \qquad OH$$
$$| \qquad |$$
$$CH_2-CH-CH_2-O-CH_2-CH=CH_2$$

Styrene-maleic anhydride copolymers are converted to autoxidizable compositions by esterification with unsaturated fatty alcohols,[3] and these polymers are formally similar to the acrylic-fatty acid esters considered above.

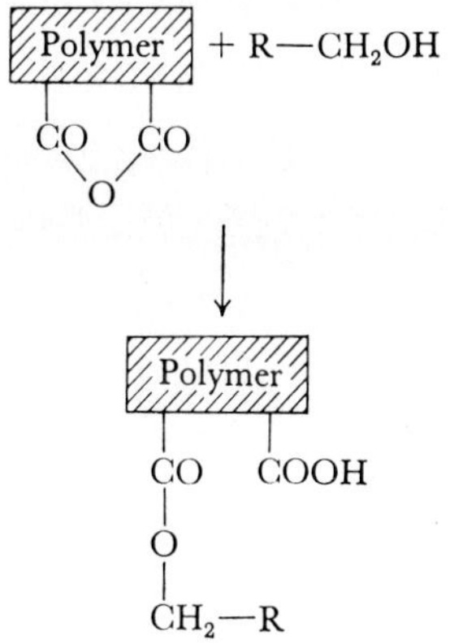

$$\boxed{Polymer} + R-CH_2OH$$
$$| \qquad |$$
$$CO \quad CO$$
$$\diagdown \diagup$$
$$O$$

$$\downarrow$$

$$\boxed{Polymer}$$
$$| \qquad |$$
$$CO \quad COOH$$
$$|$$
$$O$$
$$|$$
$$CH_2-R$$

$$(R = C_{17}H_{(35-2x)})$$

$(x =$ number of double bonds per fatty alcohol)

Base-catalyzed polymerization of allyl glycidyl ether is another possible synthetic route to an oxidizable acrylic polymer where the cross-linking reaction involves the active methylene group α- to the carbon-carbon double bond.

$$n\,CH_2\underset{\displaystyle O}{-}CH \longrightarrow \{CH_2-CH-O\}_n$$

allyl glycidyl ether

(point of oxidative attack)

Vinyl ethers of the fatty alcohols give another type of polymer that cross-links by oxidation (Chapter 4).

A formulation modification applicable in most of the above systems is the incorporation of maleate or fumarate unsaturation into the backbone polymer. By analogy with unsaturated polyester compositions, such a modification should enhance cross-linking of the polymer film by the mechanisms shown in Chapter 5.

REFERENCES

1. Canadian Industries Ltd., Brit. Pat. 955,670 (1964); Dorough, G. L., E. I. du Pont de Nemours & Co., U.S. Pat. 2,277,083 (1942); Hanford, W. E., and D. F. Holmes, E. I. du Pont de Nemours & Co., U.S. Pat. 2,284,896 (1942); Brown, W. H., and T. J. Miranda, *Offic. Dig. Federation Soc. Paint Technol.*, **36**, (**475**), 92 (1964).
2. Oster, G., *Polymer Letters*, **2**, 1181 (1964).
3. Dennis, K. S., C. E. Lyons, and D. R. Spiekerman, *Offic. Dig. Federation Soc. Paint Technol.*, **37**, 378 (1965).
4. Imperial Chemical Industries Ltd., Australian Pat. 166,197 (1955).
5. Shechter, L., and J. Wynstra, *Ind. Eng. Chem.*, **48**, 86 (1956).
6. Chitwood, H. C., and B. T. Freure, *J. Am. Chem. Soc.*, **68**, 680 (1946).

CHAPTER 11

WATER DILUTABLE POLYMER SYSTEMS

The use of organic solvents or diluents to give polymer solutions, or dispersions, with acceptable properties when used as surface coatings, presents certain problems. The volatile organic compounds are a fire hazard, they are toxic, they contribute to atmospheric pollution, and they increase the cost of the coating; yet, they do not form part of the final film. These problems are overcome when water is used as the solvent or diluent for the coating composition. In addition, water-thinnable systems offer the possibility of easier cleaning of manufacturing and application equipment. However, the use of water as a solvent presents other problems to the paint formulator; water has a high latent heat of vaporization (when compared to organic solvents) and a fixed boiling point, hence, it is not easy to control flash-off rates as is done with blends of organic solvents; water does not readily wet some surfaces, and it attacks or swells others; and conventional polymers must be modified to provide acceptable solubility, or dispersibility, in water.

The design of polymer systems that are thinnable—or dilutable—with water has proceeded along two lines. The polymer structure and other components have been chosen so that a water-soluble system results; alternatively, the polymer has been dispersed in water, by the use of suitable techniques, to give a latex. The water-soluble, and water-dispersible, compositions generally offer a different balance of properties; consequently, the tendency has been to utilize each for different types of coating compositions. The water-soluble polymers can be treated in a similar manner to existing solvent-soluble systems. Therefore, the paint formulator can directly apply many of the formulating principles developed with the solvent-based polymers. Pigment dispersion is comparatively simple, and the solution polymers can give glossy films at acceptable pigment:binder ratios.[1] The development of glossy aqueous emulsion paints has necessitated changes in formulation practice; however recent publications suggest that obtaining "high" gloss from aqueous latex systems is no longer a major problem. A contributing factor to the low gloss of some emulsion systems is the abrupt change in rheology of the film as the water evaporates.[2] On the other hand, the emulsion or latex system gives much higher "solids at application viscos-

293

ity" than a solution coating, and it is not necessary to chemically modify the polymer to make it hydrophilic. As a consequence, the development of a water-resistant film is more readily achieved from a latex than from a solution coating. The relative merits of the various water soluble and water dispersible polymers in practical formulations are discussed in reference.[3]

REFERENCES

1. Formo, M. W., and W. B. Smallwood, *Paint, Varnish, Prod.,* **49**, 39 (1959).
2. Stickle, F. J., *Offic. Dig. Federation Soc. Paint Technol.,* **31**, 93 (1959).
3. Hunt, T., *J. Oil Colour Chemists' Assoc.,* **35**, 380 (1970).

AQUEOUS DISPERSIONS OF POLYMERS

A dispersion of an organic polymer in water can be prepared by two distinct methods. A preformed polymer can be emulsified in water; alternatively, the polymer can be prepared by polymerization of an aqueous emulsion of the monomer. The method chosen to prepare an emulsion of a particular polymer system will depend on the type of polymer and on the second-order transition temperature of the polymer. In general, emulsions of polymers, which are made by stepwise reactions involving the elimination of water, are prepared by the emulsification of the preformed polymer; the vinyl and acrylic polymer emulsions are prepared by emulsion polymerization of the monomer.

Vegetable oils, polyesters, and alkyd resins are emulsified by mixing with water in equipment designed to give high shear rates. Suitable surfactants and colloidal stabilizers are added to give a stable emulsion. Some resin formulations have the surfactant as part of the polymer structure. In polyesters and alkyd resins, this is usually done by using poly(ethylene glycols)[1] as part of the polyol, and/or by forming an amine salt of the free carboxyl groups. For example a polyester which contains styrene has been emulsified by the use of triethanolamine salts of the terminal carboxyl groups in the the polyester.[5] Reactive surfactants can also be used; these offer the advantage of becoming water insensitive after the film has formed. The amine salts of unsaturated fatty acids are commonly used for this purpose.[2]

Cationic latexes can be prepared by the incorporation of basic groups into the polymer. An example of this approach is the polyurethane latex, described by Suskind;[3] an isocyanate prepolymer is reacted with an alkyl diethanolamine, then dispersed in aqueous acetic acid. The emulsified polymer undergoes cross-linking by reaction of the free isocyanate groups with water that diffuses into the polymer droplets. Alternatively the polyurethane can be crosslinked by formaldehyde which forms methylene bridges between two urethane linkages as shown below.[6]

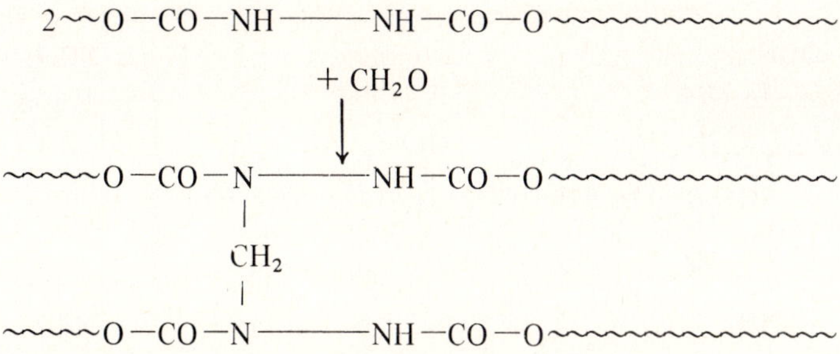

$$2 \sim\!\!\sim O-CO-NH\text{———}NH-CO-O\sim\!\!\sim\!\!\sim\!\!\sim\!\!\sim$$

$$+ CH_2O$$

$$\downarrow$$

$$\sim\!\!\sim\!\!\sim O-CO-N\text{———}NH-CO-O\sim\!\!\sim\!\!\sim\!\!\sim\!\!\sim$$

$$|$$

$$CH_2$$

$$|$$

$$\sim\!\!\sim\!\!\sim O-CO-N\text{———}NH-CO-O\sim\!\!\sim\!\!\sim\!\!\sim\!\!\sim$$

Yet another method of crosslinking the urethane latex is by bifunctional quaternising agents such as 1, 4, *bis.*-chloromethyl benzene. The crosslinked structure will contain the groups shown below.[6]

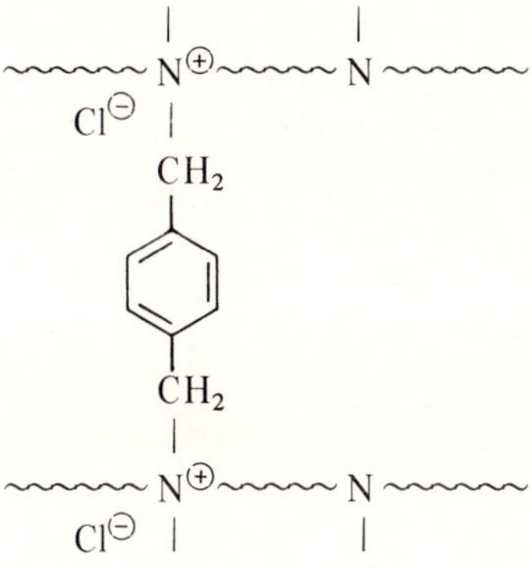

$$\sim\!\!\sim\!\!\sim N^{\oplus}\!\!\sim\!\!\sim\!\!\sim N\sim\!\!\sim \quad Cl^{\ominus}$$

$$CH_2$$

$$CH_2$$

$$\sim\!\!\sim\!\!\sim N^{\oplus}\!\!\sim\!\!\sim\!\!\sim N\sim\!\!\sim \quad Cl^{\ominus}$$

Film formation from the emulsions of oils, alkyds, etc. occurs first, by loss of water and coalescence of the polymer drops and, second, by some type of cross-linking reaction. In broad principle, these reactions follow similar courses to those discussed for solvent carried polymers; unsaturated portions of the molecules may undergo autoxidation;[4] interaction between hydroxyl and carboxyl groups of the polymer with *N*-methoylol ethers of a second polymer (often present in the aqueous phase) can occur; or poly-(epoxy) compounds can undergo self-polymerization. The yellowing sometimes encountered with films from linseed oil, or linseed oil alkyd, emulsions can be minimized by the addition to the latex, or methyl azelaaldehydate.[7]

$$4\,OCN\text{—}[Polymer]\text{—}NCO + 2\,RN(CH_2\text{—}CH_2\text{—}OH)_2$$

$$\downarrow$$

$$2\,[Polymer]\underset{OCN}{\overset{}{\big|}} \quad NH\text{—}\underset{O}{\overset{}{C}}\text{—}O\text{—}CH_2\text{—}CH_2\text{—}\underset{R}{\overset{}{N}}\text{—}CH_2\text{—}CH_2\text{—}O\text{—}\underset{O}{\overset{}{C}}\text{—}HN \quad [Polymer]\underset{NCO}{\overset{}{\big|}}$$

$$\downarrow H_2O$$

$$2\,[Polymer]\underset{NH_2}{\overset{}{\big|}} \quad NH\text{—}\underset{O}{\overset{}{C}}\text{—}O\text{—}CH_2\text{—}CH_2\text{—}\underset{R}{\overset{}{N}}\text{—}CH_2\text{—}CH_2\text{—}O\text{—}\underset{O}{\overset{}{C}}\text{—}HN \quad [Polymer]\underset{NCO}{\overset{}{\big|}}$$

$$\downarrow$$

(chemical structure of crosslinked polymer network)

REFERENCES

1. Arndt, R. P., Pittsburgh Plate Glass Co., U.S. Pat. 2,634, 245 (1953), via *Chem. Abstr.*, **47**, 90030, 1953; Armitage, F., and L. G. Trace, Berger and Sons Ltd., Brit. Pat. 845, 861 (1960); Arndt, R. P., Pittsburgh Plate Glass Co., Can. Pat. 562,913 (1958); Sherwin-Williams Co., U.S. Pat. 3,001,961 (1953); Wood, D. L., L.L. Hopper, and F. G. Dollear, *J. Am. Oil Chemists' Soc. Abstr.*, **40**, No. 9, 27 Eastman Kodak Company, Aust. Pat. Appl. 48548/68, Staley Manufacturing Co., U.S. Pat., 3, 457, 206 (1963).
2. Clarke, D. R. Shell Research Ltd., Brit. Pat 878, 139 (1960).
3. Suskind, S. P., *J. Appl. Polymer Sci.*, **9**, 2451 (1965).
4. Schwab, A. W., W. L. Kubic, J. A. Stolp, J. C. Cowan, and H. M. Teeter, *Offic. Dig. Federation Soc. Paint Technol.*, **33**, 747 (1961).
5. Hakutaro, K., M. Itaru, and H. Kazuyuki, Japanese Pat., 6,931, 831 (1969).
6. Dieterich, D., W. Keberle, and R. Wuest, *J. Oil Colour Chemists' Assoc.*, **53**, 363 (1970).
7. McManis, G. E., E. W. Swain, L. E. Gast, and J. C. Cowan, *J. Paint Technol.*, **43** (554), 53 (1971).

Vinyl and Acrylic Emulsions. The emulsion polymerization of a vinyl or acrylic monomer takes place by a free-radical mechanism. The general features of such a mechanism involve three distinct stages: initiation, propagation, and termination.

The initiators used in emulsion polymerization are water soluble, and they generate free radicals in the aqueous solution as a result of thermal or catalyzed (redox system) decomposition. This initiating free radical then attacks a monomer unit inside a micelle of emulsifier which has solubilized some of the monomer. It is also possible that a few monomer units add to the radical in the aqueous phase; this would render the radical less hydrophilic and would encourage migration to the micelle. The polymer chain then continues to grow in the micelle, with the monomer concentration being maintained by diffusion, via the aqueous phase, of monomer from other solubilized droplets of monomer. The propagation step will continue until a second initiator radical diffuses into the particle and causes termination by combination. In this manner, the latex particles are built up; stabilization results from the adsorption of emulsifier onto the latex particle (Gen. Ref. 1). In the particular case of persulphate-initiated polymerizations, the chain ends are SO_4^- groups, which contribute to the stability of the emulsified polymer drops. Kinetic studies have suggested that the molecular weight of the polymer is dependent upon the particle concentration, and is inversely proportional to the initiator concentration.[1,2,3] The details of the initial stages of the polymerization do not always follow the simple quantitative theory developed by Smith and Ewart.[1] A detailed analysis of the theoretical predictions of the Smith-Ewart theory and a discussion of experimental data in the context of this theory has been presented by Gardon.[21]

The initiators used in emulsion polymerization have been reviewed (Gen. Ref. 1); the common initiators include the water-soluble persulfates and redox systems such as hydrogen peroxide and a ferrous salt. The redox system, which generates the initiating radicals at room temperature, is of

Graft copolymer formation by hydrogen abstraction

$$
\begin{bmatrix} CH_2-CH-CH_2-CH \\ \quad\; | \qquad\qquad | \\ \quad\; OH \qquad\quad OH \end{bmatrix}_n \xrightarrow{R\cdot} \begin{bmatrix} CH_2-\overset{\cdot}{C}-CH_2-CH \\ \quad\; | \qquad\qquad\; | \\ \quad\; OH \qquad\quad OH \end{bmatrix}_n + RH
$$

$$\downarrow x\,CH_2{=}CH-O-CO-CH_3$$

$$
\begin{bmatrix} \qquad\quad OH \\ \qquad\quad\; | \\ CH_2-C-CH_2-CH \\ \qquad\; | \qquad\qquad | \\ \quad \begin{bmatrix} CH_2 \qquad\quad OH \\ \;\; | \\ \overset{\cdot}{C}H-O-CO-CH_3 \end{bmatrix}_x \end{bmatrix}_n
$$

$$\left[\!\!\begin{array}{c} \text{O—CO—CH}_3 \\ | \\ \text{CH}_2\text{—CH—CH}_2\text{—CH} \\ | \\ \text{O—CO—CH}_3 \end{array}\!\!\right]_n$$

$$\Big\downarrow\ \text{H}_2\text{SO}_4/\text{CH}_3\text{—OH}$$

$$\left[\!\!\begin{array}{c} \text{CH}\!=\!\text{CH—CH}_2\text{—CH} \\ | \\ \text{OH} \end{array}\!\!\right]_n$$

$$\Big\downarrow\ \begin{array}{l}\text{R}^{\bullet} \\ +\ x\,\text{CH}_2\!=\!\text{CH—O—CO—CH}_3\end{array}$$

$$\left[\!\!\begin{array}{c} \text{R} \\ | \\ \text{CH—CH—CH}_2\text{—CH} \\ | \qquad\qquad\ | \\ \left[\!\begin{array}{c}\text{CH}_2 \\ |\\ \overset{\bullet}{\text{CH}}\text{—O—CO—CH}_3\end{array}\!\right]_x \qquad \text{OH} \end{array}\!\!\right]_n$$

particular interest where the emulsion is unstable at elevated temperatures, or where the structure of the polymer is influenced by temperature. In the emulsion polymerization of butadiene, or butadiene copolymers, the amount of *trans* 1:4 structures formed increases as the temperature is lowered (Gen. Ref. 1). The *trans* structure influences the physical properties of the polymer and, other things being equal, would also be expected to increase the resistance to oxidative degradation. An alternative means of initiating emulsion polymerization at low temperatures, and at elevated temperatures, is by radiation using a cobalt 60 source. This method of preparing emulsions is claimed to give a vinyl acetate-butyl acrylate copolymer which gives paints with superior enamel hold to conventionally initiated latexes.[22]

Protective colloids are sometimes used in an emulsion polymerization, either in conjunction with conventional surfactants, or as the sole stabilizer. Poly(vinyl alcohol), which is often used as the colloid in the preparation of poly(vinyl acetate) emulsions, is modified chemically during the polymerization. Graft copolymers are formed by two mechanisms; a hydrogen atom can be abstracted from the poly(vinyl alcohol) chain,[4] or attack can be at double bonds in the poly(vinyl alcohol) chain.[5] The latter type of colloid is prepared by the alcoholysis of poly(vinyl acetate) by sulfuric acid in dry methanol;[5] the carbon-carbon double bonds arise from the loss of water or acetic acid from the polymer chain.

These graft copolymers are surface-active agents,[4] and they influence the course of the emulsion polymerization; the more graft copolymer, the greater is the number of growing particles, and the faster is the rate of polymerization.[4] The graft copolymer also influences the stability and film-forming properties of the emulsion; more graft chains per molecule of poly(vinyl alcohol) give an emulsion with greater stability to organic solvents.[5] This property is the result of the graft copolymer not being soluble in water or in the organic solvent. Film formation of externally plasticized latexes depends, in part, on the ability of the plasticizer to attack the stabilizing layer of graft copolymer.

The mechanism by which a dispersion of a polymer is converted to a film has been studied by a number of workers.[6–10] While some details of the drying process are still not agreed upon, the general mechanism appears to proceed by the following steps. The water or continuous phase evaporates to allow the polymer particles to compact. This stage is followed by loss of water from the capillaries in the surface. The forces created during the loss of this water cause the polymer drops to be pulled into the void spaces; film formation by particle coalescence will then take place if the polymer particles are sufficiently soft, or in other words, if their minimum film-forming temperature is approximately equal to the drying temperature. Studies on a poly (vinylidene chloride/methyl acrylate/acrylic acid) latex have emphasized the significance of Tg of the polymer to the coalescence of the latex particles. Coalescence only occurred above the Tg of the polymer.[23]

The minimum film-forming temperature of the polymer droplet is related to a number of factors. The chemical composition of the polymer influences the second-order transition temperature, which is related to the film-forming temperature. Homopolymers of vinyl acetate, methyl methacrylate, and styrene are all too hard to form continuous films from an emulsion at low temperatures, consequently, this must be plasticized or softened. Copolymerization with monomers whose polymers have a low transition temperature is one common technique used to modify the film-forming properties of an emulsion. Alternatively, an external plasticizer (such as dibutyl phthalate) can be added, and this migrates from the aqueous phase to the polymer drop. The molecular weight and chemical distribution of the copolymer is also related to the film-forming temperature. A broad distribution of molecular weights and a heterogeneous chemical structure give a wider transition range than a more uniform polymer. This is an advantage,[10] since a latex paint has not only to form a continuous film, but also to withstand substrate movement (and hence, be rubbery), and yet not pick up excessive dirt (and hence, be glassy). Chemical distribution within the polymer drop aids film formation if the softer polymer is on the outer surface of the particle. For example, a poly(vinylidene chloride) core with an alkyl acrylate sheath gives improved properties over the conventional latexes of this type.[11] Another example of the manner in which the polymer composition can be varied in the latex polymer particle is where the inner core is a rubbery crosslinked polymer (e.g. C_2-C_8 alkyl acrylate units) and subsequent layers have different degrees of crosslinking and higher Tg's.[24]

Solvents are often added to emulsion paints to improve the low temperature coalescent properties, the film levelling, and the freeze-thaw stability.[12] Typical solvents used are the higher glycols and their ethers.

During the drying of an emulsion film, considerable polymer flocculation can occur, particularly in the presence of some synthetic thickening agents (added to give acceptable viscosity characteristics). It has been predicted that polymer flocculation reduces the durability of the system[13] and, for maximum performances, a nonflocculating thickening agent is required.[13] A special silica aerogel is claimed to be a nonflocculating thickener.[14] Organic polymers which increase the viscosity of latexes with a minimum of flocculation and which result in paint films with semi-full gloss have been reported. These new thickeners have resulted from changes in the approach to latex stabilization and thickening. Conventional thickeners, such as hydroxy ethyl cellulose, polyvinyl alcohol, sodium carboxy cellulose and sodium polyacrylate, general increase the viscosity of the continuous phase (i.e. water phase) and result in some particle bridging and hence flocculation

TABLE 11.1

ENTROPIC STABILIZERS FOR LATEXES.

Stabilizer	*Reference*
Block copolymers of ethylene oxide and propylene oxide with Vinyl or acrylic esters	D. H. Napper, *J Colloid and Interface Sci.,* 29, 168 (1969)
Random copolymers vinyl acetate-Vinyl alcohol (these are unsuitable in alkali)	R. W. Kershaw and F. J. Lubbock U.S. Pat., 3,514, 421 (1970)
Random copolymers vinyl alcohol-vinyl acetal	R. W. Kershaw and F. J. Lubbock U.S. Pat., 3, 514,421 (1970)
Random copolymers vinyl alcohol-vinyl ethers	R. W. Kershaw F. J. Lubbock and L. Polgar, U.S. Pat., 3,553,152 (1971)
Vinyl alcohol-Styrene	J. J. Hopwood, R. W. Kershaw and F. J. Lubbock South Africa Pat., 6,808,407 (1970)
Vinyl alcohol-vinyl versatic ester	J. J. Hopwood, R. W. Kershaw, and F. J. Lubbock Can. Pat., 851,771 (1970)

of the latex. Using the principle of entropic stabilization, synthetic thickeners have been developed which minimize particle bridging and which are only present to a small amount in the continuous phase. These thickeners function by being absorbed onto single polymer particles and thereby increasing the hydrodynamic volume of the particles.[25] Typical entropic stabilizers for latexes are listed in Table 11.1.

The latexes used in household paints are usually of the thermoplastic type; typical compositions are (1) vinyl acetate homopolymers or copolymers with maleates, fumarates, acrylates, ethylene[26] and vinyl esters of versatic acids,[27] (2) methyl methacrylate copolymers internally plasticized with ethyl acrylate, (3) styrene-butadiene formulations. A comparatively recent development has been the use of non-linear polymers in latex formulations. It has been suggested that some crosslinking of the polymer before film application is desirable and can result in an improvement in the mechanical properties of the paint. In other formulations the crosslinking has been introduced after the film formation by promoting oxidation by the use of sensitizers, such as benzophenone or by the use of autoxidizable groups in the copolymer. Plasticizers, such as those which contain allyl ether groups, susceptable to crosslinking by oxidation have also been used. Generally these approaches have aimed at using latexes with low Tg's, to facilitate film formation, and a subsequent increase in Tg (by crosslinking) to minimize the dirt pick-up of the films.

Small amounts of monomers with groups which improve the adhesion of latex paints are often included in the polymer formulation. An example of this approach is the use of monomers which contain carboxyl, and aziridine groups.[28] In some cases, small amounts of acid monomers are included to provide a self-thickening emulsion when ammonia is added to form a salt. Alternatively, water-soluble thickeners of the ammonium poly (acrylate) type can be added.

Thermosetting emulsion polymers can be formulated by including suitable functional groups in the polymer droplet; in some cases, a water-soluble polymer is used in conjunction with the emulsion to assist in the thermosetting reactions. Styrene-butadiene emulsions undergo autoxidation in the presence of metallic driers (such as cobalt salts) at room temperature; the cross-linked films are more abrasion and solvent resistant than the thermoplastic polymer.[15] Butadiene-type emulsion polymers, which also contain carboxyl groups, can be converted to vulcanized films with alkali metal aluminates by the formation of aluminum salts.[16]

The reaction of an *N*-methylol, or *N*-methylol ether group, with either hydroxy, amino, or carboxy groups has been applied to emulsion systems. Both groups may be present in the emulsion drop,[17] or in a water-soluble formaldehyde resin [hexa(methoxymethyl)melamine] used to supply the

N-methylol ether groups.[18] The cross-linking reaction occurs after particle fusion. The use of elevated temperatures not only aids the cross-linking reactions, but also assists in giving a continuous film with less stress than that obtained from many air dried emulsions. The distribution of the reactive groups in the emulsion droplet can be used to advantage in masking the group until after particle fusion; this can assist in improving the stability of the system.

One of the more recent applications of water-dilutable polymer systems has been in the formation of coatings by electrodeposition.[19, 20] The technology and theory of this process are discussed in Chapter 13.

GENERAL REFERENCES

1. Bovey, F. A., I. M. Kolthoff, A. I. Medalia, and E. J. Meehan, *High Polymers 9, Emulsion Polymerization*, Interscience Publishers, Inc., New York, 1955.

REFERENCES

1. Parts, A. G., and D. E. Moore, *J. Oil and Colour Chemists' Assoc.*, **45**, 648 (1962).
2. Alexander, A. E., *J. Oil and Colour Chemists' Assoc.*, **45**, 12 (1962).
3. Fitch, R. M., *Offic. Dig. Federation Soc. Paint Technol.*, **37**, Oct. Part 2, 32 (1965).
4. Hartley, F. D., *J. Polymer Sci.*, **34**, 397 (1959).
5. Gregor, F., and E. Pavlacka, *Chem. Prumysl*, **12/37**, No. 12, 689 (1962).
6. Brown, G. L., *Offic. Dig. Federation Soc. Paint Technol.*, **37**, Oct. Part 2, 15 (1965).
7. Myers, R. R., and R. K. Schultz, *J. Appl. Polymer Sci.*, **8**, 755 (1964).
8. Hwa, J. C. H., *J. Polymer Sci.*, **2**, 785 (1964).
9. Brodnyan, J. G., and T. Konen, *J. Appl. Polymer Sci.*, **8**, 687 (1964).
10. Jaffe, H. L., and J. H. Fickenscher, *Offic. Dig. Federation Soc. Paint Technol.*, **35**, 169 (1963).
11. The Dow Chemical Co., Brit. Pat. 963,296 (1964).
12. Cogan, H. D., *Offic. Dig. Federation Soc. Paint Technol.*, **32**, 1216 (1960); Wilson, I. V., Monsanto Chemical Co., Can. Pat. 529,707 (1956).
13. Hahn, F. J., and J. F. Heaps, *Offic. Dig. Federation Soc. Paint Technol.*, **35**, 459 (1963).
14. Marotta, R., and H. Teicher, *Offic. Dig. Federation Soc. Paint Technol.*, **36**, (**475**), 853 (1964).
15. Schaeffer, L., *Can. Paint Varnish*, **34**, No. 4, 53 (1960).
16. Eilbeck, G. E., and E. R. Urig, B.F. Goodrich Co., U.S. Pat. 2,868,754 (1959).
17. Eilbeck, G. E., and R. Y. Garrett, B.F. Goodrich Co., Brit. Pat. 888,503 (1962); Noma, K., R. Yosmiya, and T. Fukumara, *J. Japan Soc. Col. Mat.*, **37**, No. 3, 80 (1964), via *Rev. Current Lit. Paint Allied Ind.*, **1965**, March, 206.
18. Hornibrook, W. J., E. I. du Pont de Nemours and Co., U.S. Pat. 2,918,391 (1959); Koral, J. N., and J. C. Petropoulos, Am. Chem. Soc. Abstr. 149th Meeting 2Q. (1965).
19. Sullivan, M. R., *Offic. Dig. Federation Soc. Paint Technol.*, **37**, No. 489, Part 2, 108 (1965); Gloyer, S. W., D. P. Hart, and R. E. Cutforth, *Offic. Dig. Federation Soc. Paint Technol.*, **37**, 113 (1965); O'Neill, L. A., *Rev. Current Lit. Paint Allied Ind.*, **37**, No. 262, 217 (1964); Finn, S. R., and C. C. Mell, *J. Oil and Colour Chemists' Assoc.*, **47**, 219 (1964).
20. Tasker, L., and J. R. Taylor, *J. Oil and Colour Chemists' Assoc.*, **48**, 121 (1965).

21. Gardon, J. L., *J. Polymer Sci.*, **6**, 623; **6**, 643; **6**, 665; **6**, 687 (1968).
22. Woodward, D. G., and J. A. Ransohoff, *J. Paint Technol.*, **42**, 467 (1970).
23. Bertha, S. L., and R. M. Ikeda, *J. Applied Polymer Sci.*, **15**, 105 (1971).
24. Rohm and Haas Co., Aust. Pat. Appl. 16,958/67.
25. Lubbock, F. J., and R. W. Kershaw, private communication.
26. Reynolds, G. E. J., *J. Oil Colour Chemists' Assoc.*, **53**, 399 (1970).
27. Vegter, G. C., and E. P. Grommers, *J. Oil Colour Chemists' Assoc.*, **50**, 72 (1967).
    Aten, W. C., and G. C. Yegter, *J. Oil Colour Chemists' Assoc.*, **53**, 448 (1970).
28. Kagan, G. M., and R. Roper, Esso Research and Engineering Co., U.S. Pat., 3,549,567 (1970).

WATER-SOLUBLE POLYMER SYSTEMS

It is doubtful if many of the water-soluble polymers form true solutions in water. In some cases, at least, the polymer exists as a colloidal dispersion but, since the particle size is below that visible to the eye, they are clear and are called "solutions." For example it has been shown that aqueous "solutions" of alkyd resins have a two-phase structure at low water contents and that this structure changes with increasing water content of the solutions. The outer phase of the microemulsion is the one rich in resin, the inner phase being rich in water. Even when solutions we diluted beyond the turbidity point, no phase inversion was observed.[27]

For an organic polymer to be water soluble, it must contain polar groups with a strong affinity for water; the ether, carboxyl, hydroxyl, amide, and ionized carboxyl groups are typical of those used to render a polymer water soluble. Water solubility is also enhanced if the molecular weight of the polymer is low (by comparison with the polymers used in organic solvents). Some water-miscible organic solvent is often added to reduce the viscosity of the solution, to control solvent release, and to improve the flow and wetting properties of the solution.[1,2]

Thermoplastic water-soluble polymers find very little use as surface coatings, since the film would remain permanently water sensitive. Consequently, water-soluble coating compositions are generally thermosetting, and the technical problems peculiar to these systems are, first, to make the polymer sufficiently hydrophilic to be water soluble and, second, to convert the soluble polymer to a water-resistant and durable film. The methods used to satisfy the above requirements are conveniently considered in relation to the polymer present.

Water-Soluble Vinyl and Acrylic Polymers. Low molecular weight poly(vinyl alcohols) are water soluble and give films with good oil resistance;[3] however, their water sensitivity has limited their general use. Water-resistant films are obtainable from poly(vinyl alcohol) and from other polyhydroxy materials, if the polymer is cross-linked or converted to a hydrophobic structure, by reaction of the hydroxyl groups. Various light-sensitive compounds have been used in conjunction with poly(vinyl alcohol) to give water-insoluble films. Diazonium and tetrazonium salts[4] react with the hydroxyl groups and form a cross-linked structure that contains phenyl ether bonds (Eq. 11.1). Similarly, chromium compounds (for example, potassium dichromate) form cross-linked networks by coordination with the hydroxyl groups of the polymer.[5] Other metal salts also cross-link poly(vinyl alcohol).[6]

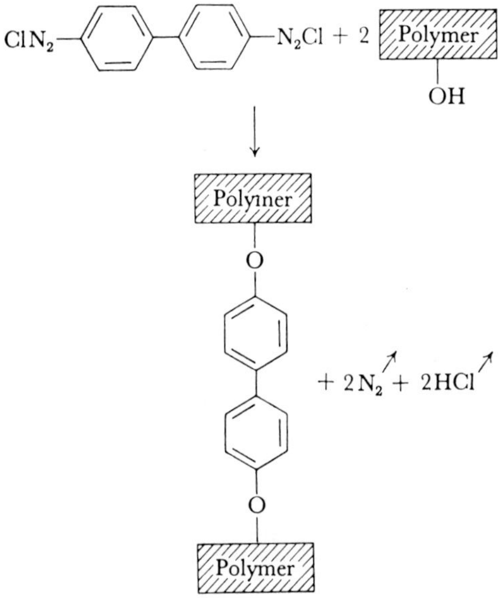

$$(11.1)$$

Boric acid forms a cyclic complex with poly(vinyl alcohol)[6] and, in so doing, removes the hydrophilic hydroxyl groups (Eq. 11.2). Aldehydes,[7] and dialdehydes,[8] also remove the hydroxyl groups by the formation of acetals (Eq. 11.3).

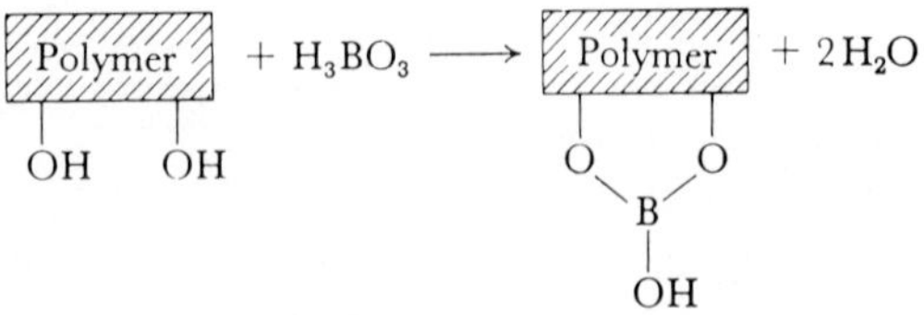

$$(11.2)$$

$$\boxed{Polymer}-CH-CH_2-CH- + RCHO \longrightarrow$$
$$\qquad\qquad\;\; | \qquad\qquad | $$
$$\qquad\qquad OH \qquad\quad OH$$

$$\boxed{Polymer}-CH-CH_2-CH- + H_2O$$
$$\qquad\qquad\qquad\qquad | \qquad\qquad\quad | $$
$$\qquad\qquad\qquad\qquad O \qquad\qquad\;\; O$$
$$\qquad\qquad\qquad\qquad\quad \diagdown\qquad\diagup$$
$$\qquad\qquad\qquad\qquad\qquad CH$$
$$\qquad\qquad\qquad\qquad\qquad\; | $$
$$\qquad\qquad\qquad\qquad\qquad\; R \qquad\qquad (11.3)$$

In the specific case of formaldehyde, dimethylol urea and ammonium chloride can be used to liberate the formaldehyde in situ and form the poly(vinyl acetal)[9] at room temperature. At elevated temperatures, dimethylol urea and other amine- or phenol-formaldehyde resins interact with the hydroxyl groups (Eq. 11.4); this reaction is catalyzed by acids.

$$\boxed{Polymer} + \boxed{Polymer} \longrightarrow \boxed{Polymer} + ROH\nearrow$$
$$\qquad | \qquad\qquad\qquad | \qquad\qquad\qquad\quad | $$
$$\qquad OH \qquad\qquad\;\; NH \qquad\qquad\qquad O$$
$$\qquad\qquad\qquad\qquad\quad | \qquad\qquad\qquad\qquad | $$
$$\qquad\qquad\qquad\qquad\; CH_2 \qquad\qquad\qquad CH_2 \qquad\qquad (11.4)$$
$$\qquad\qquad\qquad\qquad\quad | \qquad\qquad\qquad\qquad | $$
$$\qquad\qquad\qquad\qquad\; OR \qquad\qquad\qquad\;\; NH$$
$$\qquad\qquad\qquad\qquad\qquad\qquad\qquad\qquad \boxed{Polymer}$$

Other cross-linking agents also interchange with the hydroxyl groups of poly(vinyl alcohol); tetrabutyl titanite loses butanol (Eq. 11.5), and bis-(3-methoxy propylidene) pentaerythritol splits out methanol.[10]

$$2\,\boxed{Polymer}-OH + Ti(OC_4H_9)_4 \longrightarrow \boxed{Polymer} + 2\,C_4H_9OH\nearrow$$
$$\qquad\qquad\qquad\qquad\qquad\qquad\qquad\qquad\qquad | $$
$$\qquad\qquad\qquad\qquad\qquad\qquad\qquad\qquad\quad O$$
$$\qquad\qquad\qquad\qquad\qquad\qquad\qquad\qquad\quad | $$
$$\qquad\qquad\qquad\qquad\qquad\qquad\qquad\;\; Ti(OC_4H_9)_2 \qquad (11.5)$$
$$\qquad\qquad\qquad\qquad\qquad\qquad\qquad\qquad\quad | $$
$$\qquad\qquad\qquad\qquad\qquad\qquad\qquad\qquad\quad O$$
$$\qquad\qquad\qquad\qquad\qquad\qquad\qquad\qquad\quad | $$
$$\qquad\qquad\qquad\qquad\qquad\qquad\qquad\qquad \boxed{Polymer}$$

The incorporation of carboxyl groups is by far the most common method of inducing water solubility into a polymer. The acid groups are subsequently neutralized by an organic, or inorganic, base. Solubilization by salt formation generally results in a solution with a higher solids content, at a given viscosity, than is obtainable by the use of nonionizable hydrophilic groups alone. The acid groups can be incorporated in the polymer by a number of methods: an unsaturated acid (acrylic, methacrylic acids) can be used in the original polymerization; suitable groups (ester, amide, nitrile) on the polymer backbone can be hydrolyzed to the acid;[11] unsaturated groups present as a conjugated diene residue (butadiene) can be oxidized to the acid;[12] or hydroxyl groups along the polymer chain can be esterified with a polycarboxylic acid or anhydride. The acid monomer content of typical water-soluble polymers varies from 6 to 25% (by weight) of the polymer. At the lower acid limit, the distribution of the acid groups in the polymer chain is very important;[13,14] a polymer containing distributed acid groups gives better solubility, and is more resistant to added electrolytes.

The solubility of the acid-containing polymers is also influenced by other factors. As the molecular weight increases, the solubility in water decreases, and the viscosity of the solutions increases. As a consequence, it is necessary to limit the molecular weight of the polymer by the methods discussed previously (Chapter 1), particularly by the use of chain-transfer agents such as mercaptans (thioglycollic acid, butyl mercaptan)[15] and alcohols (isopropanol).[16]

The solubility of the polymer also relies on the presence of hydrophilic groups other than the ionized carboxyl; in some cases, these groups contribute as much to the polymer solubility as does the carboxyl anion.[13,17] The choice of the base used to solubilize the polymer is dictated by a number of considerations; generally, a low viscosity solution is desired, and the solution viscosity varies with the base used although, at the moment, satisfactory explanations for some of these viscosity variations are lacking. The base should also be readily lost from the film during the drying process, so as to remove the hydrophilic center of the molecule and/or to allow curing reactions to proceed. This requirement limits the general use of inorganic bases, although these are sometimes used in conjunction with volatile organic bases.[11] The solubilized polymer system must be stable; this is more likely to result when tertiary amines are used, since reactions possible with primary and secondary amines—aminolysis of ester links and/or depolymerization of added curing agents (page 245)—are avoided.

$$\boxed{\text{Polymer}}\text{—CO—OR} + R_2NH \longrightarrow \boxed{\text{Polymer}}\text{—CO—NR}_2 + ROH$$

The development of a water-resistant coating from a water-soluble acidic polymer can be achieved in a number of ways; the curing mechanism will be dictated by the drying conditions and the polymer structure. A degree of water insolubility is given by the loss of the volatile base. However, the ease with which this occurs depends on the polymer structure and the base. For example, a styrene:maleic anhydride copolymer, solubilized with ammonia, when dried at up to 105°C is resoluble in water, whereas a styrene:maleic anhydride half ester copolymer becomes insoluble.[18] These results have been taken to indicate the formation of a diamide-imide structure from the anhydride copolymer, and a monoamide from the half-ester copolymer.[18] On the other hand, acrylic acid copolymers, solubilized with ammonia, do not form an amide, but revert to the free acid.[19] The use of tertiary amines, as the solubilizing base, would be expected to favor regeneration of the acid form of the copolymer, since amide formation is precluded. In general use, the simple loss of the base is not sufficient to give films with adequate water resistance, and other chemical reactions are introduced to cross-link the polymer molecules.

Where the polymer system is required to dry or cure at room temperature, the cross-linking reactions involve salt formation or catalytic oxidation of unsaturated groups in the molecule. Salt formation can take place with divalent metal ions; these may be present in the polymer solution as a stabilized complex, or in the substrate; for example, an acid copolymer containing ammonium zirconyl carbonate, $(NH_4)_3HZrO(CO_3)_3$, dries at room temperature by the loss of carbon dioxide and ammonia, to give an insoluble zirconium-polymer salt.[20]

$$4\ \boxed{\text{Polymer}} + (NH_4)_3HZrO(CO_3)_3 \longrightarrow$$
$$\text{COO}^-\text{NH}_4^+$$

$$\left[\boxed{\text{Polymer}}-\text{COO}^-\right]_4 Zr^{4+} + 7NH_3\uparrow + 3CO_2\uparrow + 4H_2O\uparrow$$

Similarly, zinc salts can be formed from dissolved zinc oxide in ammoniacal solutions of polymers;[21] alternatively, the divalent ion may be present in the substrate (for example, calcium in plaster of paris, clay, or cement).[22] The gelation of an acid copolymer with divalent ions in solution is illustrated in Figure 11.1.[23] Salt formation is also possible between an organic polymer and the acidic copolymer; poly(ethylene oxide), or poly(ethylene glycol) as it is more commonly known, becomes insoluble by an acid-base type interaction with acid copolymers.[24]

$$\boxed{\text{Polymer}}$$
$$\text{COO}^-\quad \text{COO}^-$$

$$\text{H}\diagdown \qquad\qquad\qquad \text{H}\diagup$$
$$\text{O}^+ \qquad\qquad ^+\text{O} \qquad\qquad \text{O}$$
$$\text{CH}_2-\text{CH}_2\quad \text{CH}_2-\text{CH}_2$$

The incorporation of autoxidizable centers into an acrylic polymer can be carried out in a number of ways; these can be illustrated by reference to the specific case of an allyl alcohol-styrene copolymer. Esterification of some of the hydroxyl groups by a fatty acid, then reaction with maleic anhydride, gives a polymer that is water soluble as a salt and has air-drying potential.[25] Alternatively, allyl ethers or esters can be prepared and solubilized by half-ester formation of the remaining hydroxyl groups.

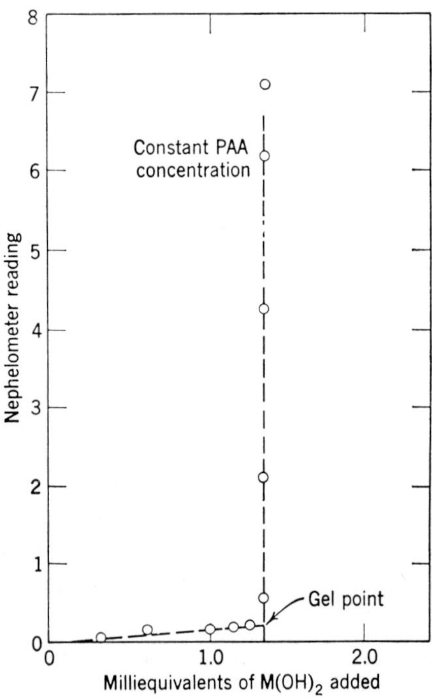

Figure 11.1.   Gelation of poly(acrylic acid) with divalent ions.[23]

The development of water-resistant films is greatly facilitated by the use of elevated temperatures for the following reasons: (1) the different types of reactions possible compared to systems that dry at room temperature; (2) the faster rate at which reactions occur at elevated temperatures; and (3) the removal of many of the hydrophilic groups in the cross-linking reactions. The most widely reported cross-linking system involves the interaction of *N*-methylol or *N*-methylol ether groups with hydroxyl and/or carboxyl groups. The chemical reactions are similar to those discussed for solvent systems (Chapter 10); however, reaction will not

$$\boxed{\text{Polymer}} + CH_3—(CH_2)_4—CH=CH—CH_2—CH=CH—(CH_2)_7—COOH$$
$$|$$
$$OH$$

$$\downarrow$$

$$\boxed{\text{Polymer}}$$
$$|$$
$$O—CO—(CH_2)_7—CH=CH—CH_2—CH=CH—(CH_2)_4—CH_3$$

$$\downarrow \quad \begin{array}{l} HC—CO \\ \quad\quad\;\;\; O \\ HC—CO \end{array}$$

$$\boxed{\text{Polymer}}$$
$$|$$
$$O—CO—(CH_2)_7—CH=CH—CH—CH=CH—(CH_2)_4—CH_3$$
$$|$$
$$HC—CO$$
$$| \quad\quad O$$
$$H_2C—CO$$

be rapid until the volatile base is lost and the acid groups are available to
catalyze and enter the reaction. The reactive groups can all be present in the
acrylic polymer, or a mixed polymer system may be used. For example,
copolymers containing acrylic acid, methyl methacrylate, and *N*-methylol
acrylamide can be water solubilized by a base, and cured to insoluble films

$$\boxed{\text{Polymer}} + CH_2=CH—CH_2—Cl$$
$$| \quad\quad |$$
$$OH \quad\; OH$$

$$\downarrow$$

$$\boxed{\text{Polymer}}$$
$$| \quad\quad\quad |$$
$$OH \quad\; O—CH_2—CH=CH_2$$

$$\downarrow$$

$$\boxed{\text{Polymer}}$$
$$| \quad\quad\quad |$$
$$O \quad\quad O—CH_2—CH=CH_2$$
$$|$$
$$CO$$
$$\quad\quad COOH$$

at elevated temperatures;[26] alternatively, an acrylic acid-methyl methacrylate copolymer can be used in conjunction with a water-soluble urea or melamine resin.

Hexakis(methoxymethyl)melamine and water-soluble phenol-formaldehyde resins are the most common cross-linking agents used with solubilized acid copolymers. The presence of hydroxyl groups in the acrylic polymers enables lower temperatures to be used for the cross-linking reaction.

REFERENCES

1. Benson, R. E., *Am. Paint J.*, **45**, 114, (1960).
2. North, A. G., *J. Oil Colour Chemists' Assoc.*, **44**, 119 (1961).
3. Schildkneckt, C. E., *Vinyl and Related Polymers,* John Wiley & Sons, Inc., New York, 1952, p. 357.
4. Tsunoda, T., and T. Yamaoka, *J. Appl. Polymer Sci.*, **8**, 1379 (1964).
5. Duncalf, B., and A. S. Dunn, *J. Applied Polymer Sci.*, **8**, 1763 (1964).
6. Deuel, H., and H. Neukom, *Makromol. Chem.*, **3**, 13 (1949).
7. Carswell, T. S., Monsanto Chemical Co., U.S. Pat. 2,441,470 (1948).
8. Robinson, W. D., E. I. du Pont de Nemours & Co., U.S. Pat. 2,510,257 (1950).
9. Plitt, K. F., *Paint Var. Prod.*, **49**, 27, 104, August (1959).
10. Hercules Powder Co., Brit. Pat. 891,441 (1962).
11. Miranda, T. J., *Offic. Dig. Federation Soc. Paint Technol.*, **37**, Oct. Part 2, 62 (1965).
12. Canterino, P. J., and L. T. Smith, Phillips Petroleum Co., U.S. Pat. 2,927,100 (1960).
13. Solomon, D. H., Unpublished observations.
14. Fordyce, D. B., J. Dupre, and W. Toy, *Offic. Dig. Federation Soc. Paint Technol.*, **31**, 284 (1959).
15. American Cyanamid Co., Brit. Pat. 824,340 (1959).
16. Canadian Industries Ltd., French Pat. 1,287,278 (1962).
17. Tawn, A. R. H., *J. Oil and Colour Chemists' Assoc.*, **47**, 794 (1964).
18. Heiligmann, R. G., and E. E. McSweeney, *Ind. Eng. Chem.*, **44**, 113 (1952).
19. Saxon, R., and J. H. Daniel, *J. Appl. Polymer Sci.*, **8**, 325 (1964).
20. Grummitt, O., and Arters, A. A., Am. Chem. Soc., Div. Org. Coatings Plastics Chem., Abstr. of Papers, 142nd Meeting, **1962**, 17R.
21. Uelzmann, H., B.F. Goodrich Co., Can. Pat. 622,140 (1961).
22. Uraneck, C. A., and R. J. Sonnenfeld, Phillips Petroleum Co., U.S. Pat. 2,894,857 (1959).
23. Wall, F. T., and J. W. Drenan, *J. Polymer Sci.*, **7**, 83 (1951).
24. Smith, K. L., Winslow, A. E., and Petersen, D. E., *Ind. Eng. Chem.*, **51**, 1361 (1959).
25. Hahn, F. J., M. R. Sullivan, and S. Steinhauer, *Offic. Dig. Federation Soc. Paint Technol.*, **37**, 1251 (1965).
26. Essig, H. J., B. F. Goodrich Co., Can. Pat. 620,439 (1961); Brit. Pat. 874,168 (1961); U.S. Pat. 3,007,887 (1961).
27. Huttman, E., K. Preim, H. Magdanz, and K. Berger, *Plaste u Kautschuk*, 17, No 3, 202 (1970).

Water-Soluble Vegetable Oils, Polyesters, and Alkyd Resins. Water-soluble oils, polyesters, and alkyd resins differ from the organic soluble resins in three main respects: they have a higher acid number so as to provide more sites for salt formation; they have more hydrophilic groups in the molecule; and they are generally of a lower molecular weight.

The maleinization of an unsaturated glyceride oil (see page 70) gives a structure that is potentially soluble in bases. Where the polymer system must dry or cure at room temperature by autoxidation, it is preferable to keep the extent of maleinization to a minimum, so as to maintain the maximum unsaturation in the oil molecule. Approximately 15 to 20% (by weight), of maleic anhydride generally gives satisfactory solubility with tung and linseed oils.[1] Some of these acid groups are lost by side reactions other than the simple maleic anhydride addition to the oil. Therefore, it is necessary to measure the available acid number, and not rely on a calculated figure.[1] Fumaric acid sometimes behaves in a different manner to maleic anhydride; with tung oil, the predicted chemical acidity is available only when fumaric acid is used but, with linseed oil, there is no noticeable difference and both fumaric acid and maleic anhydride give lower acid contents than expected.[1] Other unsaturated acids used to modify and solubilize vegetable oils in aqueous solutions are acrylic[2] and crotonic acid.[3]

The maleinized oil is often further modified to improve its solubility and drying potential. The solubility of the oil is improved by reacting the maleic adduct with a polyether, often of the poly(ethylene glycol) type.[4] The drying potential is increased if allyl ether groups are included in the molecule—for example, by reaction with glyceryl diallyl ether[5] or allyl alcohol (see overleaf).[6]

Alternatively, the maleinized oil can be further modified with a vinyl or acrylic type monomer. This results in a product that gives a short "touch free" drying time, but "through drying", and the development of water insensitivity, is prolonged.[7] The cyclopentadiene-linseed oil adduct, when reacted with maleic anhydride, gives water-soluble resins which, after air-drying overnight, become hard and water resistant.[8]

Alkyds and polyesters are modified in a number of ways in order to give polymer structures soluble in alkaline solution. The degree of polymerization is usually less than with an organic soluble polymer; this greatly assists in gaining water solubility because of the lower molecular weight, the increased number of carboxyl groups, and the greater number of free and available hydroxyl groups. A number of further formulation changes enhance the solubility of the resin in water. The use of an alcohol, which also contains ether groups, as a partial or complete replacement of the conventional polyol,

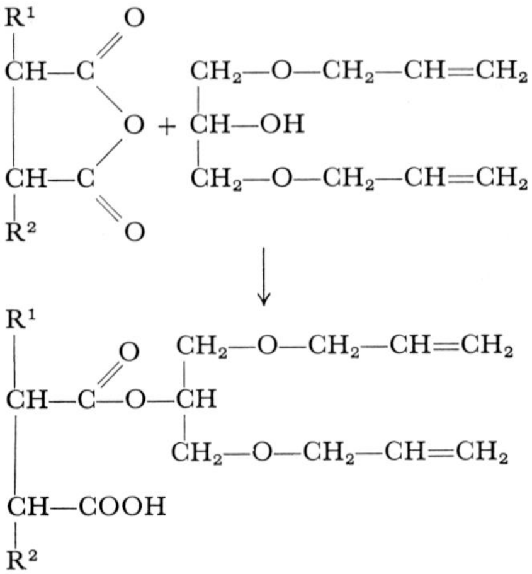

$$
\begin{array}{l}
R^1 \\
| \\
CH-C{\overset{O}{\diagup}}{\diagdown}_{O} \;+\; \begin{array}{l} CH_2-O-CH_2-CH=CH_2 \\ | \\ CH-OH \\ | \\ CH_2-O-CH_2-CH=CH_2 \end{array} \\
| \\
CH-C{\diagdown}{O} \\
| \\
R^2
\end{array}
$$

$$\downarrow$$

$$
\begin{array}{l}
R^1 \\
| \\
CH-C\overset{O}{=}O-CH{\begin{array}{l} CH_2-O-CH_2-CH=CH_2 \\ \\ CH_2-O-CH_2-CH=CH_2 \end{array}} \\
| \\
CH-COOH \\
| \\
R^2
\end{array}
$$

is one method used; poly(ethylene glycol) alkyds are either water soluble or water dispersible,[9] while tetrahydrofurfuryl alcohol[10] gives increased water solubility and air-drying potential. The latter results from the presence of the active methylene, α- to the ether oxygen.

The replacement of the dibasic phthalic anhydride, with either trimellitic (**1**) or pyromellitic anhydrides (**2**), increases the likelihood of a polymer becoming water soluble.

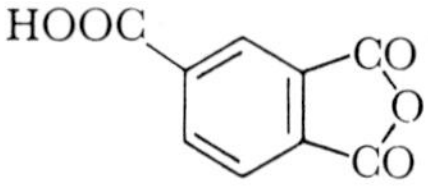

$$
\underset{(\mathbf{1})}{\text{HOOC}\!-\!\!\bigcirc\!\!-\!\!\begin{array}{l}CO \\ \;\;\diagdown O \\ CO\end{array}}
\qquad\qquad
\underset{(\mathbf{2})}{\begin{array}{l}OC \\ O\!\diagup \\ OC\end{array}\!-\!\!\bigcirc\!\!-\!\!\begin{array}{l}CO \\ \;\;\diagdown O \\ CO\end{array}}
$$

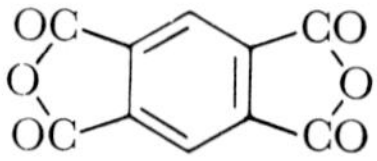

When the replacement is made on a mole basis, it is possible to maintain the polymer molecular weight at a reasonably high level, and yet gain water solubility. This is illustrated opposite for a formulation of fatty acid: glycerol:polyacid = 1:1:1.

Other differences between a trimellitic and a phthalic anhydride polyester also enhance the water solubility of the resin. The trimellitic anhydride gives a different distribution of the acid groups;[1] this effect is particularly noticeable in the high molecular weight fractions (difficult to solubilize), where the trimellic anhydride gives a much greater proportion of carboxyl groups than does phthalic anhydride. Another advantage of the trimellitic anhydride is that the remaining carboxyl group is a stronger acid than a phthalic half-ester, and this influences the solubility in water. An alternative method of preparing a polymer in which the residual carboxyl groups are a relatively strong acid, is to carry out a half-ester reaction, late in the poly-

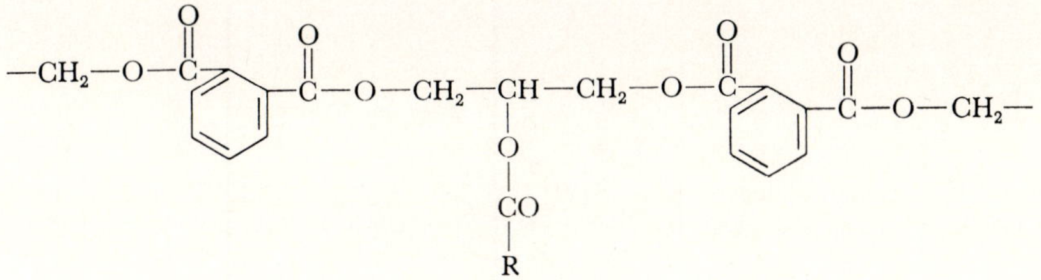

Phthalate polyester

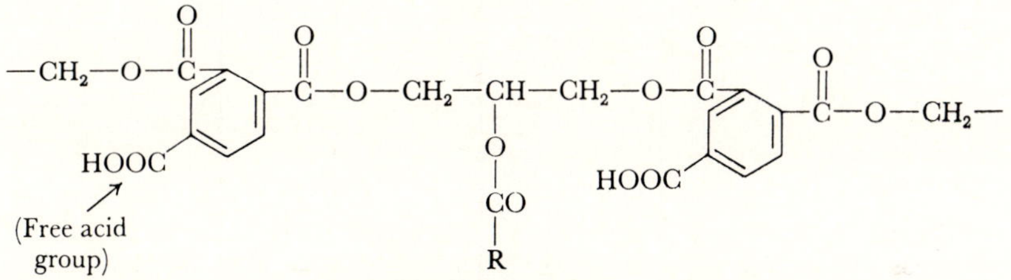

Trimellitate polyester

merization, with an appropriate anhydride—for example, tetrachloro-phthalic anhydride.[11]

The presence of nonionized hydrophilic groups in a polymer molecule influences the solubility in water. In fact, it has been claimed that these groups are of even greater importance than the ionizable carboxyls.[12] Another determining factor, as far as the water solubility of a polymer is concerned, is the availability of the hydrophilic groups, particularly the hydroxyl groups. Recently, it has been demonstrated that the acetylation of the reactive hydroxyl (approximately one third of the total hydroxyl groups) in a water-soluble alkyd renders the resin insoluble in water.[13] The availability of the hydroxyl groups, and their contribution to the water solubility of the polymer, can be measured by acetylation.[13]

Water-soluble oils, alkyds, and polyesters can be converted to films by a number of reactions. Those formulated so as to contain sufficient oxygen-sensitive groups will cure by mechanisms similar to conventional oils or alkyds. Water-soluble driers are used as catalysts. Alternatively, at elevated temperatures, the interaction of the hydroxyl and carboxyl groups with a second resin component, particularly a melamine or phenol resin, can be used to give a water-resistant film. The mechanisms by which these reactions proceed have been discussed previously for solvent-borne systems and for water-soluble acrylic polymers (Chapters 3 and page 303); however, for the water-soluble polymer mixture to react, the system must first become acid. Whether or not the amine is completely lost from the film or forms an

TABLE 11.2

NITROGEN LOSSES OF ALKYDS AND MALEINIZED OIL FILMS STOVED AT $120°C$[1]

Type of Medium	Acid Value	Viscosity (50% Solids) in Aqueous Base (Poises)	Solubilizing Base	Original Nitrogen Content (Percent)	Proportion of Nitrogen Lost (Percent)
Trimellitic/capric alkyd	57	3.5	$(CH_3)_3N$	1.64	100
Trimellitic/capric alkyd	57	4	NH_3	1.54	95
Trimellitic/adipic alkyd	57	>200	$(CH_3)_3N$	1.40	64
Trimellitic/adipic alkyd	57	>200	NH_3	1.24	61
Maleinized linseed oil + 0.1% manganese	—	2	NH_3	4.78	78

amide seems to depend on the resin structure and the amine used. Brett[1] has shown (Table 11.2) that for ammonia and trimethylamine with various resins, from 61 to 100% of the base is lost; with 2-dimethylaminoethanol, significant amounts of base remained in the film after $\frac{1}{2}$ hour at 120°C. On the other hand, with some polyesters, Wilkinson[10] has found only 3 to 5% of the original dimethylaminoethanol in the cured film.

The stability of the water-soluble resin blend depends on a number of factors; these include the following: the chemical composition of the resin, the solubilizing base, and the presence of a water-miscible co-solvent. Trimethylol propane[10] and branch chain carboxylic acids[14] improve the stability of the resin to hydrolysis. Tertiary amines reduce the likelihood of aminolysis (page 102) and depolymerization of the melamine- or phenol-formaldehyde resin (page 245); these aspects are discussed more fully in references[1,4,10] and in Chapter 3. The methyl, or lower alkyl ethers, of the methylol melamines and phenolic constituents are preferred, since they are more stable in aqueous solutions.[15] Partial reaction between the two polymer species prior to film application is suggested in a number of patents, and is of possible value in improving the solubility and stability of the resin blend.[16]

REFERENCES

1. Brett, R. A., *J. Oil and Colour Chemists' Assoc.*, **47**, 767 (1964).
2. Danzig, M. J., J. L. O'Donnell, E. W. Bell, J. C. Cowan, and H. M. Teeter, *J. Am. Oil Chemists' Soc.*, **34**, 136 (1957).
3. Teeter, H. M., C. R. Scholfield, and J. C. Cowan, *Oil Soap*, **23**, 216 (1946).
4. Miranda, T. J., *Offic. Dig. Federation Soc. Paint Technol.*, **37**, Oct. Part 2, 62 (1965).
5. Distillers Co., Ltd., Belg. Pat. 618,504 (1962).
6. Moczar, E., Dr. K. Roberts & Co., Ger. Pat. 1,109,811 (1958).
7. Hopwood, J. J., *J. Oil Colour Chemists' Assoc.*, **48**, 157 (1965); Lombardi, L. J., H. J. Wright, and P. F. Westfall, Cook Paint and Varnish Co., U.S. Pat. 3,030,321 (1962); Cook Paint and Varnish Co., Brit. Pat. 925,038 (1963).
8. Jerabek, E. D., Archer-Daniels-Midland Co., U.S. Pat. 3,098,834 (1963).
9. Arndt, R. P., Pittsburgh Plate Glass, Co., U.S. Pat. 2,634,245 (1950); Stephens, J. R., Standard Oil Company, U.S. Pat. 3,046,252 (1962); Armitage, F., and L. G. Trace, *J. Oil and Colour Chemists' Assoc.*, **40**, 849 (1957).
10. Wilkinson, R. F., *Offic. Dig. Federation Soc. Paint Technol.*, **35**, 129 (1963).
11. Archer Daniels Midland Co., Brit. Pat. 921,622 (1963); Brit. Pat. 921,623 (1963).
12. Tawn, A. R. H., *J. Oil Colour Chemists' Assoc.*, **47**, 794 (1964).
13. Solomon, D. H., and J. J. Hopwood, *J. Appl. Polymer Sci.*, **10**, 981 (1966).
14. McIntosh, A., *Proceedings, News Oil Colour Chemists' Assoc. of Australia*, **2**, No. 8, 4 (1965).
15. Stephens, J. R., *Offic. Dig. Federation Soc. Paint Technol.*, **35**, 380 (1963).
16. Honel, H., Brit. Pat. 665,195 (1952); Shelley, J. P., Rohm and Haas, Co., U.S. Pat. 2,915,486 & 7 (1959).

CHAPTER 12

RUBBER AND SILICONE RESINS

Natural and synthetic polymers with rubberlike properties are of considerable interest to the coating industry, although their use in conventional systems has necessitated chemical modification or control to give acceptable solution properties. Some of the chemical modifications have involved the introduction of compounds which contain heteroelements into the rubberlike material. Typical examples are sulfur, chloro, and silicon compounds.

NATURAL AND SYNTHETIC RUBBERS AND DERIVATIVES

Natural rubber possesses a number of desirable properties when considered as a surface-coating material, namely toughness, high tensile strength, and flexibility. However, these attractive properties are far outweighed by the limited solubility of rubber; a 5% solution of crepe rubber in benzene is too viscous for normal application procedures (Gen. Ref. 1). This high viscosity in dilute solution is a reflection of the very high molecular weight of the polymer—in the vicinity of 1,000,000 (Gen. Ref. 2). Consequently, for use as a surface-coating polymer, natural rubber is degraded to an acceptable molecular weight; chemical modification is also practiced to control certain properties.

Chemically, natural rubber is a polymer that can be regarded as being derived from isoprene, C_5H_8, units.

$$CH_2{=}C{-}CH{=}CH_2$$
$$|$$
$$CH_3$$

Isoprene

These units are linked essentially in a head to tail manner; the addition is at least 97% 1:4.[1,2] The geometrical configuration around the remaining double bonds of the isoprene residues is a characteristic of the rubber, and has a direct bearing on its properties; natural rubber is the *cis* isomer, and gutta-percha is the *trans*. The gutta-percha is a more crystalline polymer, with a higher melting point than the *cis* isomer.

The molecular weight of rubber can be reduced by mechanical and/or chemical methods; the rubber is often degraded by subjecting the polymer

317

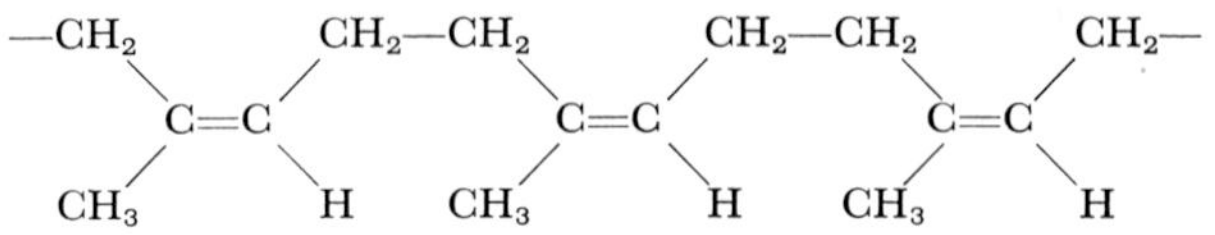

$$-CH_2\diagdownCH_2-CH_2CH_2-CH_2CH_2-$$

Natural rubber

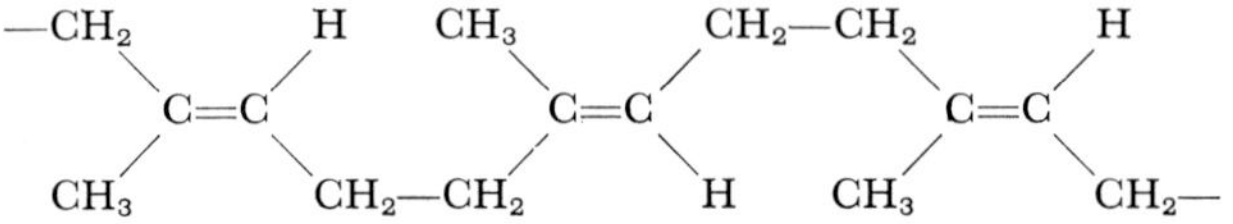

β-Gutta percha

molecules to a shearing force on hot differential rolls, then by oxidation (Gen. Ref. 1). The chemical structure of a rubber chain is similar to that of a drying oil; it contains a double bond with α-methylene and methyl groups. Therefore, oxidative degradation takes place by the formation of hydroperoxides, then by the free radical decomposition of these structures. Chain scission accounts for the reduction in molecular weight. The attack of oxygen on a rubber molecule is catalyzed by the conventional driers—for example, cobalt, lead, and manganese salts. The mechanisms by which these reactions occur have been discussed previously (Chapter 2). Oxidized, or depolymerized, rubbers (as they are sometimes termed) are used to improve the flow and resistance properties of paints, and to reduce pigment settling. Usually, the other resin component in these paint systems is a drying oil; during the film-forming stage, both the oil and rubber undergo oxidation and probably combine chemically to give a three-dimensional cross-linked structure. The mechanisms by which this drying process occurs are similar to those discussed in Chapter 2.

The susceptibility of a rubber molecule to oxidation can be reduced by two chemical modifications: the linear structure can be converted to a cyclized form with the loss of unsaturation, or the molecule can be chlorinated. Cyclized rubber is prepared by degrading the polymer molecule and then adding a catalyst to facilitate the cyclization. Typical catalysts are stannic chloride, titanium tetrachloride, and organo-metallic compounds. The crude cyclized rubber resin is purified by a variety of methods: one process involves solidification in the presence of phenol, which is subsequently removed by washing; an alternative method is to precipitate the cyclized rubber from a toluene solution with an alcohol (Gen. Ref. 1). The probable structure of cyclized rubber is given overleaf.

It should be noted that in the cyclization process, one half of the double bonds are lost. Consequently, the cyclized rubber is less susceptible to oxidation than is natural rubber; however, the residual unsaturation corresponds to an iodine value of 50 to 100, and the polymer is capable of autoxidation. This is particularly so in the presence of driers, such as cobalt and

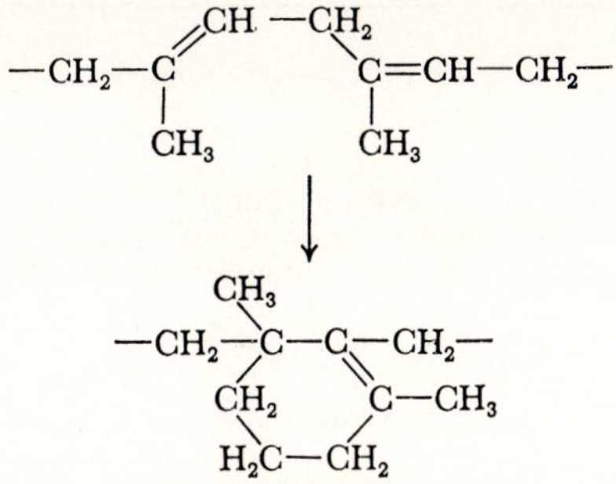

$$-CH_2-C \overset{CH-CH_2}{\underset{CH_3}{|}} \quad \overset{}{\underset{CH_3}{|}} C=CH-CH_2-$$

$$-CH_2-\overset{CH_3}{\underset{CH_2}{|}}C-C-CH_2- \quad \overset{}{\underset{H_2C-CH_2}{C-CH_3}}$$

Cyclized rubber

manganese salts. The commercially available resins have molecular weights in the range 3000 to 5000 (Gen. Ref. 1), and this, in conjunction with the structure of the polymer, results in the polymer being soluble in aliphatic hydrocarbons. The cyclized rubbers are compatible with glyceride and mineral oils, varnishes, and some alkyd resins; commercially, blends of cyclized rubber and these resins are used where application of the finish by brush is required, and where moderate chemical resistance is necessary. The film-forming reactions involve oxidation at the active methylene group α- to the double bond, and the formation and decomposition of a hydroperoxide. These reactions are analogous to the autoxidation of similar structures in the glyceride oils (Chapter 2).

Chlorinated rubbers are insoluble in aliphatic hydrocarbons, but give solutions in esters, higher ketones, and aromatic hydrocarbons. The use of these more toxic and more expensive solvents (compared with those required for cyclized rubber) is balanced by the greater chemical and corrosion resistance of the coatings (Gen. Ref. 3). Chlorinated rubber is prepared by the degradation of natural rubber, followed by chlorination that apparently involves substitution at the active methylene groups α- to the double bond (a free radical process) and addition across the double bond. To account for the 67% average chlorine content of commercial chlorinated rubbers, it is necessary to postulate a mixture of structures corresponding to three and four chlorine atoms per isoprene unit. A possible reaction sequence for the preparation of chlorinated rubber is shown in Figure 12.1.

Further modifications to cyclized or oxidized rubber are possible; graft copolymerization with a vinyl or acrylic monomer can be used to give a variety of products useful as film formers in their own right,[3] or as stabilizers for other polymer systems (page 167).

The properties of the chemically modified natural rubbers, and the availability of conjugated dienes from petroleum cracking, has resulted in the development of a range of synthetic rubberlike polymers. The conditions

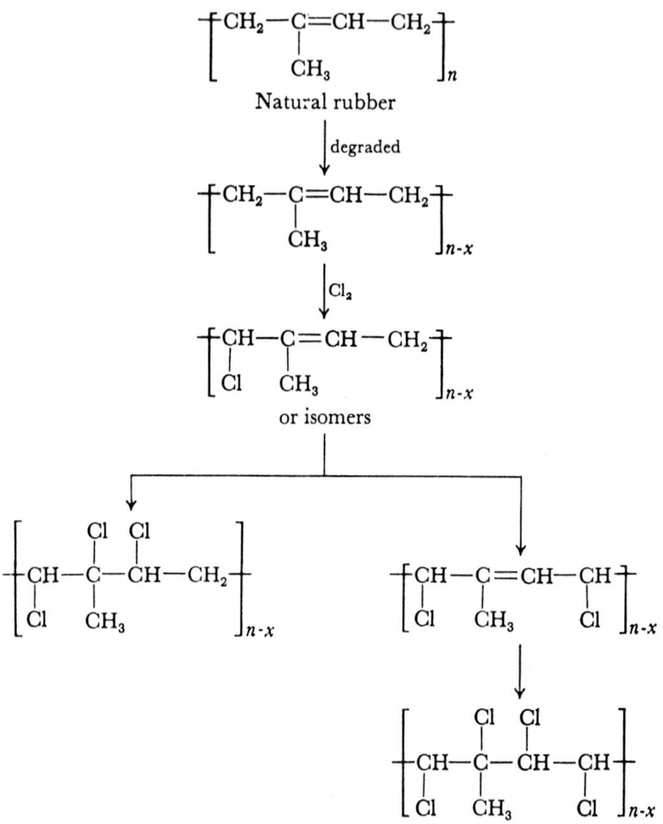

$$\begin{array}{c} \left[CH_2-\underset{\underset{CH_3}{|}}{C}=CH-CH_2 \right]_n \end{array}$$

Natural rubber

↓ degraded

$$\left[CH_2-\underset{\underset{CH_3}{|}}{C}=CH-CH_2 \right]_{n-x}$$

↓ Cl_2

$$\left[\underset{\underset{Cl}{|}}{CH}-\underset{\underset{CH_3}{|}}{C}=CH-CH_2 \right]_{n-x}$$

or isomers

$$\left[\underset{\underset{Cl}{|}}{CH}-\underset{\overset{\overset{Cl}{|}}{\underset{CH_3}{|}}}{C}-CH-CH_2 \right]_{n-x} \qquad \left[\underset{\underset{Cl}{|}}{CH}-\underset{\underset{CH_3}{|}}{C}=CH-\underset{\underset{Cl}{|}}{CH} \right]_{n-x}$$

↓

$$\left[\underset{\underset{Cl}{|}}{CH}-\underset{\overset{\overset{Cl}{|}}{\underset{CH_3}{|}}}{C}-CH-\underset{\underset{Cl}{|}}{CH} \right]_{n-x}$$

Figure 12.1. Possible reaction sequence for the preparation of chlorinated rubber.

chosen for the polymerization of the conjugated dienes influence the structure and properties of the resultant polymer. The number of polymer structures possible is greater than with a mono-olefin, since 1:2 and 1:4 additions to the conjugated system are possible. This can be illustrated by considering the polymers of butadiene; by careful selection of the Ziegler-type catalyst used to initiate the polymerization, five distinct polymer structures can be obtained (Gen. Refs. 4, 5).

Titanium trichloride with less than molar equivalents of aluminum triethyl yields *trans*-1:4-poly(butadiene); aluminum alkyls and titanium tetraiodide give *cis*-1:4-poly(butadiene); Ziegler catalysts prepared from titanium alkoxides or acetylacetonates give 1:2 poly(butadienes). Three

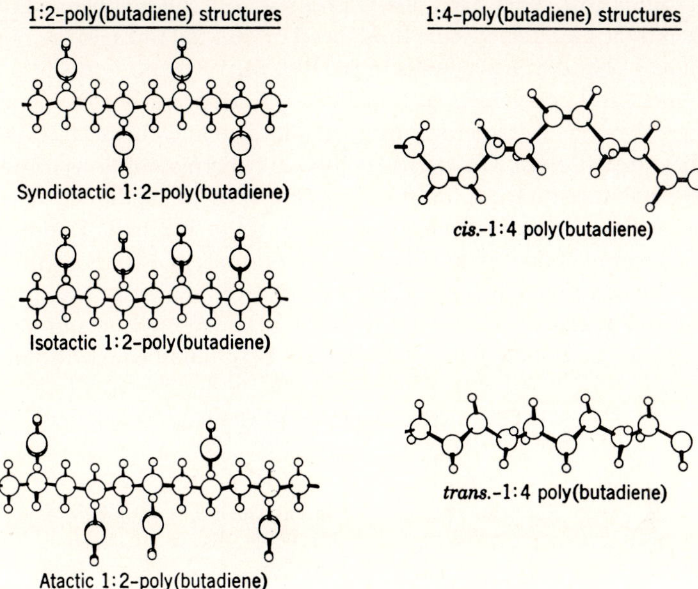

Figure 12.2.   Polymer of butadiene (Gen, Ref. 4).

polymers with a 1:2 structure are possible; these have an isotactic, syndiotactic, and atactic arrangement of the substituent groups. The polymer structures resulting from the polymerization of poly(butadiene) are shown in Figure 12.2, and the properties of these polymers in Table 12.1 (Gen. Ref. 4).

TABLE 12.1

PHYSICAL PROPERTIES OF POLY(BUTADIENE) (Gen. Ref. 4)

Polymer Structure	Density	T_m, °C	Percent Rebound at 20°C	Percent Rebound at 90°C
trans-1:4 <60°C (zigzag)	0.96–1.02	135–148	75–80	90–93
60°C (helical)		130		
cis-1:4	1.01	Dependent on degree of crystallinity	88–90	92–95
1:2-Isotactic	0.96	120–125	45–55	90–92
1:2-Syndiotactic	0.96	154–155		

Unsymmetrical conjugated dienes give rise to isomers other than those considered for butadiene; some head–head or tail–tail linkages are possible as well as 1:2 and 3:4 structures (Gen. Ref. 5).

Polymers and copolymers of butadiene are particularly suitable as can coatings; they are characterized by good adhesion and chemical resistance, and they are odorless. The polymers used are thermosetting as a result of the residual unsaturation, or from the presence of modifying chemical groups. A liquid poly(butadiene), with a ratio of 1:2 to 1:4 addition of 3:1,[4] is recommended as an interior can coating; polymers of this type require the addition of an antioxidant to improve the storage stability. The film-forming reactions involve autoxidation of the methylene and methyne groups α- to the double bond (Chapter 2). Poly(butadiene) with an even

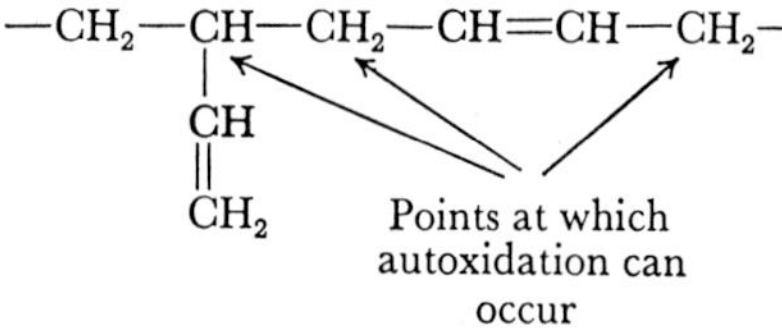

$$-CH_2-CH-CH_2-CH=CH-CH_2-$$

$$\begin{array}{c} | \\ CH \\ || \\ CH_2 \end{array} \quad \text{Points at which autoxidation can occur}$$

higher content of the 1:2 isomer (93% 1:2 and 7% *trans* 1:4) is produced by the use of titanium trichloride and phenyl magnesium bromide. As a result of oxidative cross-linking,[5] the film formed by stoving for 15 minutes at 200°C has good adhesion, and resists hot solvents and alkali. Copolymers of butadiene and styrene are also used as container linings, and as the resins for metal primers, masonry coatings, and temperature-resistant coatings; typical copolymers have a molecular weight of 2000 to 3000, and the polymer structure is represented below.[6] The film dries by autoxidation of the polymer.

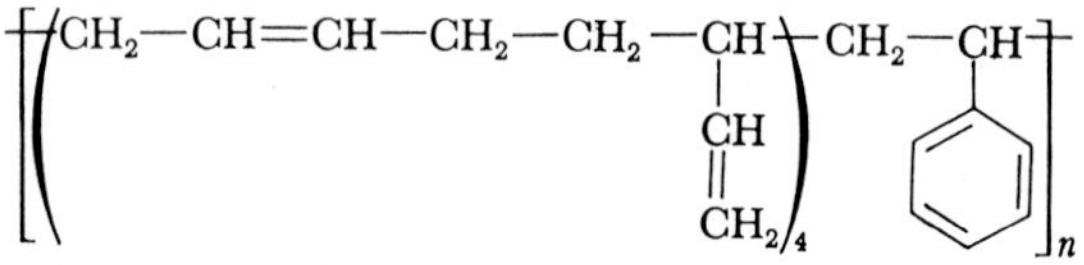

$$\left[\left(CH_2-CH=CH-CH_2-CH_2-CH \underset{\underset{CH_2}{\overset{\displaystyle CH}{||}}}{} \right)_4 CH_2-CH \underset{\langle\!\!\!\bigcirc\!\!\!\rangle}{} \right]_n$$

Both the natural and synthetic rubbers can be cross-linked by dimaleimides;[7] a number of reactions, in which the hydrogen atom of the α-methylene group in the rubber is removed by oxidation, have been proposed to account for the cross-linking; peroxides catalyze these reactions as shown in Figure 12.3.[7]

The homo- and copolymers of butadiene can be chemically modified to improve their adhesion to substrates, their pigment-wetting properties, and their film-forming potential. Polar groups are introduced by a number of

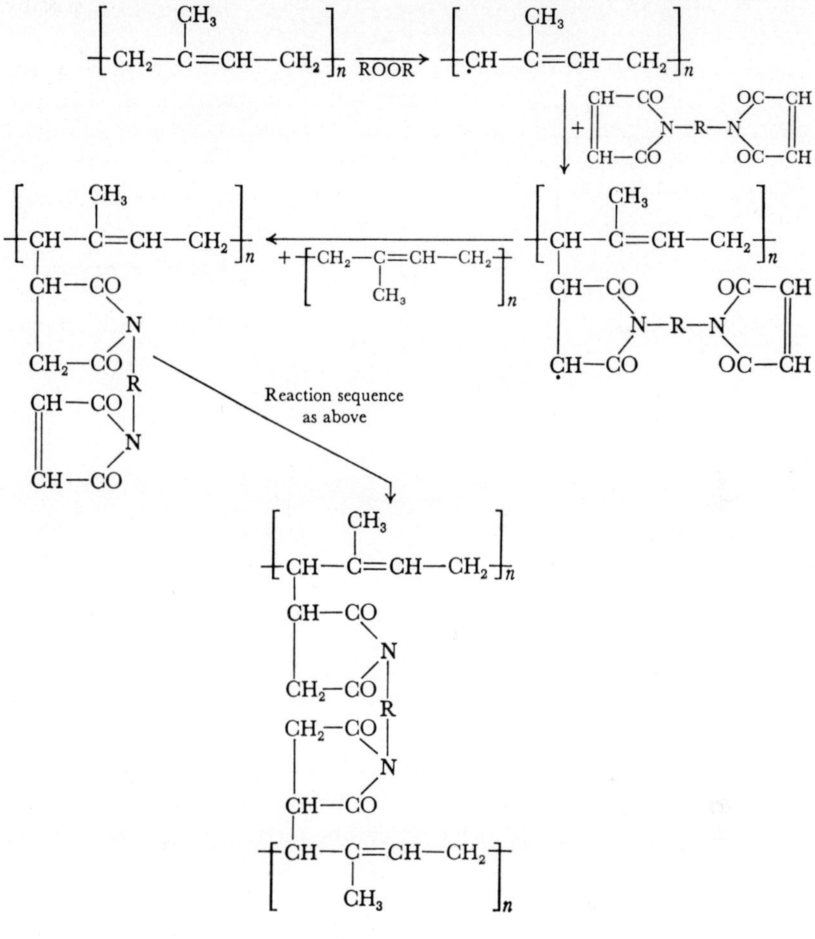

$$\left[CH_2-\underset{\underset{CH_3}{|}}{C}=CH-CH_2\right]_n \xrightarrow{ROOR} \left[\overset{\bullet}{C}H-\underset{\underset{CH_3}{|}}{C}=CH-CH_2\right]_n$$

$$+ \; \begin{array}{c} CH-CO \\ \| \\ CH-CO \end{array} N-R-N \begin{array}{c} OC-CH \\ \| \\ OC-CH \end{array}$$

Reaction sequence as above

Figure 12.3. Possible mechanisms for the cross-linking of rubbers with dimaleimides.[7]

chemical reactions. (1) Treatment with maleic anhydride[6,8] gives the typical addition product of an olefin, a substituted succinic anhydride; (2) Hydrogen peroxide and acetic acid result in the formation of an epoxide group;[9] (3) Trifluoracetic acid, and then hydrolysis, gives a residual hydroxyl group.[6] Hydrogenation of butadiene polymers and copolymers, particularly those with 1:4 structures, reduces the tendency of the polymer to undergo stress cracking; these hydrogenated polymers are useful for coating wire.[10] The reactions involved in the chemical modification of liquid butadiene polymer are shown in Figure 12.4.

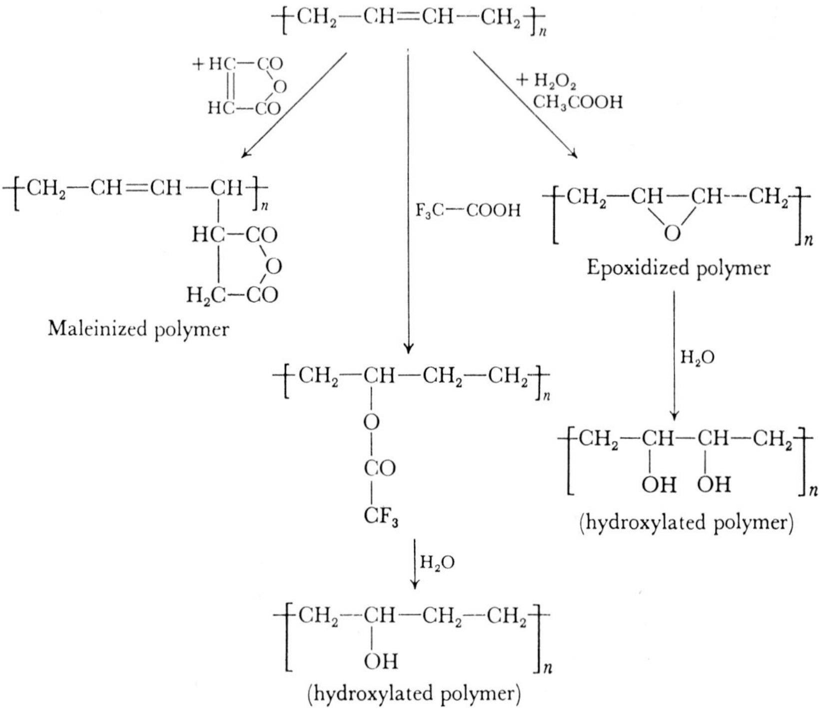

$$\{CH_2-CH=CH-CH_2\}_n$$

$+ HC-CO,\ HC-CO$ (maleic anhydride) $+ H_2O_2,\ CH_3COOH$ $F_3C-COOH$

Maleinized polymer:

$$\{CH_2-CH=CH-CH\}_n \quad \text{with} \quad HC-CO,\ H_2C-CO \ (O)$$

Epoxidized polymer:

$$\left[CH_2-CH-CH-CH_2\right]_n \quad (\text{epoxide, O bridge})$$

From $F_3C-COOH$ branch:

$$\{CH_2-CH-CH_2-CH_2\}_n \quad \text{with} \quad O-CO-CF_3$$

$\downarrow H_2O$

$$\left[CH_2-CH-CH_2-CH_2\right]_n \quad \text{with} \quad OH$$

(hydroxylated polymer)

From epoxidized polymer, $\downarrow H_2O$:

$$\{CH_2-CH-CH-CH_2\}_n \quad \text{with} \quad OH\ OH$$

(hydroxylated polymer)

Figure 12.4. Chemical modification of butadiene polymers and copolymers.

The chemically modified polymers can be used as the sole film former; when this is done, the residual unsaturation (not attacked in the modifying process) enables cross-linking to take place by autoxidation. Conventional driers accelerate the reaction, and the order of activity is similar to that found with vegetable oils (Co > Mn > Fe > Ce > Pb > Th > Ni and Zn).[11] Alternatively, cross-linking agents or resins can be added to the modified polymer of butadiene to increase the rate of cure, and to enhance the resistance properties. The hydroxy-containing polymers react at room or elevated temperature with a polyisocyanate adduct,[12] or with hexamethylene diisocyanate.[6] Urea- and melamine-formaldehyde condensates react with the hydroxylated poly(butadienes)[6] in a similar manner to other polymers which contain hydroxyl groups, such as alkyds (page 98) and acrylics (page 266). Titanium derivatives, such as orthotitanate esters or acylates, cross-link acid[13] or hydroxyl polymers of butadiene by interchange reactions.

Poly(butadienes) can be modified by fatty acids, or fatty acid esters, by chemical reaction; hydroxylated poly(butadienes) can be esterified with, for example, linseed oil fatty acids,[14] or the poly(butadiene) and glyceride oil can be thermally copolymerized,[6] by interaction of the unsaturated groups, as shown in Figure 12.5.

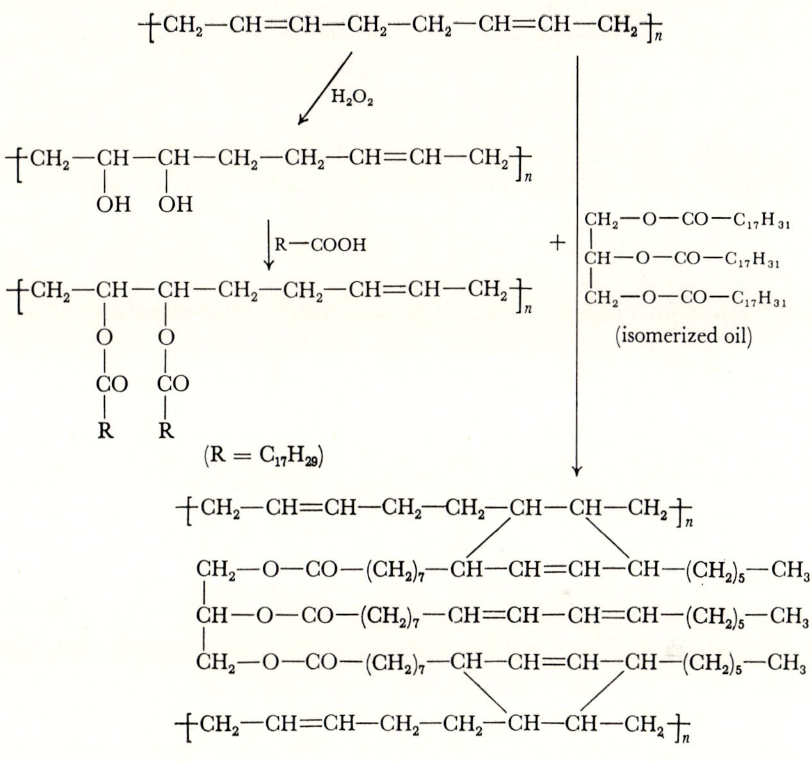

$$\left[CH_2-CH=CH-CH_2-CH_2-CH=CH-CH_2\right]_n$$

$\downarrow H_2O_2$

$$\left[CH_2-\underset{OH}{CH}-\underset{OH}{CH}-CH_2-CH_2-CH=CH-CH_2\right]_n$$

$\downarrow R-COOH$

$$\left[CH_2-\underset{\underset{R}{\overset{CO}{|}}}{\overset{O}{|}}{CH}-\underset{\underset{R}{\overset{CO}{|}}}{\overset{O}{|}}{CH}-CH_2-CH_2-CH=CH-CH_2\right]_n$$

$(R = C_{17}H_{29})$

$+$

$$CH_2-O-CO-C_{17}H_{31}$$
$$CH-O-CO-C_{17}H_{31}$$
$$CH_2-O-CO-C_{17}H_{31}$$

(isomerized oil)

$$\left[CH_2-CH=CH-CH_2-CH_2-CH-CH-CH_2\right]_n$$

$$CH_2-O-CO-(CH_2)_7-CH-CH=CH-CH-(CH_2)_5-CH_3$$
$$CH-O-CO-(CH_2)_7-CH=CH-CH=CH-(CH_2)_5-CH_3$$
$$CH_2-O-CO-(CH_2)_7-CH-CH=CH-CH-(CH_2)_5-CH_3$$

$$\left[CH_2-CH=CH-CH_2-CH_2-CH-CH-CH_2\right]_n$$

Figure 12.5.   Modification of poly(butadienes) by vegetable oils or fatty acids.

Rubber-type polymers that contain substituents other than carbon, hydrogen, and oxygen are the basis of a number of coatings. Neoprene, poly(chloroprene), is resistant to heat, sunlight, oils, and chemicals; Viton, a copolymer of vinylidene fluoride and hexafluoropropylene, gives flame-resistant coatings as a result of the fluorine groups, while Hypalon is particularly resistant to oxidizing chemicals.[15] Hypalon is prepared by treating poly(ethylene) with chlorine and sulfur dioxide to give a chlorosulfonated poly(ethylene).[16]

Film formation from these rubber-like polymers involves vulcanization which is brought about by the addition of curing agents and by the use of elevated temperatures. Viton has the structure shown in Figure 12.6[17] and,

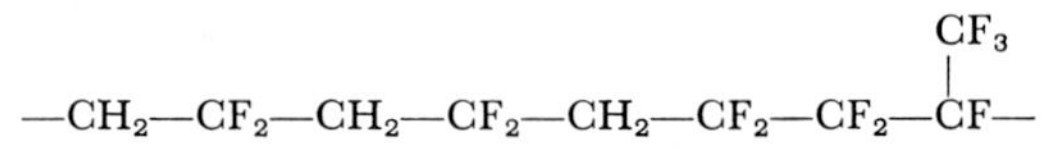

$$-CH_2-CF_2-CH_2-CF_2-CH_2-CF_2-CF_2-\overset{\overset{\textstyle CF_3}{\textstyle |}}{C}F-$$

Figure 12.6. Structure of Viton.[17]

when used as a coating composition, it is compounded with a base (usually magnesium oxide) and with a curing agent (usually a diamine or a dithiol).[17] In order to obtain the maximum physical strength from Viton coatings it is recommended that the Viton be compounded on a mill with the other ingredients, and the compound sheeted out, cut into small pieces, and then dissolved in the solvent.[23] The vulcanization proceeds in two stages. First, dehydrofluorination takes place to give a polymer with residual unsaturation. The magnesium oxide assists this reaction by acting as an acid acceptor. The

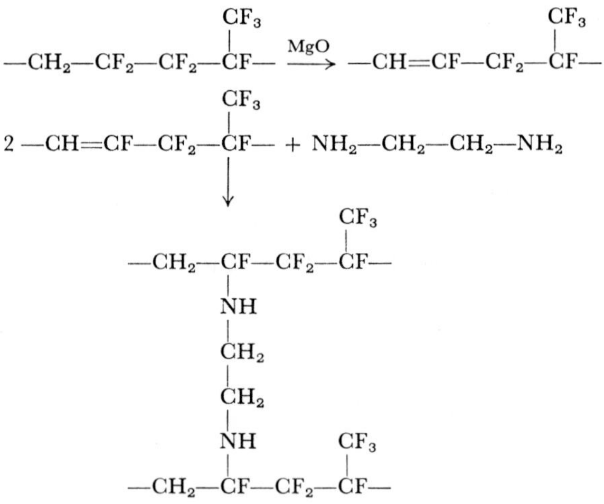

$$-CH_2-CF_2-CF_2-\overset{\overset{\textstyle CF_3}{\textstyle |}}{C}F-\ \xrightarrow{MgO}\ -CH{=}CF-CF_2-\overset{\overset{\textstyle CF_3}{\textstyle |}}{C}F-$$

$$2\ -CH{=}CF-CF_2-\overset{\overset{\textstyle CF_3}{\textstyle |}}{\underset{\downarrow}{C}}F-\ +\ NH_2-CH_2-CH_2-NH_2$$

$$-CH_2-\underset{\underset{\textstyle NH}{\textstyle |}}{C}F-CF_2-\overset{\overset{\textstyle CF_3}{\textstyle |}}{C}F-$$

$$\begin{array}{c} | \\ CH_2 \\ | \\ CH_2 \\ | \end{array}$$

$$-CH_2-\underset{\underset{\textstyle NH}{\textstyle |}}{C}F-CF_2-\overset{\overset{\textstyle CF_3}{\textstyle |}}{C}F-$$

Figure 12.7. First-stage vulcanization of Viton.[17]

difunctional amine or thiol then adds to the double bond or attacks the adjacent active methylene group; this reaction usually takes place at 120 to 130°C. This reaction sequence is shown in Figure 12.7.[17] The second stage of the vulcanization process of Viton requires higher temperatures, usually in the vicinity of 230°C. Further elimination of hydrogen fluoride takes place, and this process is autocatalytic due to the activating influence of the double bond on the substituents on the α-carbon atom. Conjugated unsaturation is developed and enables Diels-Alder type condensation between polymer molecules to take place.[18] Elimination of hydrogen fluoride yields an aromatic ring structure. The reaction sequence is shown in Figure 12.8.[18]

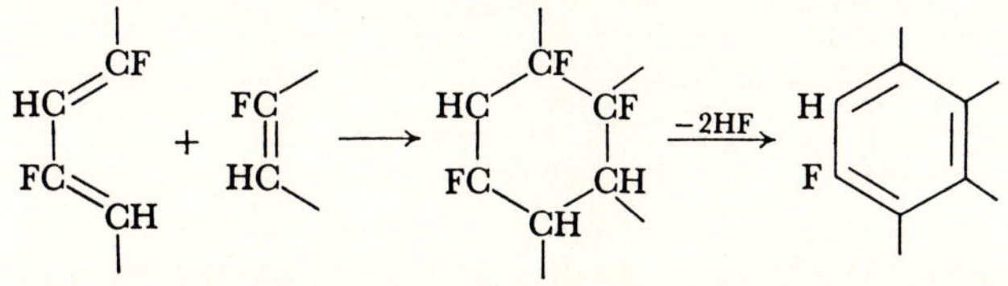

Figure 12.8.   Second-stage vulcanization of Viton.[18]

The vulcanization reactions of Hypalon and Neoprene rely on the presence of a small number of reactive sites in the rubber molecules.[17] These sites are not apparent in the generalized formula of the rubbers, and it is necessary to consider the structures in some detail in order to understand vulcanization.

Neoprene contains 2-chlorobutadiene monomer units, 98% of which are linked in the 1:4 position. However, it is the small number of units $(1.5 \pm 0.1\%)$ joined in the 1:2 positions that are responsible for the cross-linking reactions, since these contain chlorine atoms which are both tertiary and allylic; consequently, they are much more reactive than the chlorine atoms in the units linked by 1:4 addition (Figure 12.9).[17]

Neoprene can be cross-linked by heating in the presence of magnesium oxide/zinc oxide mixtures, or by using selected organic amines or sulfur

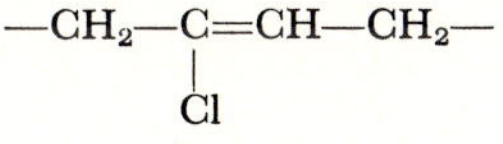

$$-CH_2-C=CH-CH_2-$$

Cl

1:4 addition
(chlorine relatively inactive)

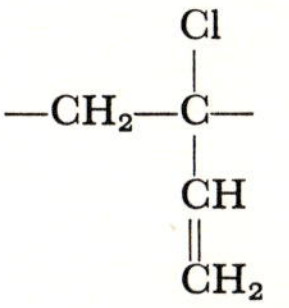

Cl

$$-CH_2-C-$$

CH

$$CH_2$$

1:2 addition
(chlorine active)

Figure 12.9.   Types of chlorine atoms in Neoprene.[17]

compounds in the presence of a base. The magnesium oxide acts as a scavenger for the chloride ions but the zinc oxide, in the absence of organic accelerators, is more directly involved in the reaction, as shown in Figure 12.10.[17]

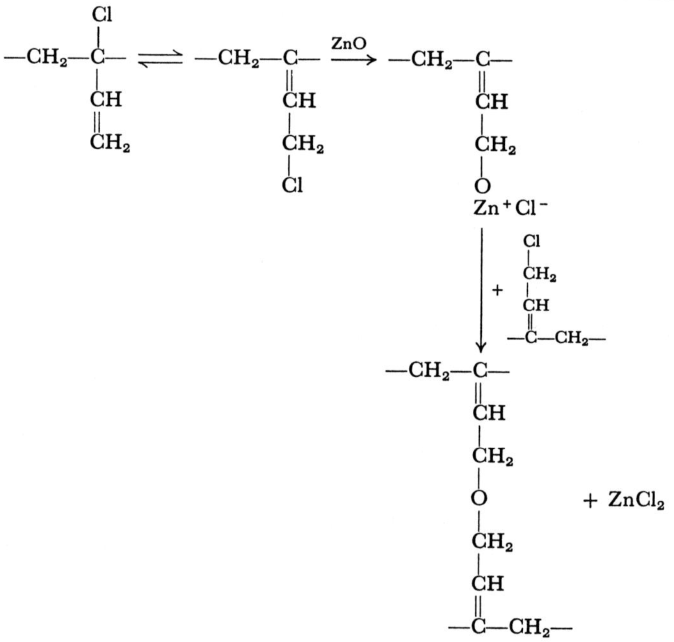

Figure 12.10. Mechanism of Neoprene cross-linking by zinc oxide.[17]

Figure 12.11. Cross-linking of Neoprene by diamines.[17]

When organic accelerators are used, the zinc oxide apparently acts as a catalyst, in a similar manner to magnesium oxide.[17] Diamines, such as piperazine, cross-link Neoprene as shown in Figure 12.11,[17] and thioureas as shown in Figure 12.12.[17]

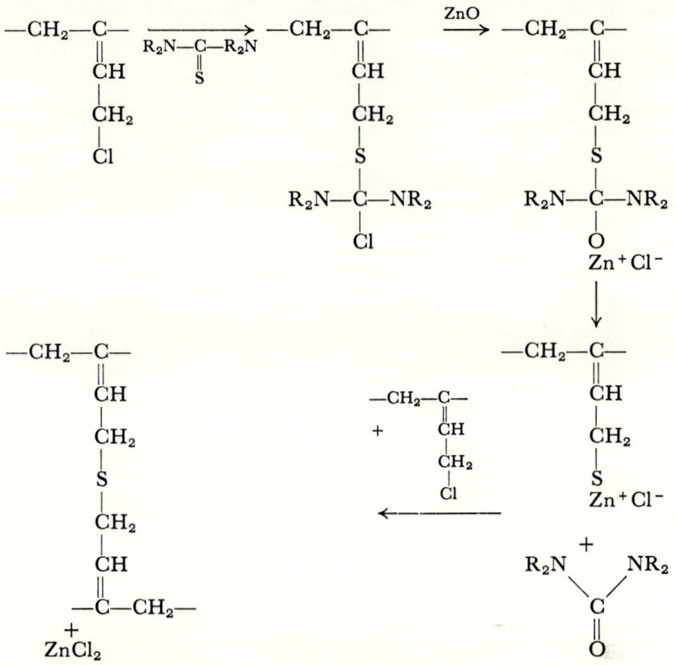

Figure 12.12. Cross-linking of Neoprene by thioureas.[17]

Hypalon is represented generally by the structure shown in Figure 12.13,[16] but it is necessary to examine the polymer in more detail to understand

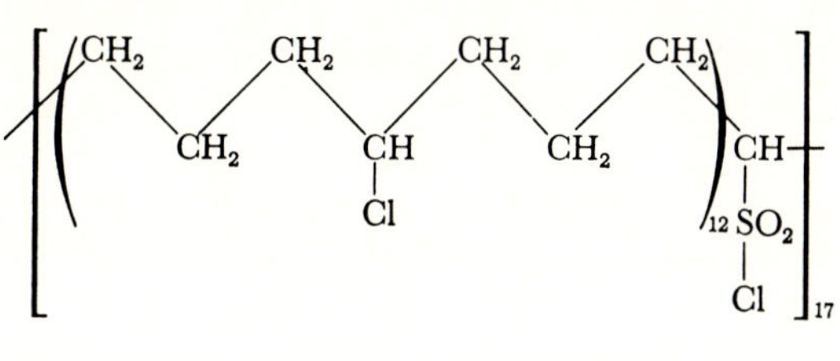

Figure 12.13. Structure of Hypalon.[16]

vulcanization.[19] The distribution of chlorine and sulfonyl chloride groups is shown in Table 12.2;[17] by far the most significant group for cross-linking

TABLE 12.2

DISTRIBUTION AND STRUCTURE OF CHLORINE ATOMS IN HYPALON[17]

Type	Structure	Total Chlorine, Weight-Percent
Primary	$R-CH_2-Cl$	2.7
Secondary	$R-CHCl-(CH_2)_n-CHCl-R$ $(n \geq 2)$	~ 71
Secondary	$R-CHCl-CHCl-R$	~ 18[a]
Secondary	$R-CHCl-CH_2-CHCl-R$	
Secondary	$R-CHCl-CH(SO_2Cl)-R$	0.5
Tertiary	$R-\underset{\underset{Cl}{\vert}}{\overset{\overset{R'}{\vert}}{C}}-R''$	2.3–3.5
Sulfonyl chloride	$R-CH_2-SO_2Cl$	0.08
	$\underset{R'}{\overset{R}{\diagdown}}CH-SO_2Cl$	4.20

[a] The 18% may also include structures of the type $R-CHCl-(CH_2)_n-CHCl-R$ where $n \geq 2$.

reactions is the secondary sulfonyl chloride substituent. The secondary chloride adjacent to a secondary sulfonyl chloride group is the next most important reactive site.

When litharge is used to cross-link Hypalon, the reactions involved are hydrolysis of the sulfonyl chloride to sulfonic acid, lead salt formation, and the formation of lead chloride. These reactions are shown in Figure 12.14.

When thiol-type accelerators are used in conjunction with the lead oxide, additional cross-links form as shown in Figure 12.15.[17]

Polyols and magnesium oxide are an alternative curing system for Hypalon. The cross-links form as a result of esterification of the hydroxyl groups by the acid chloride; the magnesium oxide neutralizes the liberated hydrochloric acid (Figure 12.16).

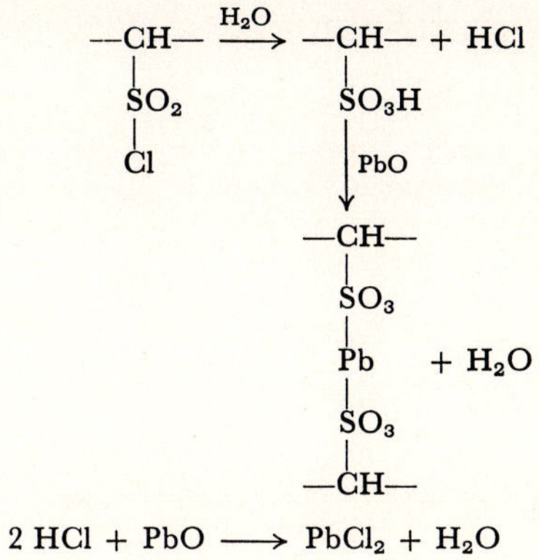

Figure 12.14. Cross-linking of Hypalon by lead oxide.

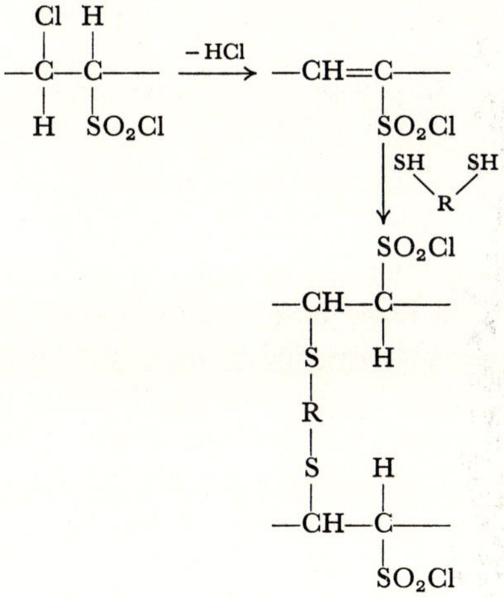

Figure 12.15. Cross-linking of Hypalon with thiol accelerators.[17]

Other polymers that contain sulfur as part of their structure are the thiokol rubbers, copolymers of sulfur dioxide, and vinyl sulphone derivatives. The thiokol polymers are prepared by the condensation of ethylene chloride with sodium polysulfide; a characteristic property is their resistance to oil (Gen. Ref. 4).

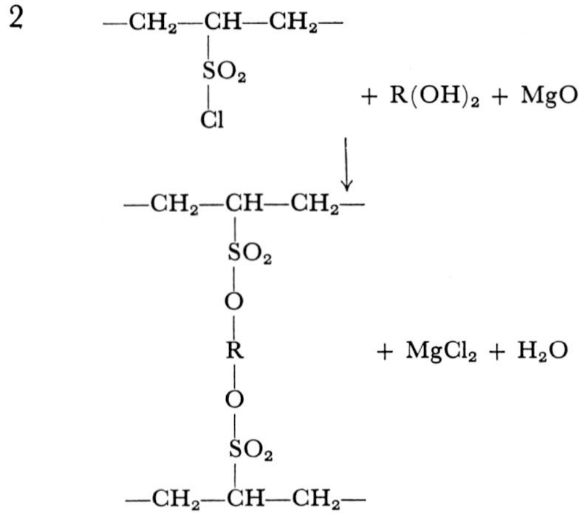

$$2 \quad \begin{array}{c} -CH_2-CH-CH_2- \\ | \\ SO_2 \\ | \\ Cl \end{array} \quad + R(OH)_2 + MgO$$

$$\downarrow$$

$$\begin{array}{c} -CH_2-CH-CH_2- \\ | \\ SO_2 \\ | \\ O \\ | \\ R \\ | \\ O \\ | \\ SO_2 \\ | \\ -CH_2-CH-CH_2- \end{array} \quad + MgCl_2 + H_2O$$

Figure 12.16. Cross-linking of Hypalon by magnesium oxide and a polyol.[17]

$$Cl-CH_2-CH_2-Cl + Na_2S_x \longrightarrow \left[CH_2-CH_2-S_x\right]_n + 2\,NaCl$$

Polymers with thiol end groups are prepared by the oxidation of alkylene dithiols.[20]

$$n\,H-S-R-S-H \xrightarrow{O_2} HS-R\left[S-S-R\right]_{n-1}S-H$$

Poly(propylene sulphide) can be polymerized by a variety of initiators to give materials which range from viscous liquids to rubbery solids. Often these polymers have reactive end groups (dithiol, diamino, dihydroxy) which can be used to react with conventional surface coating polymers such as aminoplasts, epoxides, and phenolic resins.[24] The chemical reaction between the polymers involves the mercapto end group of the polysulphide, the other end group on the polysulphide, and the functional groups of the second resin component. For example in the case of a urea formaldehyde/poly(propylene sulphide) mixture the following reaction occurs:

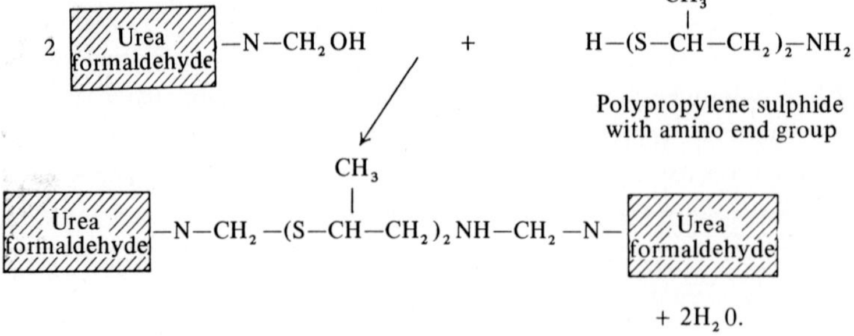

$$2\;[\text{Urea formaldehyde}]-N-CH_2OH \quad + \quad H-(S-\overset{\displaystyle CH_3}{\overset{|}{CH}}-CH_2)_{\overline{2}}-NH_2$$

Polypropylene sulphide with amino end group

$$\downarrow$$

$$[\text{Urea formaldehyde}]-N-CH_2-(S-\overset{\displaystyle CH_3}{\overset{|}{CH}}-CH_2)_2\,NH-CH_2-N-[\text{Urea formaldehyde}]$$

$$+\,2H_2O.$$

Vinyl monomers and sulfur dioxide form a copolymer when polymerization is initiated by free radicals.[21] The probable structure of the copolymer is as shown below:

$$-CH_2-CH-SO_2-CH_2-CH-SO_2-$$
$$\qquad\ \ |\qquad\qquad\qquad\ |$$
$$\qquad\ \ R\qquad\qquad\qquad\ R$$

Divinyl sulphone also undergoes copolymerization, by a free radical mechanism, with vinyl monomers.[22]

GENERAL REFERENCES

1. Worsdall, H. C. in J. B. G. Lewin, collator *Paint Technology Manual, Part 1, Non-Convertible Coatings*, Chapman and Hall, London, 1961.
2. Bateman, L., ed., *The Chemistry and Physics of Rubber-Like Substances*, John Wiley and Sons, New York, 1963.
3. Payne, H. F., *Organic Coating Technology*, John Wiley and Sons, New York, 1954.
4. Stille, J. K., *Introduction to Polymer Chemistry*, John Wiley and Sons, New York, 1962.
5. Gaylord, N. G., and H. F. Mark, *Linear and Stereoregular Addition Polymers*, Interscience Publishers, Inc., New York, 1959.

REFERENCES

1. Harries, C. D., via Gen. Ref. 1; Pummerer, R., G. Ebermeyer, and K. Gerlach, *Chem. Ber.*, **64**, 809 (1931).
2. Salomon, G., A. Chr. van der Schee, J. A. Ketelaar, and B. J. van Eyk, *Discuss. Faraday Soc.*, **9**, 291 (1950); Salomon G., and A. Chr. van der Schee, *J. Polymer Sci.*, **14**, 181 (1954).
3. Albert, Chemische Werke, East Ger. Patent 15,389.
4. Enjay Chemical Company, Technical Bulletin on Buton 152.
5. N. V. de Bataafsche Petroleum M.I.J., Brit. Pat. 837,689 (1960).
6. Enjay Chemical Company, Technical Bulletin on Buton 100.
7. Kovacic, P. and R. W. Hein, *J. Am. Chem. Soc.* **81**, 1187, 1190 (1959).
8. Henderson, L. A., E. I. du Pont de Nemours and Co., U.S. Pat. 2,876,207 (1959).
9. Greenspan, F. P., and A. E. Pepe, Food Machinery and Chemical Corp., U.S. Pat. 2,829,131 (1958).
10. Phillips Petroleum Co., Brit. Pat. 863,256 (1961).
11. Wilborn, F., and J. Morgner, *Chem. Tech. (Berlin)*, **4**, 296 (1952), via *Chem. Abstr.*, **48**, 7318 (1954).
12. Tucker, M. A., and O. C. Slotterbeck, Esso Research and Engineering Co., U.S. Pat. 2,968,648 (1961).
13. Henderson, L. A., E. I. du Pont de Nemours and Co., U.S. Pat. 2,875,919 (1959).
14. Shotton, J. A., Phillips Petroleum Co., U.S. Pat. 2,741,397 (1956).
15. E. I. du Pont de Nemours and Co., Protective Linings and Coatings.
16. E. I. du Pont de Nemours and Co., Booklet on Hypalon.
17. Stevenson, A. C., in G. Alliger and I. J. Sjothun, eds., *Vulcanization of Elastomers*, Reinhold Publishing Corp., New York, 1964.
18. Smith, J. F., *Proceedings International Rubber Conference*, Washington, D.C., November, 1959, 575, via ref. 17.
19. Nersasian, A., and D. E. Andersen, *J. Appl. Polymer Sci.*, **4**, 74 (1960).

20. Marvel, C. S., and L. E. Olson, *J. Am. Chem. Soc.*, **79**, 3089 (1957).
21. Fettes, E. M., and F. O. Davis in N. G. Gaylord, ed. *High Polymers 13, Polyethers Part 3*, Interscience Publishers, Inc., New York, 1962.
22. Union Carbide Chemical Comments, **12**, No. 8, 4 (1960).
23. Bowman, J. M., *Solution Coatings of Viton*, Viton Bulletin No 16, E.I. duPont de Nemours Co. (Inc.), 1966.
24. Smith, J. L., G. T. Williams and R. D. Singer, *J. Oil Colour Chemists' Assoc.*, **53**, 147 (1970).

SILICONE OILS AND RESINS

The presence of carbon atoms in a polymer backbone is not always desirable, particularly where the polymer is subjected to elevated temperatures. Polymers with silicon-oxygen backbones resemble the inorganic silicates, and are much more resistant to heat than similar carbon-containing polymers. To render a silicon-oxygen polymeric structure soluble in organic solvents, it is necessary to have aryl or alkyl side chains present in the molecule. Consequently, a silicone polymer has a mixture of hydrophilic and organophilic groups, and this can result in the molecule having surface-active properties. The silicones range from rubberlike polymers down to oily surface-active agents, depending on the molecular structure and molecular weight.

The silicone rubbers and oils are prepared from chlorosilanes by hydrolysis and then condensation of the resultant silane alcohol. A dichlorosilane forms linear polymers and cyclic derivatives,[1] as shown below. The cyclic product can be converted to a linear polymer by hydrolysis.[1]

$$R_2SiCl_2 \longrightarrow R_2Si(OH)_2 \longrightarrow$$

$$
\begin{array}{ccccccc}
 & R & & R & & R & \\
 & | & & | & & | & \\
-&Si&-O-&Si&-O-&Si&-O-H \\
 & | & & | & & | & \\
 & R & & R & & R & \\
\end{array}
$$

(linear polymer)

For some commercial uses, it is desirable to remove the silanol groups by reaction with a monofunctional chlorosilane, for example, trimethylchlorosilane (or its derivatives). This gives polymers of the type:

$$
(CH_3)_2Si \left[O - \underset{\underset{R}{|}}{\overset{\overset{R}{|}}{Si}} - O \right]_n Si(CH_3)_2
$$

Where three-dimensional structures are required in the resin, or in the film derived from the resin, trichlorosilanes are added to the difunctional silicone derivatives. A typical example is methyl trichlorosilane, which gives polymers with silanol groups; branching, or chemical reaction, is possible at these points in the molecule.

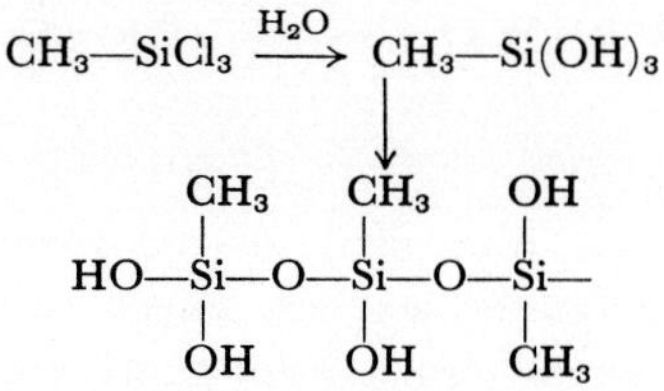

$$CH_3\text{—}SiCl_3 \xrightarrow{H_2O} CH_3\text{—}Si(OH)_3$$

$$
\begin{array}{ccccccc}
 & & CH_3 & & CH_3 & & OH \\
 & & | & & | & & | \\
HO\text{—} & Si\text{—} & O\text{—} & Si\text{—} & O\text{—} & Si\text{—} & \\
 & | & & | & & | & \\
 & OH & & OH & & CH_3 &
\end{array}
$$

Therefore, the silicone oils and resins differ in their molecular weight, molecular structure, and chemical composition. The two most commonly used organic substituents are methyl and phenyl groups.

The low molecular weight, linear silicones are used as flow control agents; the silicone oil concentrates at the inner and outer surfaces of the film.[2] The higher molecular weight and more reactive silicone resins can be used as the sole film-forming component of a coating composition. Catalysts are present in these compositions, or are added immediately prior to use, to promote the curing reaction—specifically, the further condensation of the free silanol groups.[3] Typical catalysts are acids, bases, and the salts of metals; zinc octoate is used where one-pack stability is required but, for two-pack use, the more active tin, lead or chromium octoates can be employed.[4]

The silicone resins can be blended or chemically combined with other film-forming polymers. The chemical interaction between a silicone, either in solution or a film, and other resins involves the silanol or methoxy silane groups; these groups react with water, alcohols, acids, and isocyanates, as shown in the following equations.[5,6]

(1) *Reaction with water*

$$2 \ \overset{|}{\underset{|}{Si}}\text{—}OR + H_2O \longrightarrow \ \overset{|}{\underset{|}{Si}}\text{—}O\text{—}\overset{|}{\underset{|}{Si}}\text{—} + 2\ ROH$$

(2) *Reaction with alcohols*

$$\overset{|}{\underset{|}{Si}}\text{—}OR + R'OH \longrightarrow \ \overset{|}{\underset{|}{Si}}\text{—}OR' + ROH \nearrow$$

(3) *Reaction with acids*

$$\overset{|}{\underset{|}{Si}}\text{—}OR + R'COOH \longrightarrow \ \overset{|}{\underset{|}{Si}}\text{—}O\text{—}CO\text{—}R' + ROH \nearrow$$

(4) *Reaction with isocyanates*

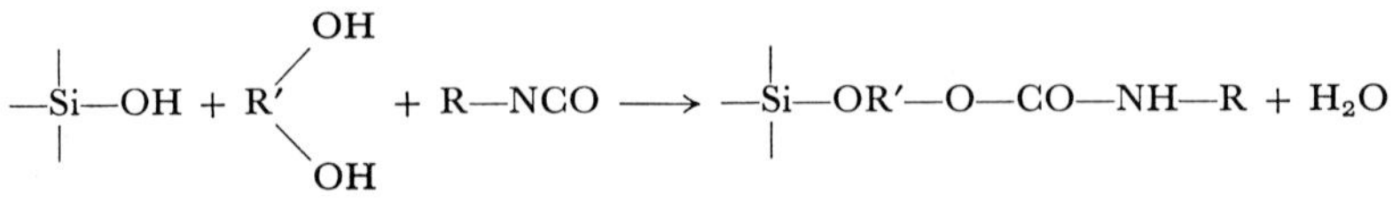

$$-\text{Si}-\text{OH} + \text{R}' \begin{smallmatrix} \text{OH} \\ \diagup \\ \diagdown \\ \text{OH} \end{smallmatrix} + \text{R}-\text{NCO} \longrightarrow -\text{Si}-\text{OR}'-\text{O}-\text{CO}-\text{NH}-\text{R} + \text{H}_2\text{O}$$

The above reactions have been used to co-condense silicone resins with the following organic polymers: acrylics, alkyds, cellulose acetate butyrate, chlorinated diphenyls, epoxy resins, ester gums, ethyl cellulose, melamine formaldehyde resins, nitrocellulose, phenolics and vinyl copolymers.[11 to 14] Often the coreaction involves hydroxyl groups on the organic polymer (i.e. Reaction 2 above).

The alkyd-silicone copolymers have given maintenance paints with outstanding gloss retention when exposed under Florida conditions; after 3 years, a 60° gloss of 70 was recorded for a baked silicone enamel compared with a figure of 18 for an alkyd-melamine finish.[7] Special silicones with low molecular weights and hydroxy functional groups have been described for the preparation of silicone-alkyd co-condensates; the silicone shown opposite is an example of this type.[8] A typical formulation and manufacturing procedure[12] is given in the Appendix on page 376.

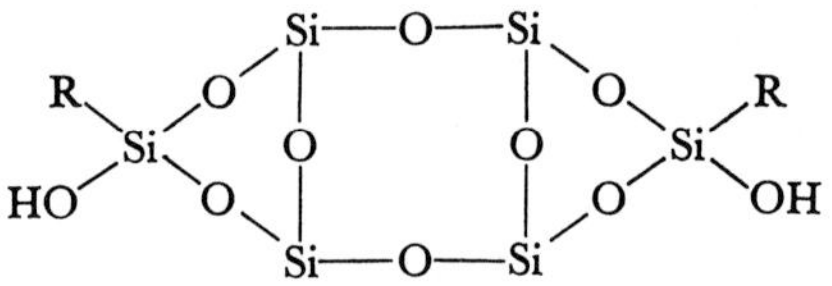

The reaction of a silicone resin with a bisphenol-A type epoxy resin involves the hydroxyl groups of the epoxy resin; the epoxide groups are available for subsequent reactions.[9] A typical formulation and manufacturing procedure[9] is given in the Appendix on page 377.

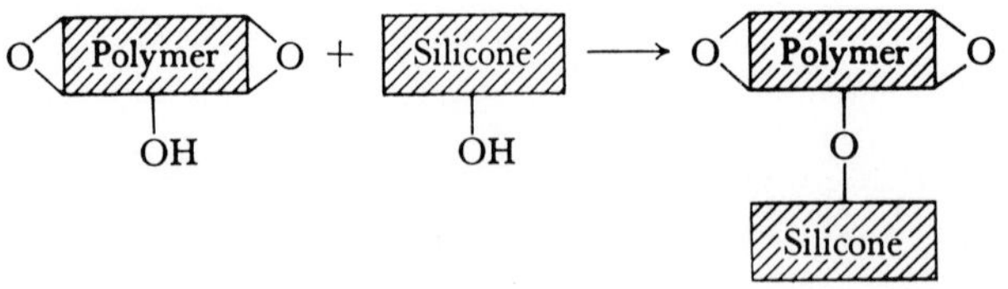

Other resins that contain epoxy groups and a silicone structure are available;

a typical example is shown below. The epoxide groups in this molecule undergo the typical reactions of epoxide groups in organic polymer molecules[10] (see Chapter 9).

$$CH_2\!-\!CH\!-\!CH_2\!-\!O\!-\!(CH_2)_3\!-\!\underset{\underset{CH_3}{|}}{\overset{\overset{CH_3}{|}}{Si}}\!-\!O\!-\!\underset{\underset{CH_3}{|}}{\overset{\overset{CH_3}{|}}{Si}}\!-\!(CH_2)_3\!-\!O\!-\!CH_2\!-\!CH\!-\!CH_2$$

1:3 bis[3(2:3-epoxypropoxy)propyl] tetramethyl disiloxane

REFERENCES

1. Stille, J. K., *Introduction to Polymer Chemistry*, John Wiley & Sons, New York, 1962, p. 109.
2. Johnson, W. J. M., *Offic. Dig. Federation Soc. Paint Technol.*, **33**, 1489 (1961).
3. Stuart, R. S. in H. W. Keenan, collator, *Paint Technology Manual, Part 3, Convertible Coatings*, Chapman and Hall, London, 1962.
4. General Electric Co., Technical Data Book, S.16.
5. Chemistry and Reactions of Sylkyd 50, Dow Corning Corp.
6. Union Carbide Corp., Bulletin No. SF-1120.
7. Thomas, R. N., *Public Works Magazine*, May 1964, via Dow Corning Corp.
8. Dow Corning Corporation, Bulletin No. 03-003 (1962).
9. Dow Corning Corporation, Bulletin No. U-7-704 (1957). Shell Chemicals Aust. Pty. Ltd., *Silicone Modified Epikote Resins*. Plastics and Resins Bulletin No. RES/29/2.
10. Dow Corning Corporation, Bulletin No. Q-2-101a (1958).
11. Cornish, J. W., *Paint Man.* 40 (4), 79 (1970).
12. Clope, R. W., M. A. Glaser, and F. D. Graziano, *Silicone Technology*, P. F. Bruins Ed., Interscience Publishers, New York (1970). Page 141.
13. Dow Corning Corp. Protective Coatings Laboratory, *Silicone Polyesters for Coil Coating Applications*.
14. Taft, D. D., and R. A. Schmidt, X *F.A.T.I.P.E.C. Congr.* 171 (1970).

CHAPTER 13

DEVELOPMENTS IN THE METHODS OF APPLYING AND CURING OF COATINGS

In the last few years there have been rapid and significant changes relating to the application and to the curing of surface coatings. These changes have been facilitated by two major requirements. The first of these has been the need for improved technical control of the coatings, particularly on irregularly shaped objects or on sensitive substrates; the second requirement has been the need to reduce environmental pollution. In some instances these two requirements have reinforced one another. For example the use of powder coatings eliminates atmospheric pollution from solvents and from a technical viewpoint allows polymers, which are insoluble in the normal solvents to be used as surface coatings. The development of powder coatings is such that it is claimed to be the most rapidly expanding area of surface coatings. Radiation curing and electrodeposition of coatings are also advances of considerable significance and both require precise understanding and control of the chemistry of the resin system.

In this chapter the chemistry of the polymers used in powder coatings, and in coatings suitable for electrodeposition or radiation curing will be briefly discussed. Previous chapters have dealt with the chemistry of most of the systems in detail.

POWDER COATINGS

Powder coatings consist of an intimate mixture of resin, pigment and suitable additives whose function is to promote flow, adhesion or chemical curing of the finish. The absence of any solvent is attractive and eliminates or minimizes problems associated with toxicity, odour, air pollution and the flamability of solution coatings. Furthermore powder coatings can be applied in films ranging from 1 to 40 mils in thickness in a single operation.[1] A further advantage of applying coatings as powders is that polymers unsuitable for application by solution can be used (e.g. nylon, polyolefines). The use of these polymers as surface coatings has brought plastics and coating technology much closer than in the past.

Two basic methods are used in the production of powder coatings. In

339

applications which are not over critical dry blending of the components is used because of the relatively low cost of this process. Fusion blending (melt mixing) followed by allowing the mixture to solidify, and grinding to a fine powder is used where greater uniformity of product is required. In some cases a combination of melt fusion and dry blending is used.[2]

Several techniques are available for applying powder coatings. These methods are usually based on either electrostatic or thermal methods of deposition and film formation. In fluidized bed application the powder coating is maintained in a fluid stage by passing a gas (often air) upward through the powder bed. The object to be coated is dipped into the bed after being suitably treated to promote adhesion of the powder particles to the object. Usually this involves pretreating the object to a temperature such that the powder particles melt and adhere to the surface when contact is made. However the use of adhesive coatings, applied before dipping the object in the fluidized bed, is an alternative process which is of interest where heat sensitive objects are to be coated.

The object usually has a residence time in the fluidized bed of only a few seconds. It is then removed and if necessary given a final curing cycle to promote optimum flow of the coating. A variation of the fluidized bed process is to use a powder spray gun to apply the coating. Flame spray methods[2,3] and Plasma Arc Guns[2,3] are also used. Electrostatic application of powder coatings is carried out either in conjunction with a fluidized bed or by special spray guns. The relative merits of the various application techniques and illustrations of the equipment used in each technique can be found in references 1-7.

Polymers suitable for powder coatings are required to fulfill the requirements of good adhesion to the substrate and good thermal flow as well as giving the required film properties (corrosion resistance, durability, etc.). Both thermoplastic and thermosetting polymer systems are used. An additional requirement of the thermosetting polymers is that the curing reaction should not proceed at an appreciable rate until after the thermal flow or levelling stage.

The major thermoplastic polymers used for powder coatings include poly(vinyl chloride), nylon, cellulose acetate butyrate, polyethylene, Penton (a chlorinated polyester) P.T.F.E. (Polytetrefluoroethylene), and P.T.F.C.E. (polytrifluorochloroethylene). Other thermoplastic polymers have been used but to a lesser extent. The properties of powder coatings based on some of the polymers are listed in Table 13.1.

Thermosetting polymers for powder coatings require careful control of the rate of the curing reaction so that the composition is stable at room temperature for long periods of time, and can be applied to the substrate and fused before significant crosslinking occurs. The epoxy resins, which

TABLE 13.1[8]

CHEMICAL RESISTANCE OF POWDER COATINGS

Chemical	Epoxy	Cellulosics	PVC	Poly-ethylene	Nylon	Chlorinated Polyether
Water (Low temperature)	6	4	6	5	5	6
Hot Water and Water Vapour	5	2	2	3	4	4
Aqueous Salt Solution	5	2	5	3	4	5
Mineral Acids	4	1	4	5	2	5
Organic Acids	2	1	3	4	1	4
Alkalis	4	1	4	4	4	5
Hydrocarbon Solvents	4	1	2	2	4	6
Chlorinated Hydrocarbons	2	0	0	1	4	4
Esters and Ketones	2	0	0	2	4	4
Overall Resistance	34	12	26	29	32	43

Ratings: 6–Excellent 4–Good 2–Fair 0–Poor

are the most common thermosetting formulation used, have a lower initial melt viscosity than the thermoplastics and hence comparatively large amounts (30-40%) of pigments and fillers can be added to them. This reduces considerably the raw material cost of the epoxies.[8]

Five chemical types of curing agents are commonly used in the manufacture of epoxy powder coatings. These are cyanamides, particularly dicyandiamide, boron trifluoride complexes, aromatic amines, aliphatic amines, and acid anhydrides.

The chemistry of the crosslinking reactions involved has been discussed in Chapter 7. In powder coatings dicyandiamide is often the preferred crosslinker and gives compositions which are extremely stable at room temperature. It also gives tough films with good color stability. The properties of epoxy powder coatings prepared from the various crosslinking agents are listed in Table 13.2.

REFERENCES

1. Kutik, L., *J. Paint Technol.*, **38**, 493 (1966).
2. Mann, A., *Paint and Varnish Production*, **Sept. 1970**, 25.
3. Brown, P., *Metal Finishing*, **June 1970**, 97.

TABLE 13.2[8]

PROPERTIES OF EPOXY POWDER COATINGS BASED ON VARIOUS CURING AGENTS

Property	Curing Agent				
	Dicyandiamide	Boron Trifluoride Complex	Aromatic Amine	Aliphatic Amine	Acid Anhydride
Application Temperature °C	150-200	120-150	120-150	100-130	150-200
Typical Cure Cycle	1 hr. at 180°C	½ hr. at 150°C	1 hr. at 120°C	½hr. at 120-130°C	1 hr. at 180°C
Storage Life at 70°F (Months)	76	6	2-5	2-3	3-6
Maximum Operating Temperature °C – continuous	100	120	150	80	150
– intermittent	200	200	250	120	200
Chemical Resistance: Water (low temperature)	6	6	6	6	6
Hot Water and Water Vapour	5	3	5	2	4
Aqueous Salt Solution	5	5	6	3	5
Acids	4	4	5	1	5
Alkalis	4	4	5	3	2
Solvents	3	3	3	2	3
Overall Chemical Resistance	27	25	30	17	25
Colour Stability	Good	Good	Poor	Good	Good

Ratings: 6–Excellent   4–Good   2–Fair   0–Poor

4. Paulson, D., and R. F. Stobel, *Metal Finishing*, **May 1970**, 53.
5. *Proceedings of the First International Symposium on Powder Coatings* Ed. L. Bilefield, Published by Mercury House, London, 1968.
6. *Proceedings of the Second International Symposium on Powder Coatings.*
7. Cole, G. E., *Industrial Finishing*, **Jan. 1970**, 47.
8. Willows, G. E., Paper 1/3 in Ref. 5.

THE USE OF RADIATION IN COATING COMPOSITIONS

The use of radiation energy to initiate chemical reactions has grown rapidly over the last ten years mainly because of the availability of lower cost linear accelerators.

Suitable radiation has been used to initiate the polymerization of monomers, to crosslink polymers or to graft onto polymers, and to cure coating compositions.

The processes of most direct interest to the surface coating industry are the use of U.V. and Electron Beam radiation to cure coatings, and the polymerization of monomers onto pigment or other surfaces using radiation to initiate the reaction.

Radiation Curing of Coating Compositions. Radiation curing is sometimes referred to as irradiation curing, electron radiation curing and electron beam curing.[1] "Electrocure" is the term used for the Ford Motor Co. process which is being used to cure coated instrument panels and "Raycron" is the trademark used by PPG Industries, Inc. for its radiation curable coating compositions.

In principal the process of radiation curing relies on the production of free radicals in the organic coating. For a surface coating to cure by electron radiation under practical conditions, the polymeric binder must incorporate chemical groups which respond efficiently to free-radical reactions. Usually this requirement is met by using solutions of polymers which contain vinyl or acrylic unsaturation, in suitable monomers.[2] Although these polymers are in effect unsaturated polyesters (Chapter 5) significant chemical differences exist between the resins used in conventional unsaturated polyesters and those used in coatings designed for electron curing.

The electron generator consists of three major components, namely a power supply, an accelerator and a control console. The power supply accepts commercial power, steps it up to a higher voltage, and then rectifies it to direct current. This is sent to the accelerator where the voltage is used to impart energy to electrons supplied by a hot filament cathode. The resultant beam may then be electromagnetically scanned and transmitted through a suitable window to the coating. The electron energies required to achieve maximum efficiency in promoting the curing reactions with the minimum of attack on the substrate and polymer molecules, have been discussed in detail by Tawn.[3] A typical installation (Figure 1) would be rated at 30 KW power, supplying 300 KV energy electrons over a 30 in. window.[5] Shielding is required to protect against scattered primary electrons and the secondary radiation.

Figure 1. Typical industrial electron processing system.[6]

Advantages claimed for electron beam curing of paint films compared with conventional catalysed methods or thermal cure are:[6-11]

1. Coatings can be cured in a very short time (at most a few seconds) at ambient temperature.

2. Less plant space is required—an electron radiation set-up would usually occupy less than 20 ft of an assembly line while drying ovens may need several hundred feet.

3. Heat sensitive materials can be coated without damage to the substrate (Ford are using electron beam curing on ABS and polypropylene).

4. Radiation curable paints have better can stability because they are not precatalysed.

5. Losses due to volatile solvent or vehicle (resin) components are small and consequently pollution is negligible.

6. In some cases multiple coats can be applied before curing.

7. The cured coating may be bonded strongly to the surface by grafting processes.

8. Line speeds are faster and this minimizes the problems associated with dust pick-up.

9. Start-up and shut-down, or variations in line speed can be dealt with simply by control of the beam intensity.

Limitations of the present method are:

1. it is difficult to handle shaped objects

2. only a limited number of paint formulations is suitable for electron curing.

3. The initial capital costs are high but when considered over a period of years the cost can be lower than for conventional curing.

4. Shielding must be provided to cope with scattered radiation.

5. if air is present, inhibition of cure may result and appreciable quantities of ozone may be formed.

The basic requirement of the polymer system used in coatings which are to be cured is that it must respond efficiently to free radical initiated reactions. The free radicals can promote crosslinking reactions, polymer degradation or polymerization depending on the chemical composition of the material. The extent of these effects is described quantitatively by G-Values; a satisfactory resin will have built in sites which respond with high G-Values (the terms used here are defined at the end of this section). Typical G-Values for a variety of chemical reactions related to polymers are listed in Table 13.3.

TABLE 13.3

G VALUES OF EVENTS IN POLYMER PROCESSES[a]

Primary events		
Radical formation	Styrene	0.2-0.8
	Polystyrene	2
	Vinyl chloride	10
	Poly(vinyl chloride)	12
	Methyl methacrylate	3-7
Crosslink formation	Natural rubber	1.2
	Polystyrene	0.05
	Silicone polymers	2.5-4.5
	Acrylate polymers	0.5
Fracture	Poly(methyl methacrylate)	1.6
Chain processes (equivalent G)		
Polymerization of	Methyl methacrylate	2,000
	Methyl acrylate	600
	Styrene	1,000
	Acrylonitrile	40,000

[a]*Charlesby (1960).*

All the primary events have values of 1 to 10. When an event initiates a chain process, the product of the G-Value for the primary event and the chain length of the reaction equals the equivalent G-Value. Therefore the equivalent G-Value for polymer formation from vinyl monomer is around 10^2 to 10^4.

As a consequence radiation curable coatings often use polymers with vinyl or acrylic unsaturation (pendant or main chain) dissolved in suitable monomers.

A number of methods have been used to synthesize polymers with pendant unsaturation. Norstrom and Hinsch[12] and Labana and McLaughlin[13] used the epoxy-acid reaction and introduced the epoxy group into the vinyl polymer by using glycidyl methacrylate as a monomer component. Acrylic or methacrylic acid was used to esterify the epoxy group.

$$
\begin{array}{c}
\boxed{\text{Vinyl or acrylic polymer}} \\
| \\
C = O \\
| \\
O \\
| \\
CH_2 \\
| \\
CH\diagdown \\
| \quad \ \ \rangle O \\
CH_2\diagup
\end{array}
\quad + \ CH_2 = \underset{\underset{COOH}{|}}{CH} \ \longrightarrow \quad
\begin{array}{c}
\boxed{\text{Vinyl or acrylic}} \\
| \\
C = O \\
| \\
O \\
| \\
CH_2 \\
| \\
CH - OH \\
| \\
CH_2 - OCO.CH = CH_2
\end{array}
$$

Similar structures have been formed by esterification of a hydroxy acrylic or styrene-allyl alcohol copolymer with acrylic or methacrylic acid.[14] Alternatively acryloyl chloride has been used to esterify the hydroxyl groups on a vinyl or acrylic backbone.[15]

The acrylic polymers with pendant unsaturation are dissolved in a suitable monomer, often methyl methacrylate which, because of the conjugated carbonyl-double bond system responds well to radiation curing.[16] Typical properties of radiation cured compositions are listed in Table 13.4.

With an increase in the crosslink density (unsaturation content of the polymer) the hardness and solvent resistance increase and the bending and impact properties decrease.

An alternative method of preparing polymers with pendant unsaturation is to react a hydroxylated polymer with the first isocyanate group of a diisocyanate monomer (to give an isocyanate terminated polymer) and then to react at least a portion of the remaining NCO groups of the diisocyanate monomer with a hydroxyalkyl acrylate or methacrylate.[17]

Unsaturated polyesters and unsaturated polyurethanes are other polymer systems studied extensively for electron beam curing.[13, 18–21] Unsaturated polyesters prepared from maleic anhydride, saturated dibasic acid, and a glycol such as ethylene or diethylene glycol and thinned in styrene have been cured by radiation curing and by peroxides and the properties compared. Radiation curing gave a higher yield of insoluble fraction and a higher con-

TABLE 13.4

VARIATION IN ACRYLIC COPOLYMER UNSATURATION

Polymer Composition, Mole %				Film Properties[a]			
MMA^a	EA^b	GMA^b	$DBPT^c$	Pencil hardness	MEK rubs[d]	Mandrel bend, inches	Reverse[e] impact, inch lb.
36	60	4	0.4	2B	11	Pass 1/8	>80
34	57	9	0.8	F	20	Pass 1/8	40
32	53	15	1.25	H	50	Pass 1/4	<10
29	48	23	1.75	H	50	Pass 3/4	<10
25	42	33	2.35	H	50	Fail 1 1/4	<10
19	32	49	3.00	H	50	Fail 1 1/4	<10

[a]Cured film thickness = 1.0 ± 0.1 mil on 24-gage bonderized steel substrate. Dose = 15 Mrad. [b]MMA methyl methacrylate, EA ethyl acrylate, GMA glycidyl methacrylate. [c]Double bonds per thousand molecular weight at total conversion. [d]Soft cloth soaked in methyl ethyl ketone rubbed across film with firm hand pressure. [e]Gardner falling dart impacter. Pass. No coating removal after taping.

$$
\boxed{Polymer} \qquad\qquad \boxed{Polymer}
$$

$$
\begin{array}{c}
| \\
OH \\
\end{array}
\qquad
+ R
\begin{array}{c}
\diagup\; NCO \\[4pt]
\diagdown\; NCO \\
\end{array}
\longrightarrow
\begin{array}{c}
| \\
O \\
| \\
CO \\
| \\
NH \\
| \\
R \\
| \\
NCO \\
\end{array}
$$

$$
\downarrow + CH_2 = \underset{\underset{CH_3}{|}}{C} - COOCH_2CH_2OH
$$

$$
\boxed{Polymer}
$$

$$
\begin{array}{c}
| \\
O \\
| \\
CO \\
| \\
NH \\
| \\
R \\
| \\
NH \\
| \\
CO \\
| \\
O \\
| \\
CH_2 - CH_2 - O.CO.\underset{\underset{CH_3}{|}}{C} = CH_2 \\
\end{array}
$$

version of the double bonds present originally.[18] Physical and chemical properties of the radiation cured polyester were superior to the peroxide cured material. When the unsaturated backbone polymer contained terminal isocyanate groups they apparently entered into the crosslinking reaction by copolymerization with either the vinyl bonds of the backbone or monomer.[18] Infrared evidence suggests that groups such as urethanes, ureas or allophanates may be formed.

Monomers other than styrene have been used in unsaturated polyesters for radiation curing.[18] The decreasing order of reactivity found was vinyl acetate > styrene > triallyl cyanurate > triallyl isocyanurate > methyl methacrylate. The vinyl acetate-unsaturated polyester copolymerization was complete at a dose of $< 1M_{RAD}$ whereas with allyl cyanurates a dose of 20 to 40 M_{RAD} was required.

In parallel studies it has been found that vinyl toluene gave higher conversions than styrene whereas chlorostyrene was less efficient than styrene. The behaviour of chlorostyrene in systems which cure by radiation contrasts with its greater reaction in peroxide cured systems. This difference in behaviour may be a consequence of the considerably higher G value for chlorostyrene than for styrene. The higher concentration of radicals formed at a given dose of ionizing radiation could produce a cage effect in which primary radicals annihilate each other.

The structure of the unsaturated polyester backbone and the ratio of polyester to styrene are of importance in coatings designed for radiation curing. At low styrene contents gel is formed mainly by polyester-polyester interaction and consequently the amount of unsaturation in the polyester is important. The rate of gel formation increases with styrene content up to a limit (around 30% styrene) and reflects the copolymerization of styrene with the unsaturated polyester. Further increases in the styrene content reduce gel formation by enhancing the termination of styrene-polyester chains of low molecular weight.[19]

Clearly then there are differences between coatings cured by peroxides and by radiation and more detailed information of the type found in references 18-21 is required to optimize coating formulations.

Unsaturated aminoplasts have been prepared from acrylamide, melamine, formaldehyde and butanol.[22] These polymers can be crosslinked alone or after mixing with monomers by electron radiation. The polymer would have a structure shown below.

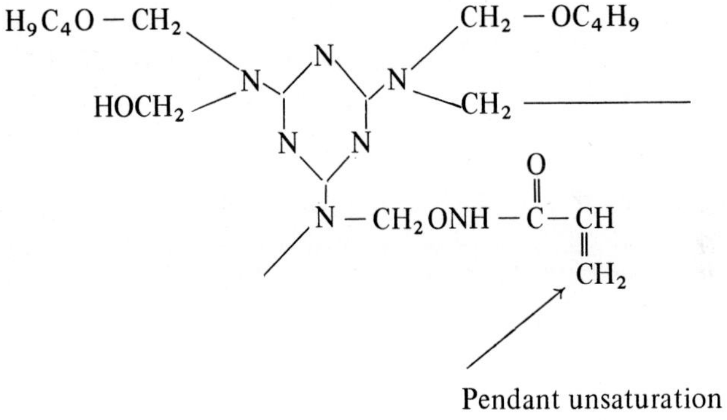

$$H_9C_4O-CH_2 \qquad\qquad CH_2-OC_4H_9$$

(triazine ring structure with substituents)

$$HOCH_2 \qquad CH_2 ————$$

$$N-CH_2ONH-\overset{O}{\overset{\|}{C}}-\overset{\|}{\underset{CH_2}{CH}}$$

Pendant unsaturation

Surface Polymerization and Encapsulation Process. An interesting extension of the radiation curing of polymers has been the direct polymerization of monomers on the substrate. For example vinyl stearate,[23] acrylamide,[23] and diacetone acrylamide[24] can be applied to a suitable surface and polymerized by ionizing radiation. A unique application is the formation of adherent films of poly(acrylonitrile) on aluminium by the electrolytic polymerization of acrylonitrile monomer, using the metal as anode, a carbon cathode, and ammonium toluenesulfate as supporting electrolyte.[25]

Polymerization onto and around pigment surfaces has been used to microencapsulate materials in fields related to coatings for many years and there is a vast literature on this subject. More recently examples of direct interest to the surface coatings industry have been reported. For example acrylic or methacrylic acids have been polymerized on the surface of TiO_2 pigments,[26] and the electron beam polymerization of monomers such as 1,3-butadiene, on a piezoelectric quartz crystal[27] has been reported. Of more direct practical interest is the polymerization of monomers around pigment and filler particles, to form microporous particles. These polymerizations can be initiated by conventional free radical initiators or by irradiation. Pigments encapsulated in these beads are more efficient opacifiers and consequently less pigment is required in a coating. This approach offers the possibility of reducing significantly the formula cost of coatings. A further extension of this technology has been to use microporous polymer beads, formed by emulsion polymerization using radiation initiation, or formed by grinding and dispersing foam sheets. These pigment-free beads give greater effective use of titanium dioxide pigments and savings of up to 50% of the titania have been reported. Usually the beads are of foamed polystyrene. Paints containing polystyrene microporous beads have been marketed successfully.

DEFINITION OF TERMS USED[1,2]

Dose – The amount of energy absorbed per unit mass of material. The unit is the rad which corresponds to an energy absorption of 100 ergs per gm. This is equivalent to 6.25×10^{13} electron volts per gm. One Megarad (M_{RAD}) equals 10^6 rads.

Dose rate – This is the dose per unit of time measured in rad sec^{-1} or Mrad min^{-1}.

Electron Volt – This is the unit of energy equal to the kinetic energy acquired by an electron when accelerated across a potential difference of 1 volt. A MeV equals 10^6 eV.

G-Value – The G-Value or energy yield is the number of individual atoms that are produced or used up in the system per 100eV of energy absorbed.

REFERENCES

1. Brushwell, W., *Amer. Paint J.*, **55** (2), 77 (Aug. 1970).
2. Nordstrom, J. D., and J. E. Hinsch, *Ind. and Eng. Chem. Prod. Res. Develop.*, **9** (2), 155 (1970).
3. Tawn, A. R. H., *J. Oil Colour Chemists' Assoc.*, **51**, 782 (1968).
4. Miranda, T. J., and T. F. Huemmer, *J. Paint Technol.*, **41**, 188 (1969).
5. Phillips, W. C. *The Australian Paint J.*, **16** (2), 7 (1970).
6. McLaren, K. G., *Proc. Royal Aust. Chem. Inst.*, **37**, 340 (1970).
7. Vanderbie, W., P. Spencer, and C. Swanholm, *Paint and Varnish Production*, **59** (3), 39 (1969).
8. Dalton, F. L., *Plastics and Polymers*, **38**, 343 (1970).
9. Cole, G. E., *Ind. Finishing*, Jan. 1970, 34.

10. Davison, W. H. T., and T. T. Greenwood, *Preprints Div. Organic Coatings and Plastics Chem., Amer. Chem. Soc.*, **29** (1), 153 (1969).
11. Mock, J. A., *Materials Engineering*, Oct. 1970, 56.
12. Norstrom, J. D., and J. E. Hinsch, *Preprints Div. Organic Coatings and Plastics Chem., Amer. Chem. Soc.*, **29** (1), 160 (1969). *Ind. Eng. Chem. Prod. Res. Develop.*, **9** (2), 155 (1970).
13. Labana, S. S., and E. O. McLaughlin, *J. Elastoplast*, **2**, 3 (1970).
14. Radlove, S. B., and A. Ravve, *Continental Can Co. Inc.*, U.S. 3,546,002 (1970).
15. Arnoff, E. J., and E. O. McLaughlin, Ford-Werke A-G, Ger. Pat., 1,932,687 (1970).
16. Huemmer, T. F., Ind. Fin., (U.S.A.) **46**, No. 5, 34 (1970).
17. Ford Motor Co., U.S. Pat., 3,509,234 (1970).
18. Omel'chenko, S. I., N. G. Videnina, V. G. Matjushova, I. N. Chervetsova, and G. N. Pyankov, *Ind. Eng. Chem. Prod. Res. Develop.*, **9** (2), 143 (1970).
19. Hoffman, A. S., J. T. Jameson, W. A. Salmon, D. E. Smith, and D. A. Trageser, *Ind. Eng. Chem. Prod. Res. Develop.*, **9** (2), 158 (1970).
20. Pietsch, G. J., *Ind. Eng. Chem. Prod. Res. Develop.*, **9** (2), 149 (1970).
21. Mettinen, J. K., *Preprints Div. Organic Coatings and Plastics Chem.*, Minneapolis Meeting, **29** (1), 182 (1969).
22. Ravve, A., Continental Can Co. Inc., U.S. Pat. 3,535,148 (1970).
23. Williams, T., M. W. Hayes, British Iron and Steel Research Assoc., Brit. Pat., 1,168,641 (1969).
24. Oliver, J., *Prod. Finishing*, (Cincinnati) **34** (7), 66 (1970).
25. Asahara, T. M., M. Seno, and M. Tsuchiya, *Kinyoku Hyomen Gijutsu*, **20** (11), 576 (1969).
26. Nollen, K., Y. Kaden, and K. Hamann, *Agnew. Makromol. Chem.*, **6**, 1 (1969).
27. Thompson, L. F., R. L. Venable, and K. G. Mayhan, *Preprints Div. Organic Coatings and Plastics Chem., Amer. Chem. Soc.*, **30** (1), 273 (1970).

Ultraviolet Curing of Coatings. Electron radiation has a sufficiently high energy to form free-radicals from the bombarded material either by ionization or excitation. Ultraviolet radiation has considerably lower quantum energy and requires the use of photo-initiators which are decomposed by the U.V. radiation to free radicals.[1,2] For example diphenyl disulfide forms radicals as shown below:

$$\text{C}_6\text{H}_5\text{—S—S—C}_6\text{H}_5 \xrightarrow{h\nu} 2\ \text{C}_6\text{H}_5\text{—S}^{\bullet}$$

Diphenyl disulfide

Phenyl thiyl radical

Opaque pigments, such as titanium dioxide, absorb U.V. radiation and consequently the process of U.V. curing is restricted to clear or semi-transparent coatings.[1] Primer coats which contain extender pigments such as asbestos, talc, and barytes however can be used.

Two major factors have contributed to the advance in U.V. curing. These are:

1. the commercial production of the sensitive catalysts which give cure times of a few seconds in polyesters and

2. the development of relatively cheap, reliable high energy U.V. lamps

The equipment for U.V. curing is very much cheaper than that required for electron beam curing.

To date most work has been done on unsaturated polyesters and a commercial application is in the production of fillers for chip board.

Cure times of 30 seconds-2 minutes are possible with precatalyzed resins. These resins are stable for long periods if protected from direct sunlight. Often it is necessary to subject the applied coating to a pre-gel stage to allow sufficient time for the wax to migrate to the surface and form a film. This is necessary to minimize air inhibition. The pre-gel is accomplished by the use of a low pressure lamp. The coating is then cured by irradiation from a high pressure lamp.

Most polymer systems used in U.V. curing are of the unsaturated polyester type, and contain a photosensitizer. Typical of the photosensitizers reported are the following: benzoin ethers of secondary alcohols[3] and a-substituted benzoin ethers of the type shown,[4]

$$Ar - \underset{\underset{O}{\overset{\|}{}}}{\overset{}{C}} - \underset{\underset{\underset{R^1}{\overset{|}{}}}{\overset{|}{O}}}{\overset{\overset{R^2}{\overset{|}{}}}{C}} - Ar$$

Ar = aromatic radical

R^1, R^2 = substituted alkyl, acryl, etc.,

other benzoin ethers,[5] various[6] halogenated and nitrogen (hydrazobenzene)[7] containing compounds. Other polymer systems that have been cured by U.V. radiation include pentaerythritol esters of unsaturated acids (i.e. acrylic acid),[8] and isocyanate terminated unsaturated polyesters.[9] Halogenated monomers have been used in U.V. initiated surface polymerizations on suitable substrates.[10] The chemistry of the polymer systems used for coatings which are to be used by U.V. radiation, does not differ significantly from the peroxide initiated systems.

REFERENCES

1. Young, S., S. Taylor and K. W. G. Butcher, *Product Finishing*, **23**, 36 (1970).
2. Deninger, W., and M. Pathiegers, *J. Oil Colour Chemists' Assoc.*, **52**, 930 (1969).
3. Farbenfabriken Bayer A.-G., Austrian Pat. 276,774 (1970); Australian Pat. Appln. 3,7283/68 (1968).
4. Farbenfabriken Bayer A.-G., Australian Pat. Appl. 57,525/69 (1969).
5. Farbenfabriken Bayer A.-G., Dutch Pat. 69.08904 (1969);
   French Demande 2,015,102 (1970); South African Pat. 6,904,724 (1970);
   Societa Italiana Resine S.P.A., Dutch Pat. 69.10753 (1969).
6. Rabek, J. F., Photochem. Photobiol., **7** (1), 5 (1968).
7. Tokyo Shibaura Electric Co., Jap. Pat. 18,075/69 (1969).
8. Bassemir, R. W., Carlick, D. J., and Sprenger, G. E., Sun Chemical Corp., U.S. Pat. 3,551,246 (1970); U.S. Pat. 3,551,311 (1970).
9. Asahi Kasei Kogyo Kabushiki Kaisha, Aust. Pat. Appl. 57,220/69 (1969).
10. General Electric Company, Aust. Pat. Appl. 34,064/68 (1968).

ELECTRODEPOSITION OF COATINGS

The electrodeposition of coatings involves the passage of an electric current across a tank of a water soluble or water dispersable paint system. The object to be coated is generally made the anode. Paint deposits on the anode and the film thickness of the coating builds up until the resistance of the coating prevents further deposition. The process offers many advantages over conventional spray or dipping processes. Foremost among the advantages are the ability to coat complex shapes (particularly the inner surfaces of rocker panels etc.), the reduced fire risk (compared with solvent containing dip tanks), the coated object can be handled immediately it is removed from the tank, and the process is well suited to automatic production lines. (See Gen. Ref.).

During electrodeposition the composition of the bath changes as a result of preferential deposition of some components. The pH usually rises and soluble salts can accumulate. Analytical techniques have been developed to measure the changes[1] and techniques are available for controlling bath composition.[2] These are discussed fully in the literature and will not be dealt with here.

Polymers suitable for electrodeposition must be readily soluble or dispersible in water, and they must be stable under the conditions of the bath (usually alkaline pH).

The methods used in synthesising these polymers have been discussed in previous chapters, particularly Chapter 5. Therefore only a brief indication of the types of resins used in electrodeposition will be given.

Vegetable oils modified by reaction with maleic anhydride (see Chapter 2 for the reactions involved) can be used as such or reacted with hydrophilic alcohols and neutralized by amine.

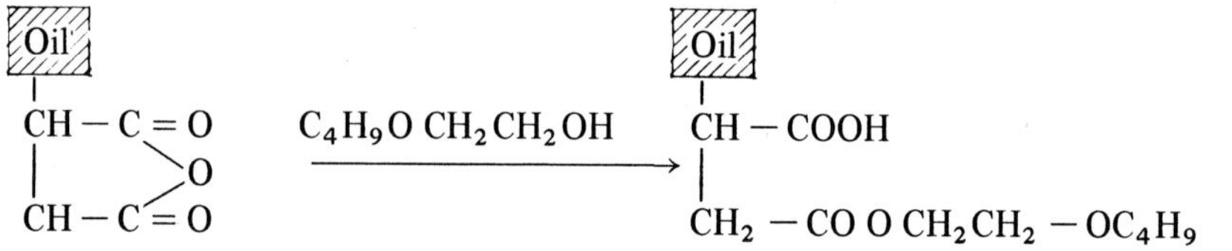

$$
\begin{array}{ccc}
\boxed{\text{Oil}} & & \boxed{\text{Oil}} \\
| & & | \\
CH-C=O & & CH-COOH \\
| \quad \diagdown O & \xrightarrow{\quad C_4H_9O\,CH_2CH_2OH \quad} & | \\
CH-C=O & & CH_2-CO\,O\,CH_2CH_2-OC_4H_9
\end{array}
$$

Similarly alkyd resins, generally of relatively low molecular weights and high acid numbers can be solubilized and used in electrodeposition processes. Often a tri-functional acid is used to give an alkyd with the required acid number/molecular weight balance; maleinized fatty acid or trimellitic anhydride[3] are used for this purpose. Branched chain fatty acids improve the hydrolytic stability of the alkyds in the electrodeposition bath. The modification of suitable alkyds with methoxymethyl isocyanate is reported to improve the crosslinking between the alkyd and a phenolic resin.[4]

A variety of acrylic resins has been applied by electrodeposition. For example, a copolymer of ethyl acrylate, glycidyl methacrylate and methacrylic acid has been used in conjunction with a phenol formaldehyde resin.[5]

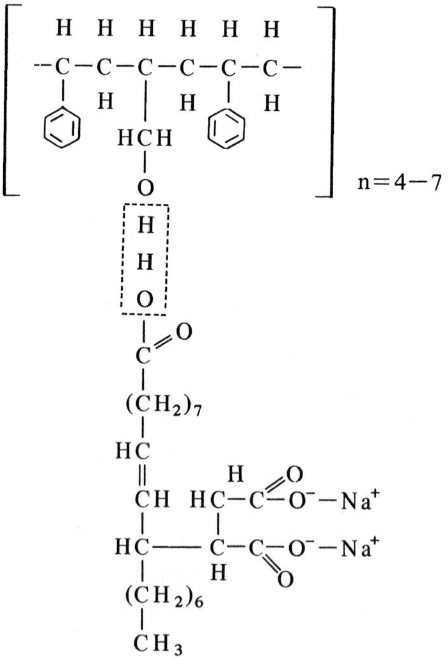

$$n = 4-7$$

−RJ-100-oleate-adduct salt

Acrylic resins which contain N-methylol or N-methylol ether groups are of particular interest because of the crosslinking potential of these groups during stoving. Acrylic modified oils formed by copolymerization of suitable oil/monomer mixtures, or by co-reaction (Chapter 4) have been suggested as vehicles for electrodeposited paints. The product obtained by esterifying RJ-100^R (a styrene-allyl alcohol copolymer) with maleinized oleic acid has been described in detail.[6]

Epoxy resins in various forms are of interest in electrodeposition because of the hydrolytic stability of the polymer backbone. Esters of epoxy resins with maleinized fatty acids have been studied in detail.[7,8] A novel application has been a room temperature curing system of polyamide-epoxy adduct.[9]

In mixed resin systems and in pigmented systems preferential migration of some components may occur. Titania pigments vary in throwing power; this is related to the coating on the titania and to the surface area of the pigment.[10] Often the pigments deposit preferentially and give a higher concentration of pigment in the coating than was present in the bath.

All polymers are mixtures of chains of different length and polarity and it is not surprising that preferential migration of some molecules occurs in electrodeposition. Similarly in blends of acid polymers, solubilized by amines, and amine resins different migration rates have been noted.[11] Chemically reacted alkyd/melamine resins reduce the tendency toward preferential migration [12] (see also page 236).

GENERAL REFERENCES

1. Brewer, G. E. F., and L. W. Wiedmayer, *J. Paint Technol.*, **42**, 588 (1970).
2. Hays, D. R., and C. S. White, *J. Paint Technol.*, **41**, 461 (1969).
3. Brushwell, W., *Amer. Paint J.*, **54**, 71 (1970).
4. Ellinger, M. L. *X F.A.T.I.P.E.C. Congr.*, 479 (1970) *J. Paint Technol.*, **41**, 202 (1969).
5. Brewer, G. E. F., and R. F. Hines, *J. Paint Technol.*, **43** (554), 71 (1971).
6. Belliveau, J. F., E. G. Bobalek, and G. L. Simard, *J. Paint Technol.*, **42**, 377 (1970).
7. Goboz, J., and J. Lahaye, *J. Paint Technol.*, **42**, 371 (1970).
8. Sato, T., Y. Motoyama, and O. Ohu, *J. Paint Technol.*, **41**, 438 (1969).
9. Nakamura, Y., K. Komata, and H. Nozaki, *Bull. Soc. Chem. Japan*, **43**, 663 (1969).

REFERENCE

1. Anderson, D. G., and D. J. Tessari, *J. Paint Technol.*, 42, 119 (1970).
2. Milne, D. G., *Metal Finishing*, 16, 185 (1970).
3. Gentles, J. K., and J. B. Harrison, Goodlass, Wall & Co. Ltd. Brit. Pat. 1,027,813 (1966); U. S. Pat. 3, 378,477 (1968).

4. Farbenfabriken Bayer A-G. Dutch Pat. 69.08670 (1969); German Pat. 1,769,545 (1970).
5. Nayarova, I. V., I. A. Krylova, V. I. Eliseeva, V. A. Stazov, and P. J. Zubov, Lakrokras. Mat. (1970) (3), 4.
6. Sullivan, M. R., *J. Paint Technol.*, **38**, 424 (1966).
7. May, C. A., *J. Paint Technol.*, **43** (552), 43 (1971).
8. Van Westrenen, W. J., and L. A. Tysall, *J. Oil Colour Chemists' Assoc.* 51, 108 (1968).
9. North, A. G., *J. Oil Colour Chemists' Assoc.*, **53**, 353 (1970).
10. Seivard, L. L., and W. W. Douney, *J. Paint Technol.*, **43** (552) 65 (1971).
11. Robinson, F. D., and B. J. Tear, *J. Oil Colour Chemists' Assoc.*, **53**, 265 (1970).
12. Vianova Kunsthary A-G French Pat. 1,540,118 (1968).

APPENDIX

TYPICAL FORMULATIONS AND MANUFACTURING METHODS FOR THE PREPARATION OF SURFACE-COATING POLYMERS

PREPARATION OF A 60% LINSEED OIL PENTAERYTHRITOL-PHTHALATE ALKYD (FUSION PROCESS)

Formulation

Linseed oil	43.0 parts
Lead naphthenate solution (14% lead)	4.3×10^{-2} parts
Pentaerythritol	9.8 parts
Phthalic anhydride	18.2 parts
Mineral spirits	29.0 parts

Method

The linseed oil was heated to 100°C and the lead naphthenate catalyst was added. After heating this mixture to 220°C, the pentaerythritol was added slowly with stirring. The mixture was maintained at 220°C until a sample (1 part) was soluble in 2 parts ethanol, and then this temperature was maintained for an additional 30 min. The phthalic anhydride was added and the mixture was condensed at 230°C until an acid number of 10 to 14 and a viscosity of X–Z, at 70% solids in mineral spirits, was reached. The resin was cooled and thinned in mineral spirits.

Equipment

A stainless steel resin kettle equipped with a stirrer.

Applications

Used in air-drying enamels.

PREPARATION OF A 33% COCONUT OIL ALKYD
(MONOGLYCERIDE PROCESS)

Formulation

A	{ Coconut oil	30.0 parts
	{ Glycerine	8.5 parts
B	Catalyst solution*	0.6 parts
C	Glycerine	13.5 parts
D	Phthalic anhydride	43.4 parts
	Xylene	4.0 parts

Method

A was heated to 115°C, B was added and the temperature raised to 240°C. When a sample of the reaction mixture (1 part) was soluble in methanol (2 parts), C was added. After 30 min., D was added and the mixture heated under reflux at 200°C until the acid number was 10 to 15 mg KOH/g and the viscosity was $X\text{–}Y$ (at 55% solids in xylene). The resin was thinned to 55% with xylene.

Equipment

A stainless steel resin kettle equipped with a stirrer, a reflux condenser, and a water separator.

Applications

Used in conjunction with a melamine-formaldehyde resin in high quality alkyd/melamine formaldehyde automotive and industrial enamels.

* The catalyst solution is prepared by dissolving sodium hydroxide (16.2 g) in water (13.8 g) and glycerine (70 g).

PREPARATION OF A 40% COCONUT OIL ALKYD
(FATTY ACID PROCESS)

Formulation

Coconut oil fatty acid	31.0 parts
Phthalic anhydride	38.5 parts
Glycerol	17.0 parts
Pentaerythritol	9.5 parts
Xylene	4.0 parts

Method

The fatty acid was heated to 180°C and then the glycerol and pentaerythritol were added slowly. After addition of the xylene, the temperature was gradually raised to 220°C and maintained at this temperature until an acid value of 15 to 22 and a viscosity of Z_1-Z_3 (at 50% solids in xylene) were obtained. The resin was cooled and then thinned in xylene.

Equipment

A stainless steel resin kettle equipped with a stirrer, a reflux condenser, and a water separator.

Applications

Used in conjunction with a melamine-formaldehyde resin in high quality alkyd/melamine formaldehyde automotive and industrial enamels.

PREPARATION OF AN ISOPHTHALIC ACID BASED ALKYD
BY THE ACIDOLYSIS PROCESS[*]

Formulation

Soya bean oil (alkali refined)	47.0 parts
Isophthalic acid	35.0 parts
Glycerine	18.0 parts

Method

The soya bean oil and the isophthalic acid were heated to 280°C in $1\frac{1}{2}$ to 2 hrs. An inert gas flow rate of twice the normal was used until the temperature reached 230°C and then the normal rate was used. The mixture was maintained at 280°C for 40 min., then cooled to 215 to 240°C. The glycerine was added and the mixture heated to 230°C in 2 hrs. The mixture was then maintained at 230°C until the acid number reached 8.8 and a viscosity at 50% solids in xylene of *W–X*.

Equipment

A stainless steel resin kettle equipped with a stirrer.

Applications

Used for air-drying industrial enamels.

[*] Amoco Chemicals Corporation Bulletin IP 5, October, 1960.

PREPARATION OF A METHACRYLATED ALKYD
(ESTERIFICATION PROCESS)

1. Preparation of Methyl Methacrylate Copolymer

Formulation

Methyl methacrylate	48.0 parts
Methacrylic acid	2.0 parts
Benzoyl peroxide	1.0 parts
Xylene	49.0 parts

Method

The monomer mixture was added at a constant rate over a period of 3 hrs. to the xylene heated under reflux. The mixture was heated under reflux for a further 2 hrs. The polymer solution had a viscosity of Y and a solids content of 47%.

Equipment

A stainless steel resin kettle, preferably steam heated, fitted with a stirrer and a reflux condenser.

2. Esterification of Copolymers with the Alkyd Components

Formulation

A	Soya bean oil	26.3 parts
	Pentaerythritol	6.05 parts
	Lead naphthenate solution (12% lead)	0.088 parts
B	Copolymer solution (from above)	21.2 parts
C	Phthalic anhydride	11.1 parts
D	White spirit	35.2 parts

Method

A was heated, with stirring, to 240°C and maintained at this temperature until a sample of the product (1 part) was soluble in ethanol (3 parts). The mixture was cooled to 150°C and *B* added. The temperature was raised to 220°C by removal of solvent and the mixture heated until the acid number dropped to 2. The solution was cooled, *C* added, and the condensation carried out at 220°C until the viscosity was V at 60% solids in white spirit and an acid number of 20.0. The resin was then thinned with *D*.

Equipment

A stainless steel resin kettle equipped with a stirrer, a reflux condenser and a water separator.

Applications

Used in fast air-drying enamels. Often, enamels based on this type of resin have advantages over those that use the conventional process. However, manufacturing costs by the esterification process may be higher than with the conventional process.

PREPARATION OF A METHACRYLATED ALKYD
(COPOLYMERIZATION PROCESS)

Formulation

A	60% Soya bean oil alkyd*	16.0 parts
	White spirit	48.0 parts
B	Methyl methacrylate	12.0 parts
	Ditertiary butyl peroxide	0.48 parts
C	White spirit	24.0 parts

Method

A was heated to 140°C and then B added at a constant rate over 2 hrs. The solution was maintained at 140°C for a further 3 hrs., then cooled and thinned with C to 60% solids content. The resin had a viscosity of V and an acid number 8.

Equipment

A stainless steel resin kettle equipped with a stirrer, reflux condenser, and a water separator.

Applications

Used in fast air-drying enamels.

* The alkyd was prepared by heating to 240°C, with stirring, a mixture of soya bean oil (1000 parts), pentaerythritol (230 parts), and lead naphthenate catalyst (0.1% lead based on the weight of oil), until a sample of product (1 part) was soluble in ethanol (3 parts). Phthalic anhydride (433 parts) and xylene (64 parts) were added and the mixture condensed at 220°C to an acid value of 11 and a viscosity of J at 75% solids in xylene.

PREPARATION OF AN UNSATURATED POLYESTER

Formulation

A	Maleic anhydride	16 parts
	Phthalic anhydride	24 parts
	Propylene glycol	29 parts
	Xylene	5 parts
B	Hydroquinone	0.01 parts
C	Styrene	26 parts

Method

A was heated under total reflux for 1 hr. and then the water of reaction removed (the resin temperature was 210°C). The reaction was continued until the acid value dropped to 35 to 40. The resin was cooled to 120°C, *B* added and the resin cooled further to 105°C, when *C* was added.

Equipment

A stainless steel resin kettle equipped with a stirrer, a reflux condenser, and a water separator.

Applications

Used in wax-type polyester formulations.

PREPARATION OF A THERMOPLASTIC POLY(METHYL METHACRYLATE)

Formulation

A	Toluene	50.0 parts
B	{Methyl methacrylate	50.0 parts
	{Benzoyl peroxide	0.2 parts

Method

A was heated to reflux and then *B* added at a constant rate over 3 hrs. The solution was heated under reflux for a further 4 hrs. The polymer that was formed had a molecular weight of 80,000, calculated using the equation:

$$M = 1.47 \times 10^6 (\eta_r - 1 - \ln\eta_r)^{0.65}$$

where η_r is the relative viscosity of a 0.5% polymer solution in ethylene dichloride.

Equipment

A stainless steel resin kettle (no copper or brass fittings), preferably steam heated, equipped with a stirrer and a reflux condenser.

Applications

Used in acrylic lacquers where gloss retention on exterior exposure is a prime requirement.

PREPARATION OF AN EPOXY ESTER

Formulation

Epikote 1004	45.0 parts
Safflower oil fatty acids	55.0 parts

Method

The mixture of ingredients was heated at 240°C until the acid value reached 8 to 12 and the viscosity was $S - V$ at 50% solids in white spirits (usually requires 5 to 6 hrs.). The resin was cooled and thinned with white spirit.

Equipment

A stainless steel resin kettle equipped with a stirrer.

Applications

Used in air-drying, brushing, industrial maintenance paints where protection against polluted atmospheres is of importance.

PREPARATION OF A URALKYD (60% OIL LENGTH)

Formulation

A	Soya bean oil	20.0 parts
B	Pentaerythritol	8.9 parts
C	Calcium oxide	1.5×10^{-2} parts
D	Mineral spirits	26.0 parts
	Tolylene diisocyanate*	19.0 parts
E	Mineral spirits	26.0 parts
	Lead naphthenate solution (14% lead)	6.0×10^{-2} parts

Method

A was stirred and heated to 150°C. B was added and the temperature raised to 200°C, followed by the addition of C and a further raising of the temperature to 245°C. The mixture was maintained at this temperature until a sample (1 part) was soluble in ethanol (2 parts). After a further hour at 245°C, D was added and the mixture cooled to 50°C. E was added and the mixture heated at 75 to 80°C for 2 hrs. The uralkyd formed had a viscosity of $Y - Z$.

Equipment

A stainless steel resin kettle equipped with a stirrer, a reflux condenser, and a water separator.

Applications

One pack air-drying polyurethane finishes.

* Tolylene diisocyanate is toxic.

PREPARATION OF A BUTYLATED MELAMINE
FORMALDEHYDE RESIN

Formulation

	Formalin	33.2 parts
A	Butanol	41.9 parts
	Toluene	5.2 parts
B	Formic acid	0.017 parts
C	Melamine	10.0 parts
D	Toluene	5.6 parts
E	Shell X3B-type solvent	3.7 parts

Method

The pH of *A* was adjusted to 4.0 ± 0.1 with *B*. *C* was added and after heating under reflux for 1 hr., and then cooling to 80°C, *D* was added. The mixture was heated under reflux until 32 parts of water had been removed in the water separator (4 to 5 hrs.), followed by removal of the total solvent distillate (33 parts), (¾ to 1 hr.). The resin was then cooled and thinned with *E* to give a final vixcosity of approximately *V*.

Equipment

A stainless steel resin kettle (no copper or brass fittings), preferably steam heated, equipped with a stirrer, a reflux condensor, and the facilities for separating the water from the solvent/water azeotrope.

Applications

This is a relatively fast curing melamine-formaldehyde resin and can be used in conjunction with suitable alkyds or thermosetting acrylic polymers to formulate stoving enamels.

PREPARATION OF A BUTYLATED
UREA-FORMALDEHYDE RESIN

Formulation

A	⎧ Formalin	34.9 parts
	⎨ Sodium hydroxide	0.017 parts
	⎩ Disodium hydrogen phosphate	0.024 parts
B	Urea	11.6 parts
C	Butanol	21.8 parts
D	Phthalic anhydride	0.35 parts
E	⎧ Toluene	3.8 parts
	⎩ Butanol	13.6 parts
F	Hydrocarbon solvent*	11.5 parts
G	⎧ Butanol	1.15 parts
	⎩ Hydrocarbon solvent*	1.15 parts

Method

The pH of the formalin was adjusted to 8.5 with the sodium hydroxide and the disodium hydrogen phosphate buffer was added to give A. B was added and the mixture stirred until all B dissolved. After the addition of C, the solution was heated under reflux for 1 hr. The mixture was cooled (70 to 80°C), the components of E added, and the mixture reheated to boiling point. Water was removed until the ratio of water to total distillate was 13:70 (4 to 5 hrs.). The total distillate was then removed until the viscosity of the resin reached Z_6 ($\frac{3}{4}$ to 1 hr.). The resin was cooled and thinned with F and then G. The solvent distillate that is removed during the manufacture of the resin can be used in following batches.

Equipment

A stainless steel resin kettle (no copper or brass fittings), preferably steam heated, equipped with a stirrer, reflux condenser, and the facilities for separating the water from the solvent/water azeotrope.

Applications

This is a relatively fast curing urea-formaldehyde resin and can be used in conjunction with suitable alkyds or thermosetting acrylic polymers to formulate stoving enamels.

* Type X3B, marketed by Shell Chemical (Aust.), Ltd.

PREPARATION OF A THERMOSETTING ACRYLIC POLYMER
(FOR USE WITH A MELAMINE RESIN)

Formulation

A	Xylene	25.0 parts
	Butanol	25.0 parts
B	Methyl methacrylate	28.0 parts
	Butyl acrylate	14.0 parts
	Hydroxyethyl methacrylate	6.3 parts
	Acrylic acid	0.7 parts
	Cumene hydroperoxide	1.0 parts

Method

A was heated under reflux and then B added at a constant rate over 2 hrs. The mixture was heated under reflux for a further 4 hrs. and then cooled. The copolymer solution had a solids content of 48% and the molecular weight of the copolymer, calculated from relative viscosity data, was 30,000.

Equipment

A stainless steel resin kettle (no copper or brass fittings), preferably steam heated, equipped with a stirrer and a reflux condenser.

Applications

Used in conjunction with a melamine-formaldehyde resin in exterior enamels.

PREPARATION OF A COPOLYMER BY EMULSION POLYMERIZATION*

Formulation

A	Water	46.5 parts
	Sodium bisulphite	0.0425 parts
	Sodium laurylsulphate	0.00425 parts
B	Styrene	23.4 parts
	Methacrylic acid	2.12 parts
	Ethyl acrylate	17.0 parts
	Potassium persulphate	0.170 parts
	Water	7.45 parts
C	Sodium laurylsulphate	0.212 parts
	Water	3.19 parts

Method

A was heated to 80°C and then the ingredients of B added in the order given above. The mixture was heated at 66°C for 30 min., C added, and the reaction continued for an additional $2\frac{1}{4}$ hrs. at 65 to 66°C.

Equipment

A stainless steel resin kettle (no copper or brass fittings) equipped with a stirrer and reflux condenser.

Applications

Used in conjunction with a water-dilutable heat-reactive formaldehyde condensation resin (for example, urea, melamine or phenol formaldehyde) to produce thermosetting water-dilutable enamels. A typical stoving schedule would be 1 hr. at 120°C.

* Hornibrook, W. J., E. I. du Pont de Nemours & Co., U.S. Pat. 2,918,391 (1959).

PREPARATION OF A WATER SOLUBLE ALKYD RESIN CAPABLE OF DRYING BY AUTOXIDATION*

Formulation

A	Safflower oil	43.90 parts
B	Trimethylol propane	16.44 parts
C	Litharge	0.02 parts
D	Tetrahydrofurfuryl alcohol	12.54 parts
E	Trimellitic anhydride	23.60 parts
F	Phthalic anhydride	3.50 parts

Method

A was heated to 150°C and then B added. C was added when the temperature of the mixture reached 220°C. This temperature was maintained until the product (1 part) was soluble in methanol (5 parts). The product was cooled (166°C), D added, and the mixture stirred until uniform distribution of D had taken place. After addition of E, the reaction mixture was reheated to 166°C, kept at this temperature for 15 to 30 min. and then maintained at 190°C for 3 to 4 hrs. When the acid number reached 35 to 40, the temperature was lowered to 170°C, F was added and the reaction continued to an acid number of 45 to 50. The resin (100 parts) was thinned in a mixture of water (101.2 parts), isopropanol (11 parts), and 28% ammonia (10 parts) to give a solids content of 45%. The viscosity of the solution was Z_4 and the pH was 8 to 9.

Equipment

A stainless steel resin kettle equipped with a stirrer and a reflux condenser.

Applications

Used as a water soluble air-drying resin.

* Wilkinson, R. F., *Offic. Dig. Federation Soc. Paint Technol.*, **35**, 129 (1963).

PREPARATION OF A WATER-SOLUBLE THERMOSETTING ACRYLIC COPOLYMER*

Formulation

Acrylic acid	3.55 parts
N-Methylol acrylamide	7.53 parts
Ethyl acrylate	16.6 parts
Methyl methacrylate	16.6 parts
Isopropanol	55.5 parts
Caprylyl peroxide	0.22 parts

Method

The ingredients were heated under reflux for 16 hrs. An equal volume of water was added together with slightly more ammonium hydroxide than required to neutralize the carboxyl groups present in the formulation mixture. The water-isopropanol azeotrope was distilled from the mixture and additional water was added to the reaction vessel from time to time so that the final isopropanol solution had a solids content of 50%.

Equipment

A glass-lined reaction vessel fitted with a stirrer, a reflux condenser, and facilities for solvent removal.

Applications

Used as the vehicle for thermosetting water-soluble enamels; it gives films which, after baking 30 min. at 160°C, are water insoluble.

* Essig, H. J., B. F. Goodrich Co., Can. Pat. 620,439 (1961).

PREPARATION OF HEXAMETHYLOL MELAMINE

Formulation

A	Paraformaldehyde	450 parts
	Water	500 parts
	Sodium Carbonate	2 parts
B	Melamine	189 parts

Method

A was charged to a flask equipped with stirrer, thermometer, nitrogen purge, and reflux condenser. The termperature was slowly raised to 80°C with stirring. On obtaining a clear solution the mixture was slowly cooled and the pH was adjusted to 8. *B* was added over 60 mins. to avoid an exotherm. A white flocculant precipitate was obtained. It was allowed to cool to room temperature and filtered. The precipitate was repeatedly redispersed in water, washed, and filtered until no odour of formaldehyde was detected.

ETHERIFICATION OF METHOYLATED MELAMINE [*]

Formulation

A	Hexamethylol melamine	30.6 parts
	Butanol	741.2 parts

Method

A was charged to a flask equipped with stirrer, thermometer, and reflux condenser. The pH of the reaction mixture was adjusted to 1.5 using sulphuric acid. The temperature was slowly raised to 40°C and was kept there for 2hrs. then allowed to cool slowly to room temperature. The solution was then adjusted to pH 8 using potassium hydroxide. The salt was removed by filtration and the unreacted alcohol removed at 0.1mm pressure and room temperature.

[*] Anderson, D. G., D. A. Netzel, and D. J. Tessari, *J. Appl. Polymer Sci.*, **14**, 3021 (1970)

PREPARATION OF ACRYLATED ALKYD RESINS[*]

Formulation

A	Riamene fatty acid	364 parts
	Pentaerythritol	113 parts
	Ethylene glycol	78 parts
	Phthalic anhydride	238 parts
	Maleic anhydride	13 parts
B	Xylene	34 parts
C	Xylene	53 parts
D	Trimethylolpropanediallyl ether	107 parts
E	Methyl methacrylate	339 parts
	Butyl methacrylate	85 parts
F	Benzoyl peroxide	45 parts

Method

A was heated at 180°C for 3 hrs. *B* was then added and heated at 250°C until the acid value was < 20. *C* was added and the mixture cooled to 150°C before *D* was added. The reaction was taken to completion at 180-220°C. The acid value was < 10. The resin was diluted with xylene, to a suitable viscosity.

Subsequently 127 parts alkyd resin (60% solids in xylene) was mixed with *E* and diluted with 191 parts toluene. *F* was added and the mixture was polymerised for 2 hrs at 110°C while CO_2 was bubbled through. The resultant polymer was adjusted to a suitable viscosity with toluene; the acid value was then 1-5.

[*] Reichhold-Albert Chemie, Ger. Pat. 1,295,816

SILICONE MODIFIED ALKYD[*]

The long oil alkyd required for this process can be prepared by the Fatty acid process. (See Appendix, page 359.)

Formulation

Long oil alkyd	70 parts
Silicone resin eg (M.S. 650)	18 parts
Mineral spirits	12 parts

Method

The formulation was charged to a resin kettle equipped with reflux condenser and trap. It was heated to a steady reflux of 170°C and held there until the resins co-polymerized. The reaction can be followed by drying a drop of resin on a glass plate. At the end point a clear film is formed. The prepared resin was diluted to 60% NVM with mineral spirits. It then had a Gardner-Holdt viscosity of V-X and an acid value of 4-5.

* Clope, R. W., M. A. Glaser, and F. D. Graziano, *Silicone Technology,* Applied Polymer Synopsia No. 14, Interscience Publishers, Inc., New York, 1970, 151.

PREPARATION OF SILICONE MODIFIED EPOXY RESIN[*]

Formulation

Epikote 1001	100 parts
Silicone Resin: MS.650	20 parts
Dodecyl Benzene Sulphonic Acid	0.12 parts

Method

The formulation was charged to a resin kettle and melted. At this stage stirring was commenced and the temperature raised gradually to 225°C. in 40-45 minutes. A typical heating program is given below:

Temperature 10 minutes after stirring commenced	190°C
Temperature 20 minutes after stirring commenced	210°C
Temperature 30 minutes after stirring commenced	220°C
Temperature 40 minutes after stirring commenced	225°C

At this stage the heat was removed and the contents of the kettle were poured into trays and allowed to solidify.

N.B. It is essential to control the process carefully since if the reaction is allowed to proceed too far, gelation results.

Equipment

A resin kettle equipped with stirrer, Dean and Stark trap (for liberated methanol) and condenser.

---

[*] Shell Chemical (Aust.) Pty. Ltd., Plastics and Resins Bulletin, RES/29/2, *Silicone Modified Epoxy Resins*, 3.

AUTHOR AND PATENT INDEX

*The page numbers in italics refer to references at the end of sections.

SUBJECT INDEX